FOOD SOVEREIGNTY, AGROECOLOGY AND BIOCULTURAL DIVERSITY

Contestations over knowledge – and who controls its production – are a key focus of social movements and other actors that promote food sovereignty, agroecology and biocultural diversity. This book critically examines the kinds of knowledge and ways of knowing needed for food sovereignty, agroecology and biocultural diversity.

'Food sovereignty' is understood here as a transformative process that seeks to recreate the democratic realm and regenerate a diversity of autonomous food systems based on agroecology, biocultural diversity, equity, social justice and ecological sustainability. It is shown that alternatives to the current model of development require radically different knowledges and epistemologies from those on offer today in mainstream institutions (including universities, policy think tanks and donor organizations). To achieve food sovereignty, agroecology and biocultural diversity, there is a need to re-imagine and construct knowledge for diversity, decentralisation, dynamic adaptation and democracy.

The authors critically explore the changes in organizations, research paradigms and professional practice that could help transform and co-create knowledge for a new modernity based on plural definitions of wellbeing. Particular attention is given to institutional, pedagogical and methodological innovations that can enhance cognitive justice by giving hitherto excluded citizens more power and agency in the construction of knowledge. The book thus contributes to the democratization of knowledge and power in the domain of food, environment and society.

Michel P. Pimbert is Professor of Agroecology and Food Politics and Director of the Centre for Agroecology, Water and Resilience at Coventry University, UK.

ROUTLEDGE STUDIES IN FOOD, SOCIETY AND THE ENVIRONMENT

For further details please visit the series page on the Routledge website: http://www.routledge.com/books/series/RSFSE/

'This important book shows both how agroecology can democratize knowledge, and how democratizing knowledge in turn is a condition for agroecology to develop. We tend to reduce agroecology to a set of agronomic techniques that reduce the need for external inputs, that de-link food production from energy consumption, and that restore soil health. But it is, more fundamentally, about the direction of knowledge: agroecology operates the shift from top-down "extension" of knowledge by experts delegated by ministries, to a bottom-up approach prioritizing the local knowledge developed by farmers. It is empowering, horizontal, based on trial and error—but it is also, as this volume shows, another way of conceiving science.'

—Olivier De Schutter, former UN Special Rapporteur on the Right to Food (2008–2014), Co-Chair, International Panel of Experts on Sustainable Food Systems (IPES-Food), Belgium

'The surge of industrial farming, mega-use of pesticides, and chronic commercialisation which appeared to "grow" the world also created a cascade of painful issues relating to poisoning "Mother Earth", generating inequities, and destroying biodiversity and cultural heritage. This book provides not just insights into those issues but, more importantly, it explores the knowledge and transformative ways of knowing we now need to re-enchant the world. Deepening knowledge democracy is key for reclaiming food sovereignty, rooting agroecology, and promoting biocultural diversity. This book shows another world is indeed possible. All it needs is action.'

—Anwar Fazal, Recipient of the Right Livelihood Award, popularly called the 'Alternative Nobel Prize', Malaysia

'Michel Pimbert is a rare combination of syncretic visionary and on-the-ground change maker. He's put together a unique volume showing that healthy farming systems and life-serving human communities emerge together. They are inseparable and they require diverse ways of knowing to free themselves from the deadening grip of dominant knowledge. In coming to appreciate this process, we learn to see the democratization of knowledge-creation as key to our future.'

—Frances Moore Lappé, Director of the Small Planet Institute, USA

'This book tells us how we can still make peace with nature and with ourselves by constructing a radically different knowledge that is ecologically wise and based on epistemic justice. This decolonisation of knowledge depends on respectful engagements with diverse ways of life and in particular of indigenous peoples and other traditional local communities.'

—Ashish Kothari, Kalpavriksh and ICCA Consortium, India

'At a time of climate extremes, unpredictability and complexities that the dominant food regime's limited understandings can't respond to, this book is an outstanding contribution to the transformation of knowledge construction for diversity. It describes the inter-dependence of food sovereignty, agroecology and biocultural diversity. And it illuminates pathways for them to flourish through knowledge justice grounded in cultural pluralism and context, including the place-based relationships vital to indigenous peoples' knowledge systems. By suggesting deep changes to empower marginalized knowledge holders, this book lays groundwork for achieving and sustaining genuine "well-being" in its diverse meanings.'

—Carol Kalafatic (Quechua), Vice-Chair of the UN High Level Panel of Experts on Food Security and Nutrition

'Family farmers, pastoralists, fishers and small food processors continue to be neglected and marginalized by the dominant agricultural research system. As this book makes abundantly clear, the exclusion of peasant farmers from the co-construction of knowledge for food and farming is not only an enduring injustice, it is also a huge wasted opportunity for the development of socially just and ecologically sustainable food systems everywhere. Achieving food sovereignty, amplifying agroecology and regenerating biocultural diversity all directly depend on peasant farmers and other citizens being centrally involved in deciding the priorities for research and innovation. This book is both timely and courageous because it clearly shows how the construction of knowledge can indeed be democratized and re-invented for the common good.'

—Mamadou Goita, Director of IRPAD and former Executive Director of ROPPA. Founding member of the Alliance for Food Sovereignty in Africa, Mali

FOOD SOVEREIGNTY, AGROECOLOGY AND BIOCULTURAL DIVERSITY

Constructing and Contesting Knowledge

Edited by Michel P. Pimbert

Routledge
Taylor & Francis Group

LONDON AND NEW YORK

earthscan
from Routledge

First published 2018
by Routledge
2 Park Square, Milton Park, Abingdon, Oxon OX14 4RN

and by Routledge
711 Third Avenue, New York, NY 10017

Routledge is an imprint of the Taylor & Francis Group, an informa business

British Library Cataloguing-in-Publication Data
A catalogue record for this book is available from the British Library

Library of Congress Cataloging-in-Publication Data
A catalog record for this book has been requested

ISBN: 978-1-138-95535-6 (hbk)
ISBN: 978-1-138-95536-3 (pbk)
ISBN: 978-1-315-66639-6 (ebk)

Typeset in Bembo
by Deanta Global Publishing Services, Chennai, India

CONTENTS

FIGURES

TABLES

BOXES

ACKNOWLEDGEMENTS

This book reflects numerous conversations over several decades with peasant farmers, indigenous communities, pastoralists, fishers, farm workers, activist scholars and other citizens around the world. They are too numerous to mention by name, but I hope they know who they are.

I particularly want to thank my longtime companions in participatory action research on food sovereignty, agroecology and biocultural diversity: Alejandro Argumedo in Peru, Boukary Barry in Mali, Mansour Fakih in Indonesia, Taghi Farvar in Iran and P.V. Satheesh in India. I owe them special thanks for their trust, courage, generosity, inspiration and joyful friendship.

Ideas for this book crystalized during my sabbatical Fellowship at the Rachel Carson Centre for Environment and Society in Munich. I thank its co-directors, Christof Mauch and Helmuth Trischler, for giving me the opportunity to be part of their stimulating community of critical scholars.

My thinking on the scope of this book significantly benefitted from the rich conversations I had with peasant farmers and activist scholars during the *St Ulrich Workshop on Democratising Agricultural Research for Food Sovereignty and Peasant Agrarian Cultures,* which was held in the Black Forest region of Germany in September 2013. I acknowledge the generous intellectual spirit of all workshop participants.

I also wish to acknowledge colleagues in the Centre for Agroecology, Water and Resilience at Coventry University for their collegiality and critical insights on the politics of knowledge: Colin Anderson, Josh Brem-Wilson, Jahi Chappell, Priscilla Claeys, Csilla Kiss, Deepa Joshi, Stefanie Lemke, Iain MacKinnon, Jasber Singh and Tom Wakeford.

I sincerely thank Fiona Hinchcliffe for her thoroughly professional editing of the manuscript. I also acknowledge the advice and patient support of Tim Hardwick and Amy Louise Johnston at Routledge.

Last, but not least, I lovingly thank my companion Nathalie Whitfield-Pimbert, and dedicate this work to our two children, Ilona and Matthias. May the ideas shared in this book inspire them to continue co-creating a more just and sustainable world.

Michel Pimbert
Oxford

ABBREVIATIONS

ABS	access and benefit sharing
AIDS	acquired immune deficiency syndrome
ARS	agricultural research systems
ASHIN	*Asociación de Shamanes Indígenas del Napo*
CBC	community-based conservation
EU	European Union
FAO	Food and Agriculture Organization of the United Nations
GDP	gross domestic product
ha	hectare
IAASTD	International Assessment of Agricultural Science, Knowledge and Technology for Development
ICCA	Indigenous Peoples' and Community Conserved Territories and Areas
IF	impact factor
IMF	International Monetary Fund
IPM	integrated pest management
IPR	intellectual property rights
ISDS	investor-state dispute settlement
LGBT	lesbian, gay, bisexual and transgender
LVC	La Via Campesina
n.d.	no date
OECD	Organisation for Economic Co-operation and Development
PES	payments for ecosystem services
PPP	public-private partnership
PSRE	public-sector research establishment
R&D	research and development
RSP	*Réseau Semences Paysannes*
SI	systems of innovation

S&T   science and technology
VIPP   Visualisation in Participatory Programmes
WHO   World Health Organization
WTO   World Trade Organization

CONTRIBUTORS

Philippe V. Baret is a tropical agronomist and quantitative geneticist by training. He is Professor at the Catholic University of Louvain in Belgium, where he coordinates research on agroecology, genetic diversity and sustainable development. He is co-founder of the Belgian Interdisciplinary Agroecology Research Group of the FNRS (the Belgian Scientific Research Foundation).

Kristen Blann is a freshwater ecologist with The Nature Conservancy in Minnesota, where she provides technical leadership for freshwater and watershed conservation planning within the Minnesota, North Dakota, South Dakota and neighboring chapters. She holds Masters and PhD degrees in conservation biology from the University of Minnesota, Department of Fisheries, Wildlife, and Conservation Biology. She lives with her husband on a small farm in central Minnesota where they raise vegetables, sheep and chickens.

Katherine Homewood studied zoology at Oxford University and gained her PhD in anthropology at the University of London. After working at the University of Dar es Salaam in Tanzania, she joined University College London (UCL) as lecturer and tutor in human sciences, an interdisciplinary and interdepartmental degree. She is now Professor in Anthropology at UCL. Her work centres on the interaction of conservation and development in sub-Saharan Africa, with a special focus on pastoralist peoples in drylands, among other groups and ecosystems. She researches the implications of natural resource policies and management for local people's livelihoods and welfare, and the implications of changing land use for environment and biodiversity. Her Human Ecology Research Group integrates natural and social sciences approaches to interactions of environment and development around the Global South.

Stephen S. Light obtained his Masters in parks and wildlands management at Pennsylvania State University and his PhD in natural resources at the University of Michigan. He was a founding member of the Resilience Alliance, and the journal *Ecology and Society*. Recently serving as adjunct faculty, McGill University, Light has engaged in some of the most fundamental of reforms in water resources management in the past century as a thought leader and practitioner of science-based approaches to management and ecological restoration. He was one of a core group of scientists that guided the development of guidelines for the Everglades' restoration (1993). Since then he has had projects in the Everglades, Upper Mississippi River, Red River of the North, Coastal Louisiana, Rio Grande and the Missouri River. His books include *Barriers and Bridges to Renewing Ecosystems and Institutions* and *Conservation of Biodiversity in Rural Sustainability in Central Europe*.

Nina Isabella Moeller's academic background is in philosophy, sociology and anthropology. In 2010, she finished her PhD thesis in which she analyzed processes of protecting traditional knowledge in the Ecuadorian Amazon as a form of capital expansion. She has worked in Latin America and Europe – amongst other things as a consultant to indigenous organizations, NGOs and the UN's Food and Agriculture Organization. Her current research continues her explorations of the 'local' complexities of 'global' socio-ecological change, and of the 'local' manifestations of 'global' discourses. It is entangled with her quest to understand how non-market relations and practices are fostered or undermined in today's world. Nina is a Research Fellow at the University of Oxford and is also a keen gardener and wild plant user.

Michel P. Pimbert is Professor of Agroecology and Food Politics and Director of the Centre for Agroecology, Water and Resilience at Coventry University, UK. An agricultural ecologist by training, he previously worked at the UK-based International Institute for Environment and Development (IIED), the International Crops Research Institute for the Semi-Arid Tropics (ICRISAT) in India, the University François Rabelais de Tours in France, and the World Wide Fund for Nature in Switzerland. He is currently a member of the High Level Panel of Experts on Food Security and Nutrition (HLPE) of the Committee on World Food Security (CFS) at the UN Food and Agriculture Organization. He is the author of several books including *Social Change and Conservation: Environmental Politics and Impacts of National Parks and Protected Areas* (with Kléber B. Ghimire) and *Sharing Power: A Global Guide to Collaborative Management of Natural Resources* (with Grazia Borrini-Feyerabend).

Gilbert Rist is Professor Emeritus at the Graduate Institute for International and Development Studies in Geneva, Switzerland. He first taught at the University of Tunis, became Director of the Europe-Third World Centre in Geneva and, later on, Senior Researcher on a United Nations University Project. Afterwards, he joined the Graduate Institute of Development Studies, where he taught intercultural relations and social anthropology. His main interest is in an anthropological approach to our contemporary society. He is the author of *The History of Development: From Western Origins to Global Faith* (Zed Books, London, 3rd Edition 2008) and *The Delusions of Economics: The Misguided Certainties of a Hazardous Science* (Zed books, 2011).

Eric B. Ross, retired Associate Professor, International Institute of Social Studies, Erasmus University, The Hague, is presently Professorial Lecturer in Anthropology at George Washington University in Washington DC, USA. His principal interests are the history of anthropology, agrarian justice and the political ecology of the world food system. Among his books are *Beyond the Myths of Culture: Essays in Cultural Materialism* (ed.), *Food and Evolution: Toward a Theory of Human Food Habits* (edited with Marvin Harris), *Death, Sex and Fertility: Population Regulation in Preindustrial and Developing Societies* (written with M. Harris) and *The Malthus Factor: Poverty, Politics and Population in Capitalist Development*.

Sian Sullivan's contribution in this volume builds on research conducted since the early 1990s in the drylands of west Namibia, southern Africa, work that continues today through a UK Arts and Humanities Research Council project called *Future Pasts* (www.futurepasts.net). Sian has published in the areas of culture-nature relationships, the financialization of nature, biodiversity offsetting, embodiment practices, social movements and alternative media, contributing to emerging fields of study in political ecology, non-equilibrium ecology, neoliberal conservation, ethnographic interpretations of rock art and ecocultural ethics. Currently Sian is Professor of Environment and Culture at Bath Spa University in Somerset, UK.

Gaëtan Vanloqueren is an agroeconomist by training. Between 2008 and 2013 he co-ordinated the support team of Olivier De Schutter, the United Nations Special Rapporteur on the Right to Food, and advised him on thematic and strategic issues. He holds a PhD from the University of Louvain (2007), and is a Guest Lecturer at ICHEC-Brussels Management School, where he teaches development economics and policies. He is currently working with the Minister of Economics and Employment of the Brussels-Capital Region, shaping policies supporting social enterprises and the circular economy.

1

CONSTRUCTING KNOWLEDGE FOR FOOD SOVEREIGNTY, AGROECOLOGY AND BIOCULTURAL DIVERSITY

An overview

Michel P. Pimbert

Introduction

Much of the knowledge produced by mainstream research is inappropriate or directly harmful to local communities and the environments on which they depend for their food security, livelihoods and culture. Narrow-lens, universal and reductionist explanatory models have generated a crisis in agriculture and natural resource management through their inability to come to terms with the dynamic complexity and variation within and among ecosystems. Similarly, the science of economics and mainstream accounts of human demography embody several reductionist biases, unproven assumptions and narrow historical perspectives that legitimize the dominant food regime and current land uses.

As a result, global narratives on people-environment interactions blame the poor, women and ethnic minorities for social and environmental ills. Despite the fact they represent by far the vast majority of the world's food and agricultural producers today (Lowder *et al.*, 2016), small and family farmers, nomadic pastoralists and agro-pastoralists, indigenous peoples and forest dwellers, artisanal fishers and urban food producers are largely excluded from participation in research and policy debates on the future of food, farming, environment and development (Chambers *et al.*, 1989; Chambers, 2008) – with women being the most excluded everywhere. Instead, representatives of large farmers and agri-food corporations are usually centre-stage in these debates. The consequence is a socially and ecologically destructive and increasingly globalized agri-food system.

Food sovereignty is an alternative paradigm for food, fisheries, agriculture, pastoralism and forest use that is emerging in response to this democratic deficit and the many environmental and social crises of food and farming. It aims to guarantee and protect people's space, ability and right to define their own models of production, food distribution and consumption patterns. It emphasizes the science and practice of *agroecology* to design sustainable agricultures that reduce carbon and ecological footprints in rural and urban areas (Altieri *et al.*, 2015; IPES-Food, 2016). It also encompasses the concept of *biocultural diversity*: the interrelated biological, cultural and linguistic diversity as well as the local knowledge, institutions and practices which are vitally important in allowing societies to adaptively manage their farming systems, natural resources, landscapes and social life.

To achieve food sovereignty, agroecology and biocultural diversity there is a need to transform and construct knowledge for diversity, decentralization, dynamic adaptation and democracy. This is the central thesis of this book.

About this volume

Conflicts and contestations over knowledge – and who controls its production – are a key focus of social movements and other actors that promote food sovereignty, agroecology and biocultural diversity. Alternatives to the current model of agriculture, conservation and development require radically different knowledges and epistemologies from those on offer today in mainstream institutions (universities, policy think tanks, donor organizations, trade unions, etc.).

Going beyond facile critique, this collection of papers critically explores new knowledge and reforms in research, technological paradigms, organizations and professional practice that could help transform and construct knowledge for food sovereignty, agroecology and biocultural diversity. In sum, the purpose of this book is to contribute to the democratization of knowledge and power. It draws on a series of conversations with peasant farmers[1] and key scholars on food, environment and society – some of whom have been invited to contribute to this book.

This introductory chapter summarizes the main issues and concepts related to transforming knowledge for food sovereignty, agroecology and biocultural diversity. The origins, history and main features of the concepts are first briefly described, together with a vision for a radically new modernity. It highlights how these concepts and issues are dealt with in the various chapters, as well as the central arguments that run through this book.

Overall, this volume offers critical reflections on the nature and politics of knowledge(s) that are centrally involved in the governance[2] and management of food systems[3] and biocultural diversity. It argues that a fundamentally new paradigm for science and knowledge is required to achieve food sovereignty and amplify agroecological solutions along with biocultural diversity. And while a paradigm shift has many dimensions (Kuhn, 2012; Lincoln *et al.*, 2011), this volume mainly focuses on transformations in the nature of knowledge (epistemology) and in ways of knowing (the nature of human inquiry). It emphasizes in particular how the production of knowledge might be democratized. The book is careful to avoid simplistic recommendations. Instead, it argues for a re-imagining and radical reconstruction of knowledge and ways of knowing for food sovereignty, agroecology and biocultural diversity.

A brief introduction to food sovereignty, agroecology and biocultural diversity

Food sovereignty

After several years of development by peasant movements and citizens,[4] the concept of 'food sovereignty' was first put forward internationally by La Vía Campesina[5] at the UN Food and Agriculture Organization's World Food Summit in 1996 (Desmarais, 2007; Desmarais and Nicholson, 2013). During this summit, La Vía Campesina (LVC) presented a set of mutually supportive principles that offered an alternative to world trade policies and would also realize the human right to food. Their statement, *Food Sovereignty: A Future without Hunger* (1996), declared that 'Food Sovereignty is a precondition to genuine food security'.[6] At its heart, this alternative paradigm for food, agriculture and land use aims to guarantee and protect people's space, ability and right to define their own models of production, food distribution and consumption patterns in rural and urban contexts (Box 1.1).

BOX 1.1 FOOD SOVEREIGNTY: A FUTURE WITHOUT HUNGER

During the 1996 World Food Summit, La Vía Campesina presented seven mutually supportive principles that define an alternative paradigm for food, agriculture and human wellbeing, summarized here:

1 Food – a basic human right

Food is a basic human right. Everyone must have access to safe, nutritious and culturally appropriate food in sufficient quantity and quality to sustain a healthy life with full human dignity. Each nation should declare that access to food is a constitutional right and guarantee the development of the primary sector to ensure the concrete realization of this fundamental right.

2 Agrarian reform

A genuine agrarian reform is necessary which gives landless and farming people – especially women – ownership and control of the land they work and which returns territories to indigenous peoples. The right to land must be free of discrimination on the basis of gender, religion, race, social class or ideology; the land belongs to those who work it. Smallholder farmer families, especially women, must have access to productive land, credit, technology, markets and extension services. Governments must establish and support decentralized rural credit systems that prioritize the production of food for domestic consumption. [...]

3 Protecting natural resources

Food sovereignty entails the sustainable care and use of natural resources, especially land, water, seeds and livestock breeds. The people who work the land must have the right to practice sustainable management of natural resources and to preserve biological diversity. This can only be done from a sound economic basis with security of tenure, healthy soils and reduced use of agro-chemicals. Long-term sustainability demands a shift away from dependence on chemical inputs, on cash-crop monocultures and intensive, industrialized production models. Balanced and diversified natural systems are required. [....] Farming communities have the right to freely use and protect the diverse genetic resources, including seeds and livestock breeds, which have been developed by them throughout history. This is the basis for food sovereignty.

4 Reorganizing food trade

Food is first and foremost a source of nutrition and only secondarily an item of trade. National agricultural policies must prioritize production for domestic consumption and food self-sufficiency. Food imports must not displace local production nor depress prices. This means that export dumping or subsidized exports must cease. Smallholder farmers have the right to produce essential food staples for their countries and to control the marketing of their products. Food prices in domestic and international markets must be regulated and reflect the true costs of producing that food. This would ensure that smallholder farmer families have adequate incomes. [....]

5 Ending the globalization of hunger

Food sovereignty is undermined by multilateral institutions and by speculative capital. The growing control of multinational corporations over agricultural policies has been facilitated by the economic policies of multilateral organizations such as the World Trade Organization (WTO), World Bank and International Monetary Fund (IMF). Regulation and taxation of speculative capital and a strictly enforced code of conduct for transnational corporations is therefore needed.

6 Social peace

Everyone has the right to be free from violence. Food must not be used as a weapon. Increasing levels of poverty and marginalization in the countryside, along with the growing oppression of ethnic minorities and indigenous populations, aggravate situations of injustice and hopelessness. The ongoing displacement, forced urbanization, repression and increasing incidence of racism of smallholder farmers cannot be tolerated.

7 Democratic control

Smallholder farmers must have direct input into formulating agricultural policies at all levels. The United Nations and related organizations will have to undergo a process of democratization to enable this to become a reality. Everyone has the right to honest, accurate information and open and democratic decision-making. These rights form the basis of good governance, accountability and equal participation in economic, political and social life, free from all forms of discrimination. Rural women, in particular, must be granted direct and active decision-making on food and rural issues.

Subsequent declarations and documents by La Vía Campesina have built on these core food sovereignty principles.

(La Vía Campesina, 1996; www.viacampesina.org)

As the largest transnational agrarian movement today, LVC is mainly recognized for championing and developing the food sovereignty paradigm (Desmarais, 2007; Wittman et al., 2010). It does so by actively building alliances with other social movements trying to respond to the impacts of capitalist development in food, agriculture and land use. For example, LVC was one of the seven organizations[7] that planned and facilitated the 2007 International Forum on Food Sovereignty in Nyéléni (Mali), where over 600 participants from 80 countries further developed the political, economic, social and ecological dimensions of this alternative policy framework. The Nyéléni Declaration affirms the centrality and primacy of 'peoples' – rather than governments of nation states – in framing policies and practices for food, agriculture, environment and human wellbeing:

Food sovereignty is the right of peoples to healthy and culturally appropriate food produced through ecologically sound and sustainable methods, and their right to define their own food and agriculture systems. It puts those who produce, distribute and consume food at the heart of food systems and policies rather than the demands of markets and corporations. It defends the interests and inclusion of the next generation. It offers a strategy to resist and dismantle the current corporate trade and food regime, and directions for food, farming, pastoral and fisheries systems determined by local producers. Food sovereignty prioritizes local and national economies and markets and empowers peasant and family farmer-driven agriculture, artisanal fishing, pastoralist-led grazing, and food production, distribution and consumption based on environmental, social and economic sustainability. Food sovereignty promotes transparent trade that guarantees just incomes to all peoples as well as the rights of consumers to control their food and nutrition. It ensures that the rights to use and manage lands, territories, waters, seeds, livestock and biodiversity are in the hands of those of us who produce food. Food sovereignty implies new social relations free of oppression and inequality between men and women, peoples, racial groups, social and economic classes and generations.

(La Vía Campesina, 2007)

The organizers of Nyéléni 2007 aimed to expand the food sovereignty debate outside producer groups of small and family farmers (Box 1.2) to include consumer groups and workers' trade unions, as well as the youth and women (see www.nyeleni2007.org). As the LVC has globalized the struggle for food sovereignty, many other organizations, social movements, indigenous peoples' networks and citizen-consumers have adopted and further developed this progressive framework, generating major statements on food sovereignty over the last ten years in particular (Desmarais and Nicholson, 2013; and see www. viacampesina.org).

LVC strongly emphasizes the importance of women's rights and their knowledge for the further development of food sovereignty (Wiebe, 2013). As stated in the Declaration of Maputo: 'If we do not eradicate violence towards women within the movement, we will not advance in our struggles, and if we do not create new gender relations, we will not be able to build a new society' (La Vía Campesina, 2008a).

BOX 1.2 FAMILY FARMS FOR AUTONOMY AND COMMUNITY RESILIENCE

Worldwide, over 72% of the total number of farms are family farms which are smaller than one hectare in size (Lowder *et al.*, 2016). A family farm can be seen as an association composed of two or more members linked by family relations or customary ties. Production factors are harnessed in common to generate resources for social reproduction as well as financial, material and moral benefits in both rural and urban areas (Sourisseau, 2015). The autonomy, economic viability and resilience of family farms are enhanced when farmers control their resource base, including land, water, seeds and labour as well as knowledge, skills, social networks, local organizations and institutions (EAFF, PROPAC and ROPPA, 2013).

The broader historical context of peasant agricultures and various traditions of agrarian social thought have also influenced the emergence, theory and practice of food sovereignty – and continue to do so today. These influences include:

- agrarian collectivism, as well as social anarchism and libertarian socialist thought – all of which view peasants as progressive agents of change (Bakunin, 1987; Herzen, 1992; Kropotkin, 1892, 1898);

- Marx's view that capitalism induces a fundamental metabolic rift between society and nature (1981);
- heterodox Marxism (Chayanov, 1989);
- peasant studies (Hernández Xolocotzi, 1985,1987; Polanyi, 1957; Shanin, 1987; van der Ploeg, 2013);
- centre-periphery and dependency theory (Amin, 1976; Gunder Franck, 1978);
- post-development theory (Escobar, 1996; Esteva and Prakash, 1998; Partant, 1999, 2002); and
- agrarian social theory and thinking on environment and radical ecology (Bookchin, 1989; Friedmann, 2005; McMichael, 2009; Gonzales de Molina, 2010; Martinez-Alier, 2002).

Some of these traditions of radical thought have deeply influenced peasant struggles for self-determination and the right to food sovereignty. For example, Bakunin's proposals on collectivist anarchism (Bakunin, 1982) and Proudhon's 'principle of federation' (Proudhon, 1979) informed the consciousness and agency of an impoverished peasantry in Spain. During the Spanish civil war (1936–1939), the peasants of Andalusia and Aragon established communal systems of land tenure, in some cases abolishing the use of money for internal transactions, setting up free systems of production and distribution and creating a decision-making procedure based on popular assemblies and direct, face-to-face democracy. In those parts of Spain not overrun by Franco's troops, about three million men, women and children were living in collectivized communes over large areas (Bookchin, 1998; Leval, 1975).

Kropotkin's ideas on agrarian and industrial mutualism (Kropotkin, 1898) influenced Mahatma Gandhi's views on self-rule *(Swaraj)* and progress based on economic self-reliance *(Sarvodaya)* to end poverty through improved agriculture and small-scale cottage industries in every village of India (Bhatt, 1982). And today, newly emerging thinking on the convergences between food sovereignty and the right to the city[8] (Lefébvre, 1968; Harvey, 2012; Purcell, 2013), with an emphasis on urban agroecology and new garden cities (Bliss, 2011; Ross and Cabannes, 2014; Tornaghi, 2016) as well as libertarian municipalism and the use of eco-technologies in locally controlled circular economies (Bookchin, 1986, 1995; Jones *et al.*, 2012), all build on, or echo, the vision of direct democracy which Peter Kropotkin described over a century ago in *Fields, Factories and Workshops* (Kropotkin, 1898).

Similarly, the enduring struggles of indigenous peoples[9] for self-determination, control over their ancestral territories and their right to protect their knowledge systems and lifeways (see Chapters 4 and 6) all strongly amplify the vision of food sovereignty put forward by peasant organizations. Adopted after decades of negotiations between representatives of indigenous peoples and governments, the landmark United Nations Declaration on the Rights of Indigenous Peoples contains many statements that overlap with the intent and practice of food sovereignty (UNDRIP, 2007). Indigenous peoples' affirmation of their rights to self-determination and cultural revitalization are driving their food sovereignty agendas. Indeed, many indigenous peoples' movements, such as the Zapatistas in the Chiapas of Mexico, have embraced food sovereignty as part of their struggles for self-determination, decolonization, cultural affirmation, autonomy and gender equity (Gahman, 2016; Collier and Quaratiello, 2004).

More generally, the food sovereignty movement draws extensively on human rights-based frameworks. Virtually everywhere in the global North (Brent *et al.*, 2015) and South (Pimbert, 2008) the movement emphasizes the right to food, farmers' and workers' rights, rights to culture and livelihoods as well as the right to self-determination. This reflects a radical conception of rights: rights that are claimed by citizens rather than granted by the state. As Priscilla Claeys writes: 'The [food sovereignty] movement's assertion of new rights contributes to shaping a cosmopolitan, multicultural, and anti-hegemonic conception of human rights' (Claeys, 2012).

Increasing visibility and influence

As a concept and framework, the term 'food sovereignty' has moved from the margins and gained more visibility during the last ten years in particular (Desmarais and Nicholson, 2013; McKeon, 2015). The term is increasingly recognized by parts of the international community, some of the United Nations organizations, governments and a growing number of academic research centres and universities. For example, several recent international reports on the state of the world's food and agriculture mention 'food sovereignty' as a possible pathway to more sustainable agricultural development (e.g. IAASTD, 2009; HLPE, 2016).

The reform of the UN's Committee on World Food Security (CFS) and the creation of the Civil Society Mechanism (CSM) in 2010 have given representatives of small-scale producers and civil society supporters of food sovereignty a unique opportunity to engage with and influence governments at the international level (Brem-Wilson, 2015; CFS, 2009; McKeon, 2015). Like any other policy framework, food sovereignty implies a purposeful course of action to advance specific objectives based on national and international policies as well as an enabling global multilateralism (see Pimbert, 2008; Windfuhr and Jonsén, 2005; www.viacampesina. org). However, rather than presenting a fixed menu of policy instruments, advocates of food sovereignty use their interventions in the CFS to propose a range of policy shifts and directions for governments who seek to implement (or harm) food sovereignty. High-level policy dialogues with governments allow the food sovereignty movement to further develop its broad policy discourse[10] and create space for change.

Countries like Mali and Senegal have included food sovereignty thinking in their national policies, and constitutional recognition of the right to food sovereignty has been achieved in Ecuador, Bolivia and Nepal (Beuchelt and Virchow, 2012). For example, the government of Mali undertook a consultation process with farmers to draft its new agricultural framework law (LOA, 2005). After more than a year of work, this law has enshrined food sovereignty as a priority for improving rural and urban living standards. While the government may use the term as a rhetorical device, Malian farmer organizations continue to discuss ways of implementing the food sovereignty framework throughout the country.[11] Other countries such as Peru, Argentina, Guatemala, Brazil, El Salvador and Indonesia have legislation supportive of food sovereignty efforts (Wittman, 2015).

In academia too, interest is growing in the critical analysis and study of food sovereignty (Agarwal, 2014; Andrée *et al.*, 2014; Bernstein, 2014; Brem-Wilson, 2015; Burnett and Murphy, 2014; Claeys, 2015; Desmarais and Wittman, 2014; Edelman, 2014; Grey and Patel, 2015; Henderson, 2017; Kloppenburg, 2014; Masson *et al.*, 2017; Martínez-Torres and Rosset, 2014; McMichael, 2014; Schiavoni, 2017; Tilzey, 2017; Trauger, 2015; van der Ploeg, 2014). *The Journal of Peasant Studies* has published a selection of academic papers that were presented at two well-attended international conferences on food sovereignty: the first in 2013 at Yale University (USA) and the second in 2014 at the Institute for Social Studies in The Hague (The Netherlands) (*The Journal of Peasant Studies*, 2014).

Anticipating the risk of co-option by powerful actors, social movements and critical scholars alike affirm that 'food sovereignty' is not, and cannot be, a piecemeal approach to change. It entails a fundamental transformation of the industrial capitalist food system by working towards autonomy and democracy (Pimbert, 2010). The proponents of food sovereignty are not working for 'inclusion' in existing political structures and the dominant culture. Instead they strive to 'transform the very political order in which they operate' (Alvarez *et al.*, 1998).

As Patel puts it, the food sovereignty movement argues 'for a mass re-politicization of food politics, through a call for people to figure out for themselves what they want the right to food to mean in their communities, bearing in mind the community's needs, climate, geography, food preferences, social mix and history' (Patel, 2007).

Contesting, re-imagining, and constructing knowledge for food sovereignty ultimately depends on putting into practice these forms of radical democracy and active citizenship in the governance of research and the production of knowledge (see Chapter 8).

Agroecology and food sovereignty: a brief history

Agroecology focuses on the ecological relations in farming systems, and it seeks to understand the dynamics, the form and the functions of these relations. Its underlying analytical framework owes much to systems theory and approaches that aim to integrate the numerous factors – environmental and social – that influence agriculture and land use (Altieri, 1987; Carrol *et al.*, 1990; Lowrance *et al.*, 1984). This agroecological knowledge can then be used as the basis for the design of more sustainable, diverse and resilient agricultures (Altieri, 1987; Vandermeer, 2010).

At the heart of agroecology is the idea that agroecosystems should mimic the biodiversity levels and functioning of natural ecosystems. Such agricultural mimics, like their natural models, can be productive, pest-resistant, nutrient-conserving and relatively resilient to stresses such as climate change. The goals of sustainability and productivity are met through agroecosystem designs that enhance functional diversity at the genetic, species, ecosystem and landscape levels. They are also met through the use of agroeco-logical methods such as genetic mixtures, crop rotations, intercropping, polycultures, mulching, terracing, the management of diverse micro-environments for nutrient con-centration and water harvesting, agro-pastoral systems and agroforestry. The design of biodiverse, energy-efficient, resource-conserving and resilient farming systems is based on mutually reinforcing agroecological principles (Box 1.3). These modern principles of agroecology have their roots in the collective knowledge, practices and ecological rationale of indigenous and peasant agriculture(s) throughout the world (Altieri, 1987).

BOX 1.3 MODERN PRINCIPLES OF AGROECOLOGY ORIGINATE IN THE KNOWLEDGE AND PRACTICES OF INDIGENOUS AND PEASANT FARMERS

- Adapting to the local environment and its diverse micro-environments
- Creating favourable soil conditions for plant growth and recycling nutri-ents, particularly by managing organic matter and by enhancing soil bio-logical activity
- Minimizing losses of energy, water, nutrients and genetic resources by enhancing the conservation and regeneration of soil, water and agro-biodiversity on the farm and neighboring landscape and watershed
- Diversifying species, crop varieties and livestock breeds in the agroecosys-tem over time and space – including integrating crops, trees and livestock at the field and wider landscape levels
- Strengthening the 'immune system' of agricultural systems through the enhancement of functional biodiversity – natural enemies of pests, allelopathy and antagonists etc., by creating appropriate habitats and through adaptive management in time and space
- Enhancing beneficial biological interactions and synergies throughout the system and among the components of agro-biodiversity, thereby promot-ing key ecological processes for sustainable production and resilience to stresses and shocks.

(Adapted from Altieri, 1995; Gliessman, 1998)

Within academic circles, the term 'agroecology' was coined in 1928 by Bensin (1928, 1930, cited in Wezel and Soldat, 2009); a number of pre-Second World War scientists such as Klages (1928) had already begun to merge the sciences of agron-omy and ecology together (Gliessman, 1990). Mexican scientists and practitioners played an important role in the development of agroecology, arguing for an eco-logical approach to food production as early as 1926 in the First Agroecological Congress in Meoqui, Mexico (Rosado-May, 2015).

Initially, agroecology strongly focused on ecological science as a basis for the design of sustainable agriculture. However, the importance of farmers' knowledge for agroecological innovation also became increasingly recognized and championed by these early pioneers of agroecology. Among Mexican scholars for example, the work of Efraim Hernández Xolocotzi between the 1940s and late 1970s is particularly noteworthy for its emphasis on intercultural processes for constructing agroecological knowledge that combines ecological science with peoples' knowledge (Hernández Xolocotzi, 1977, 1985, 1987).

However, it was the increasing awareness of the environmental impacts and pollution caused by industrial farming and Green Revolution agriculture that really encouraged closer links between agronomy and ecology in the search for a more sustainable agriculture (Herber, 1962; Merrill, 1976). For example, as part of the growing movement to resist the introduction of Green Revolution agriculture in Mexico, several 'International Courses on Tropical Ecology with an Agroecological Approach' were organized between 1979 and 1981 at the College for Tropical Agriculture in Tabasco (Mexico). These courses allowed Mexican students to interact with scientists from Mexico and the USA, and helped seed resistance to industrialized food and farming (Gliessman, 2015). In the USA, the pioneering work of Miguel Altieri (1987) and Stephen Gliessman (1990) helped put agroecology on the map in the early 1980s as a credible alternative to industrial monocultures. Around the same time, Pierre Rabhi championed agroecological approaches in France and in West Africa where he ran training courses in agricultural ecology at the CEFRA (*Centre d'études et de formation rurales appliquées*) and the Gorom Gorom Agroecology Centre in Burkina Faso, which he set up in 1985 (Rabhi, 1989; Rabhi and Caplat, 2015).

The conceptual foundations of Altieri and Gliessman's agroecology are firmly rooted in the science of ecology and agroecosystem analysis. Hernàndez Xolocotzi's understanding of intercultural agroecology embraced a broad conceptualization that included social, economic, cultural, political, ethical, ecological and technological factors (Rosado-May, 2015). And Rabhi's approach is built on ecology and is explicitly grounded in the tradition of 'anthroposophy' (Steiner, 1974) and indigenous cosmovisions, emphasizing a life-affirming ethic with a central focus on the Earth rather than only the agroecosystem. In their uniquely different ways, these pioneering agroecologists and their early followers have helped to frame the foundations of today's transdisciplinary agroecology.

More recent debates in peasant studies have further enriched our understanding of the origins of agroecology and its transdisciplinary history. For example, Sevilla Guzmán and Woodgate (2015) have traced the origins of agroecology – and its links to food sovereignty – to neo-Narodnism, heterodox Marxism and different strands of libertarian thought, including social anarchism (see also Sevilla Guzmán, 2011). Building on the seminal thinking of Chayanov (1989), van der Ploeg has also analyzed agroecological praxis as a form of resistance to capitalist modernization by agrarian social movements and peasants struggling for autonomy (van der Ploeg, 2009, 2013).

Unlike most conventional agricultural research and development (Chapter 2), agro-ecological approaches consciously seek to combine the experiential knowledge of peasant farmers and indigenous peoples with the latest insights from the science of ecology. Local knowledge and indigenous management systems are usually effective responses to place and site-specific challenges and opportunities. They are, after all, based on literally hundreds of years of collective observation, experimentation and adaptive management of diversity and dynamic complexity across time and space. The historical record shows that this vernacular science has been remarkably innovative across the world. Farmers, pastoralists, forest dwellers, fisherfolk and indigenous peoples collectively harnessed their knowledge to generate sophisticated agricultural and land-use systems in Africa, the Americas and Asia before the arrival of the Europeans. The Incas, the Mayas and the Aztecs all developed systems capable of feeding large and concentrated populations in the Americas (Gómez-Pompa and Kaus, 1992; Gliessman, 1991; Fedick and Morrison, 2004; Ford and Nigh, 2015). European explorers and travellers to Africa and Asia in the sixteenth, seventeenth and eighteenth centuries chronicled the ingenuity and sustainability of highly diverse local agricultures, and the prosperous agrarian life they allowed (Dharampal, 1983; King, 1911; Jones, 1936).

In exploring the epistemological basis of agroecological thought, Hecht (1995) has highlighted the importance of indigenous and peasant knowledge in the construction of modern agroecology. Good agroecologists value and respectfully build on peoples' knowledge and farmer-led experimentation to develop locally appropriate farming practices and agroecological solutions (Box 1.4). Agroecological solutions for sustainable food systems are not delivered top-down. They are developed through respectful intercultural dialogue between scientists and farmers/citizens, building on peoples' local priorities, knowledge and capacity to innovate. Farmer-led and people-centred agroecological research thus rejects the transfer-of-technology model of research and development (R&D) in favour of a decentralized, bottom-up and participatory process of knowledge creation tailored to unique local contexts (Méndez et al., 2016; Levidow et al., 2014; Rosset et al., 2011). Agroecology's interest in indigenous and peasant knowledge thus converges with other approaches that emphasize the importance of 'ethno science' and 'peoples' knowledge' in meeting fundamental human needs in culturally and environmentally appropriate ways (Brokensha et al., 1980; Richards, 1985; Chambers et al., 1989; Posey, 1999).

BOX 1.4 AGROECOLOGY BUILDS ON THE KNOWLEDGE OF FARMERS, INDIGENOUS PEOPLES, FISHERFOLK, PASTORALISTS AND FOREST DWELLERS

Four areas of farmer and peoples' knowledge are particularly important for agroecologists:

1. Local taxonomies – wo/men's detailed knowledge and classification of different types of soils, plants, animals and ecosystems

2. Ecological knowledge

    a. climate, winds, topography, minerals, micro-climates, plant communities and local ecology
    b. knowledge of not only structures but also of processes and dynamic relations e.g. influence of the moon and other planets on growth cycles of crops and livestock

3. Knowledge of farming practices

    a. functional biodiversity e.g. the intentional mixing of different crop and livestock species and varieties to stabilize yields, reduce the incidence of diseases and pests and enhance resilience to shocks and stresses
    b. optimal use of resources and space
    c. recycling of nutrients
    d. water conservation and management

4. Experimental knowledge that stems from:

    a. wo/men farmers' careful observations of dynamic processes over time and space
    b. active experimentation. For example, farmers' seed selection as well as their animal and plant breeding has generated myriads of locally adapted crop varieties and animal breeds. Indeed, most of the world's crop and livestock genetic diversity we still see today is an embodiment of the knowledge and creative work of previous generations of wo/men farmers across the world.

All this collective knowledge reflects the multi-use strategies of men and women farmers, indigenous peoples, pastoralists, fisherfolk and forest dwellers deriving their food and livelihoods in culturally specific ways in highly diverse contexts.

In the 1990s, 'agroecology as a scientific discipline went through a strong change, moving beyond the field or agroecosystems scales towards a larger focus on the whole food system, defined as a global network of food production, distribution and consumption' (Wezel and Soldat, 2009; Wezel et al., 2009). This led to a new and more comprehensive definition of agroecology as: 'the integrative study of the ecology of the entire food system, encompassing ecological, economic and social dimensions, or more simply the ecology of food systems' (Francis et al., 2003).

Agroecological research thus widened its focus to critically analyze the global food system and explore alternative food networks that re-localize production and consumption (e.g. short food chains and webs, local food procurement schemes…). This approach seeks to reinforce connections between producers and consumers and integrate agroecological practices with alternative market relationships within specific territories (Gliessman, 2014; CSM, 2016). Increasingly too, a transformative agroecology aims to facilitate a shift from linear food systems to circular ones that mimic natural cycles and reduce carbon and ecological footprints – ensuring that circular systems are designed to replace specialized and centralized linear supply chains with decentralized webs of food and energy systems that are integrated with water and waste management in sustainable rural and urban circular economies (Jones *et al.*, 2012; Pimbert, 2015b).

This broader perspective has also encouraged closer links with farmer organizations, consumer-citizen groups and social movements supporting alternatives to industrial food systems and Green Revolution agriculture. For many social movements and farmer organizations, agroecology became explicitly linked with food sovereignty. Most notably, ecologically sound and sustainable methods of farming and food provisioning are projected as an integral part of the vision for food sovereignty (La Vía Campesina, 2007).

Agroecology: in danger of co-option

Barely recognized within official circles only six years ago, agroecology is now more visible in policy discourses on food and farming. For example, in its third Foresight Report the European Union's Standing Committee on Agricultural Research calls for research that gives a high priority to approaches that 'integrate historical knowledge and agroecological principles' to create 'radically new farming systems' that must 'differ in significant respects from current mainstream production systems' (EU SCAR, 2012). Similarly, the report of the International Assessment of Agricultural Knowledge, Science and Technology for Development (IAASTD, 2009) argues that the vulnerabilities of the global food system can be reduced through locally based innovations and agroecological approaches. The UN Special Rapporteur on the Right to Food – in his report on *Agroecology and the Right to Food* presented at the United Nations Human Rights Council in 2011 – has also helped put agroecology on the map of the international community and policy makers (De Schutter, 2010). And both civil society groups and scientists continue to marshal the latest evidence on the multiple benefits of agroecological models of production, including reduction in carbon footprints, adaptation to climate change, reversals in the loss of biodiversity, reductions in agri-chemical pollution (chemical fertilizers and pesticides, antibiotics and growth hormones), reduced water footprints of crops and farm enterprises, multiple yields and often higher profit margins than industrial monoculture farms, creation of employment and new sources of livelihoods and fewer public health hazards for wider society.[12]

However, agroecological knowledge and practices are increasingly contested and interpreted to mean different things to different people. The term 'agroecology' is now used by different actors as part of a normative vision of the future that either seeks to conform to the dominant industrial food and farming system, or to radically transform it (Levidow *et al.*, 2014; Pimbert, 2015a). For instance, the National Institute of Research in Agriculture (INRA) in France introduced agroecology in its 2010-2020 strategic research plan (INRA, 2010). In 2012, the Minister of Agriculture declared that France aims to be 'the champion of agroecology' in

Europe. However, civil society organizations and farmer networks argue that the French government proposes a 'form of agroecology very distant from what they hope to see promoted for our agriculture' because it encourages, for example, no-till methods with herbicide sprays in the context of so called 'sustainable agricultural intensification' (Garnett *et al.*, 2013; Royal Society, 2009) and 'climate smart agriculture' (Campbell *et al.*, 2014; Pimbert, 2015). This coalition of citizens and small farmers ask that the French government promote instead an agrarian reform that strongly favours a diversified organic agriculture on a human scale. For them: 'Agroecology is synonymous with greater producer-consumer proximity, employment creation, a solidarity economy and diverse food products for citizens' (Fédération Nature & Progrès, 2012).

Subsequently, the European Coordination of La Vía Campesina stated that: 'Agroecology as understood by social movements is complementary and inseparable from the food sovereignty we want to build' (ECVC, 2013). More recently, representatives of indigenous and peasant communities from across the world reaffirmed this idea:

> Agroecology is the answer to how to transform and repair our material reality in a food system and rural world that has been devastated by industrial food production and its so-called Green and Blue Revolutions. We see Agroecology as a key form of resistance to an economic system that puts profit before life. [...] Our diverse forms of smallholder food production based on Agroecology generate local knowledge, promote social justice, nurture identity and culture, and strengthen the economic viability of rural areas. As smallholders, we defend our dignity when we choose to produce in an agroecological way.
>
> *(Nyéléni, 2015)*

Speaking at the 2015 International Forum on Agroecology in Nyéléni (Mali), Ibrahima Coulibaly went further in saying that: 'There is no food sovereignty without agroecology. And certainly, agroecology will not last without a food sovereignty policy that backs it up' (www.agroecologynow.com/video/ag/). A similar opinion was strongly expressed by a South Korean delegate from La Vía Campesina: 'Agroecology without food sovereignty is a mere technicism. And food sovereignty without agroecology is hollow discourse' (cited by Rosset and Martínez-Torres, 2014).

Throughout the world, transnational social movements are mobilizing to build, defend and strengthen agroecology as the pathway towards more just, sustainable, resilient and viable food and agricultural systems. These social movements are claiming agroecology as a bottom-up construction of knowledge and practice that needs to be supported – rather than led – by science and policy. They reject an agroecology which promotes 'input substitution' approaches that maintain dependency on suppliers of external inputs and commodity markets, and which leave untouched the ecological, economic and social vulnerabilities of genetically uniform monocultures and linear food chains (McRae *et al.*, 1990; Rosset and Altieri, 1997). Instead, these social movements favour a transformative agroecology based on a redesign and functional diversification of the agroecosystem as well as its integration with re-territorialized local and regional markets (CSM, 2016; Pimbert, 2015; Rosset and Altieri, 1997). And they clearly emphasize the indivisibility of agroecology as a science, a practice and social movement. Today's more transformative visions of agroecology for food sovereignty thus integrate peoples' knowledge and transdisciplinarity, farmers' practices and social movements – while recognizing their mutual dependence and agency (Anderson *et al.*, 2015; Méndez *et al.*, 2016; Nyéléni, 2015; Pimbert, 2007).

It is this transformative agroecology which the contributions to this volume seek to advance, rather than one that conforms to the dominant agri-food regime.

Biocultural diversity: the intimate link with agroecology and food sovereignty

'Biocultural diversity' describes the diversity of life in all its manifestations: biological, cultural and linguistic. This concept encompasses biological diversity at all levels (genetic, species, ecosystem and landscape) as well as cultural diversity in all its forms.

Languages are a major indicator of cultural diversity and linguists estimate that some 5,000 to 7,000 languages are spoken today on all five continents (Maffi, 2001). Many of the areas of the planet with the highest biological diversity are inhabited by indigenous and traditional peoples (Maffi and Woodley, 2010). Out of the nine countries which account for 60% of the world's languages, six of these centres of linguistic/cultural diversity are also mega-diversity countries harbouring remarkably high numbers of unique plant and animal species (Durning, 1993). This strong overlap between biodiversity 'hotspots' and indigenous peoples' territories is what the 1988 Declaration of Belém calls the inextricable link between biological diversity and cultural diversity.[13]

Biocultural diversity emerges from the countless ways in which people have interacted with their natural surroundings. This co-evolution of people and nature has generated local knowledge and practices which are vitally important in allowing societies to adaptively manage their farming systems, natural resources, biocultural landcapes and social life. In turn, culture, memory and identity have become embedded in the land and waters, with human agency co-creating biodiversity with nature – from crop genetic diversity to the species composition of humanized ecosystems and biocultural landscapes (Gómez-Pompa and Kaus, 1992; Posey, 1999).

Diverse worldviews and unique knowledge systems have emerged through this reciprocal interplay between biological and cultural diversity. Toledo's description of the intimate relationship between nature and culture in Mexico is relevant for other countries where cultural diversity among rural inhabitants is high:

> Each species of plant, group of animals, type of soil and landscape nearly always has a corresponding linguistic expression, a category of knowledge, a practical use, a religious meaning, a role in ritual, an individual or collective vitality. To safeguard the natural heritage of the country without safeguarding the cultures which have given it feeling is to reduce Nature to something beyond recognition; static, distant, nearly dead.
>
> *(Toledo, 1988)*

Languages embody the knowledge, intellectual heritage and frameworks for each society's unique understanding of life. This is one of the main reasons why the disappearance of languages is a major concern. It is estimated that half the world's languages will disappear within a century (UNESCO, 2010). And an even higher number of languages are losing the 'ecological contexts' that sustain them as vibrant languages (Mühlhäusler, 1996; Posey, 1999).

Biocultural diversity also offers a normative vision for the future – an antithesis to the controllable uniformity favoured by corporations and governments committed to the relentless pursuit of economic growth and a singular idea of modernity. It is therefore not surprising that there are convergences and overlaps between initiatives that seek to enhance biocultural diversity and the struggles for agroecology and food sovereignty.

For example, many indigenous peoples and local communities are on the frontline in the struggle to sustain, protect, restore and defend the commons and its biocultural diversity. Communities focus in particular on the territories and areas they live in and collectively conserve on the basis of their traditional knowledge and customary practices, law and institutions. The term Indigenous Peoples' and Community Conserved Territories and Areas (ICCAs) is used to describe these grassroots efforts to enhance biocultural diversity (Borrini-Feyerabend and ICCA Consortium, 2012). The International Union for Conservation of Nature (IUCN) defines ICCAs as 'natural and/or modified ecosystems, containing significant biodiversity values, ecological benefits and cultural values, voluntarily conserved by indigenous peoples and local communities, both sedentary and mobile, through customary laws or other effective means' (IUCN-CEESP, 2008).

In its connection to a well-defined territory, the community is the major actor in making decisions about the local adaptive management of the territory's biocultural diversity. This decentralized governance implies that local institutions have – *de facto* and/or *de jure* – the capacity to develop and enforce decisions (Borrini-Feyerabend *et al.*, 2007). Other actors may collaborate as partners, especially when the land is owned by the state. But local decisions and self-determination in the management of ICCAs are predominant.

The decentralized and distributed network of ICCAs helps to conserve critical ecosystems and threatened species as well as maintain essential ecosystem functions (e.g. carbon sequestration in soils, the renewed availability and quality of water, keystone species for pollination). The scale of these community-led efforts is significant: the global coverage of ICCAs has been estimated to be comparable to that of the official network of protected areas, which now covers 15.4% of the world's total land area (Kothari *et al.*, 2012; Juffe-Bignoli *et al.*, 2014). However, the conscious objective of community management is usually different than conservation *per se*. Unlike conventional conservation that excludes people from protected areas and restricts resource use by local populations (Colchester, 2004; Dowie, 2009; Ghimire and Pimbert, 1997a; Vidal, 2016), in ICCAs community decisions and efforts are usually geared to sustaining the material basis of livelihoods as well as safeguarding the diversity of cultural and spiritual places. Moreover, networks of ICCAs that rely on agroecological farming actually create a mosaic or matrix in which the landscape is shared by agriculture and biodiversity conservation (Perfecto and Vandermeer, 2017). This 'nature's matrix' model does not assume that agriculture is necessarily harmful to biodiversity. It is the *kind* of farming and land use that matters along with an approach that simultaneously embraces biodiversity conservation, food production and food sovereignty as interrelated and mutually supportive goals (Perfecto *et al.*, 2009). By helping to reconcile conservation with the satisfaction of fundamental human needs, ICCAs provide the basis of economic livelihoods and culture for millions of people (Kothari *et al.*, 2012; www.iccaconsortium.org). Locally available resources (food, water, fodder and energy) are valued and used to generate the means of life and income, as well as plural definitions of human wellbeing and spirituality (Pimbert and Pretty, 1998; Posey, 1999).

There are also some noteworthy international initiatives that link biocultural diversity with agroecological models of production and land use. For example, the Program on Globally Important Agricultural Heritage Systems (GIAHS) has championed agroecology and biocultural landscapes from within the UN Food and Agriculture Organization (Koohafkan and Altieri, 2012). Agricultural heritage systems are complex, highly diverse and locally adapted agricultural systems and biocultural landscapes. They have emerged over centuries of cultural and biological co-evolution and embody the accumulated experiences and traditional – but evolving – knowledge, practices and technologies of rural peoples. According to the FAO, today agricultural heritage systems cover about five million hectares and generate multiple social, cultural, ecological and economic benefits to communities throughout the world.[14]

Sustaining biocultural diversity is essential for the future of agroecology, food sovereignty and of a plural world. Advancing knowledge on the links between biological and cultural diversity can help ensure that diversity finds a central place in policy agendas for food, agriculture, environment and human wellbeing. Several chapters in this book address some of the critical gaps in knowledge identified at a workshop on biocultural diversity organized by UNESCO and The Christensen Fund (UNESCO, 2008): biocultural landscapes – including their dynamics and management (Chapters 3 and 4); the recognition and protection of knowledge systems (Chapters 4 and 6); and the values of diversity (Chapters 4, 6 and 7).

Contesting and constructing knowledge for food sovereignty, agroecology and biocultural diversity

Battlefields of knowledge and contested modernity

Food and agriculture development projects can be seen as 'social arenas' in which numerous actors with different interests, knowledge and values interact. Norman Long has emphasized the centrality of knowledge and power in contesting, defining and shaping the outcomes of these interactions between different social actors (Long and Long, 1992). Far from being innocent intellectual games, discussions and disputes over what knowledge counts – and whose knowledge should prevail – are usually intense, often brutal, sometimes violent and ultimately have huge consequences for the wellbeing of people and the land. Indeed, Long has aptly described the development and modernization process as a 'battlefield of knowledge' (Long and Long, 1992).

The dominant development paradigm envisions having fewer people farming and depending on local food systems and biodiversity-rich landscapes. It encourages an exodus of people from rural areas to work in industry and urban-based trade and services (Delgado Wise and Veltmeyer, 2016; Perez-Vitoria, 2015; Pimbert *et al.*, 2006). Many development policies are indeed based on the belief that family farmers and other small-scale producers who continue to farm, fish, harvest forests and rear livestock on common property lands should 'modernize' as quickly as possible. The policies imply they should become fully commercial producers by applying industrial food and farming technologies that allow for economies of scale (Perez-Vitoria, 2005, 2015; Desmarais, 2007). Those who cannot make this transition should move out of farming and rural areas to seek alternative livelihoods.

This modernization agenda is seen as desirable and inevitable by most governments today. The ideologies of both Marxist and capitalist nation states have similar views on the future of peasants in modern industrial society. As Walden Bello says:

> The two dominant modernist ideologies of our time give short shrift to the peasantry. In classical socialism, peasants were viewed as relics of an obsolete mode of production and designated for transformation into a rural working class producing on collective farms owned and managed by the state. In the different varieties of capitalist ideology, efficiency in agricultural production could only be brought about with the radical reduction of the numbers of peasants and the substitution of labour by machines. In both visions, the peasant had no future.
>
> *(Bello, in Desmarais, 2007)*

As a vision for universal progress, this idea of modernity also facilitates and legitimizes the global restructuring of agri-food systems in which a few transnational corporations gain monopoly control over various links in the food chain (Clapp and Fuchs, 2009; ETC, 2016), thereby extending capitalism in the web of life (Moore, 2015). An important part of this process of corporate control is what Ivan

Illich has termed 'radical monopoly': 'the substitution of an industrial product or a professional service for a useful activity in which people engage or would like to engage', leading to the deterioration of autonomous systems and modes of production (Illich, 1973). Radical monopolies replace non-marketable use-values with commodities by reshaping the social and physical environment. They do so by appropriating the components and means of life that enable people to cope on their own, thus undermining freedom and cultural diversity (Illich, 1973).

The scale and impacts of the rural transformations associated with this vision of modernity are staggering for human security and the environment. This is mainly because modernizing development primarily targets for displacement *the vast majority of small-scale producers on which much of the world's food security and environmental care depends.* According to the latest data available, there are more than 570 million farms worldwide, most of which are small and family-operated (FAO, 2014; Lowder *et al.*, 2016). Of these 74% are located in Asia, with China alone representing 35% and India 24% of all farms. Moreover, 72% of the world's farms are smaller than one hectare (ha) in size (family farms – Box 1.2); 12% are one to two ha in size (small farms); and 10% are between two and five ha.[15] Only 6% of the world's farms are larger than five ha. Small farms manage around 12% and family farms about 75% of the world's agricultural land. In their careful analysis of the number, size and distribution of the world's farms, Lowder *et al.* (2016) conclude that family farms work 75% of the world's agricultural land and are responsible for most of the world's food and agricultural production.[16] Family farmers are also responsible for the largest share of investment made in agriculture on a day-to-day basis on their farms (Lowder *et al.*, 2015) – public sector support is relatively small and outside investment from aid programmes and private enterprises is marginal (Figure 1.1).

In Chapter 5, Eric Ross analyzes the enduring influence of Thomas Malthus's *Essay on the Principle of Population* and how it has enabled the continuous process of enclosure, displacement of subsistence producers and environmental degradation that continues today. He shows how the Malthusian paradigm[17] in contemporary development thinking continues to legitimize the spread of commercial agriculture that massively displaces small and family farming as a source of food security, local innovation, as well as biocultural and ecological diversity.

Eric Ross's work offers a careful analysis of past history that illuminates the present: from the enclosures and the consolidation of capitalism in eighteenth- and nineteenth-century Britain, the Irish potato famine in the middle of the nineteenth century, the spread of colonialism and modernity outside of Europe, to current development practice. The methods of critical anthropology and comparative historical analysis used by Ross are also relevant for understanding current 'battlefields of knowledge' on the future of food and agriculture, including the vision of modernization and productivism promoted by G8 governments and corporations that are part of the New Alliance for Food Security and Nutrition in Africa (McKeon, 2014; European Parliament, 2016). Chapter 5 thus offers intellectual resources for citizens to counter the recurrence of Malthusian thinking that blames peasants for their backwardness and inability to control their sexual urges, causing

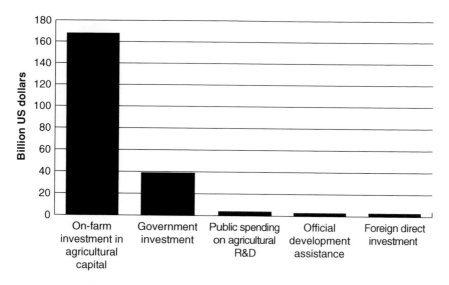

FIGURE 1.1 Who invests in agriculture?

Source: Lowder *et al.*, 2015.

them to have too many children. Ross's contribution also shows how misrepresentations of the life of peasant family farmers and pastoralists are selectively useful in furthering the interests of powerful actors in programmes designed to combat HIV/AIDS in Africa.

More generally, Malthusian thought has been closely associated with the construction of an economic science committed to a modernity based on the commodification of nature and social relations (Rist, 2011, 2013). In Chapter 7, Rist explores some of the circumstances that prevailed when neo-classical economics was invented as a science that claims to be valid for all people and places.

As enduring belief systems, both modern economics and variants of Malthusian thought continue to play a key role in the long history of blaming peasants for economic and social backwardness. Jim Handy has shown how the 'almost idiotic wretchedness' of peasants has been imagined and institutionalized in eighteenth-century Britain and in the global South during colonialism and twentieth-century development (Handy, 2009). Apart from blaming them for their propensity to have too many children, political and economic elites have also dismissed peasants for not being sufficiently enamoured with consumption – they stifle economic development because their needs are far too easily met. Peasants everywhere tend also to be viewed as inefficient because they do not use land and capital effectively – they are seen as lazy and should be compelled to work harder. Peasants need to be swept off the land because they delay economic growth and the necessary process of capital intensification of the land. And last but not least, peasants and other small-scale food producers are dangerous and unruly – they need to be incorporated into nation states as lawful and responsible citizens (Berger, 1978; Handy, 2009; Thompson, 1991).

In the current 'battlefields of knowledge' for conservation and development, these 'faulty imaginings' (Handy, 2009) of peasants serve to promote social, economic and environmental policies that are designed to expel them from the land and turn them into wage labourers and migrant workers (Delgado Wise and Veltmeyer, 2016; Ghimire and Pimbert, 1997a; Handy and Fehr, 2010). Launched in 1843, *The Economist* magazine was an influential voice in providing the intellectual rationale for what has become an enduring vision of modernity based on free trade in agriculture, capital-intensive 'scientific' agriculture and the accumulation of land in private property – land not held in private property was 'inconsistent with a civilised state' (*The Economist*, 1851, cited in Handy, 2009).

However, the idea that peasant farmers and indigenous peoples as a group are bound to disappear reflects just one vision of the future – it is a political choice that relies on specific theories of change that are increasingly disputed and rejected by social movements. For example, in response to a development model geared to ensuring the extinction of family farmers and other small-scale food providers, La Vía Campesina is redefining what it means to be a 'peasant'.[18] A process of 're-peasantization' is slowly unfolding as more people proudly embrace the term 'peasant' to describe themselves, projecting an alternative identity rich in meaning and hope for the future (Desmarais, 2007; Perez-Vitoria, 2015). People's knowledge is being mobilized from below (Chapter 8) to challenge the inevitability of progress and the end of the peasantry.

Central to this process of 're-peasantization' is the constant struggle for autonomy (Pimbert, 2008; van der Ploeg, 2009).[19] Embraced by a growing number of youth who seek to regenerate autonomous 'life worlds' (Habermas, 1984) in rural and urban spaces, the construction of this alternative modernity also looks to other definitions of 'the good life' – including *Buen Vivir* or *Sumak Kausai* in Latin America, De-growth in Europe and *Ecological Swaraj* in India (Chapter 7; Latouche, 2011; Kothari *et al.*, 2014).

Institutionalized bias in research: lessons from the uneven development of agroecology and genetic engineering

The development of ideas in the sciences and humanities has had its own complex intellectual history and sociology, in which certain theories were able to rise and prevail. Following the sociologist Karin Knorr Cetina, it is essential to ask how do 'the machineries of knowledge construction' arise and selectively favour the production of certain types of knowledge whilst actively excluding others? How do particular framings[20] of the problem, research technologies, social configuration of scientists' funding contexts, as well as laboratory and research settings combine to form 'epistemic cultures', which Knorr Cetina describes as: 'amalgams of arrangements and mechanisms – bonded through affinity, necessity, and historical coincidence – which, in a given field, make up how we know what we know' (Knorr Cetina, 1999).

As a story of the 'history of the present' (cf. Foucault, 1991), the example of agroecology is particularly instructive in this regard. Despite the growing recognition of the many benefits of agroecology, it remains largely unsupported and marginalized in public research and development (R&D):

- According to a recent analysis done by the Union of Concerned Scientists and partners, just 15% of funding granted in 2014 by the US Department of Agriculture for research and education incorporated any element of agroecology (Union of Concerned Scientists, 2015). The agency allocated even less funding to projects emphasizing agroecological research or implementation. Systems-based projects that included both agroecological farming practices and connections between producers and consumers to support a socio-ecological transformation of the food system were particularly poorly funded (4%), as were agroecology R&D projects that included complex rotations (3%), spatially diversified farms (3%), integrated crop-livestock systems (1%), rotational or regenerative grazing (1%), or agroforestry (<1%) (DeLonge *et al.*, 2016; Union of Concerned Scientists, 2015).
- Funding for agroecological research in the UK represents a tiny 1.5% of the total UK budget for agricultural R&D. The percentage of funds for the development of agroecological solutions is even lower in the UK's official aid programme for Africa, Asia and Latin America. Agroecological research projects receive less than 0.1% of the UK's Department for International Development's budget for official aid on food and farming (Moeller and Pimbert, unpublished data based on publicly available DFID reports from 1995 to 2016).

It is noteworthy that both genetic engineering and agroecology were insignificant or non-existent scientific branches before the early 1970s. They were both seen as two fields of research with the potential to improve food and agricultural systems. However, from the 1930s onwards influential actors such as The Rockefeller Foundation selectively favoured research in molecular biology and a strategic re-organization of the life sciences (Kay, 1997) that made it possible for genetic engineering to harness 'life as a productive force' (Yoxen, 1981). Since the 1960s, the science of genetic engineering has received by far the largest share of research funding and government support, and continues to do so throughout the world.[21] After the first product of genetic engineering became commercially available in 1995, powerful corporate actors, such as Monsanto[22] and Syngenta, have worked to promote and legitimize the rapid – albeit contested – proliferation of the science and products of genetic engineering. A recent study of the legitimation strategies of Monsanto shows that for nearly two decades, this powerful corporation has systematically used discursive resources that have concealed details about actors and actions, including in scientific research and risk assessments (Lamphere and East, 2016). Monsanto's own documents published over an 18-year period have fundamentally reshaped narratives to promote the company, its products and genetic engineering as the solution to hunger and sustainability problems (Lamphere and East, 2016; see also The Monsanto Tribunal, 2016).

But the failure of research to produce knowledge for agroecological solutions cannot be fully explained by funding decisions and corporate influence alone – important as they are. In Chapter 2, Gaëtan Vanloqueren and Philippe Baret describe the institutionalized bias and some of the systemic reasons that help explain why research favours scientific and technological regimes that develop genetic engineering and 'lock out' agroecological innovations. The authors identify the determinants of innovation and factors that influence choices within agricultural research systems. They show how the interactions between funding priorities, public-private sector partnerships, the assumptions and cognitive routines of scientists and other key determinants of innovation, construct a technological regime that favours genetic engineering and hinders the development of more holistic agroecological approaches to farming and land use. In turn, this systemic bias in research acts in combination with wider political structures and agricultural markets to 'lock out' agroecology from society and keep industrial agriculture in place (see IPES-Food, 2016).

In describing the politics of knowledge involved, Vanloqueran and Baret offer a comprehensive analytical framework that helps better understand why, when and how agroecological knowledge can be constructed. The authors also provide valuable insights into the constraints that need to be addressed to develop agroecological knowledge and innovations.

More generally, the methodology used in Vanloqueran and Baret's analysis of 'lock-in' and institutionalized bias in research is highly relevant for social movements and scholars who seek to contest, construct and transform other areas of knowledge for food sovereignty and biocultural diversity.

Disabling knowledge and bureaucracies undermine indigenous and peasant tenure rights

Secure tenure rights and equitable access to land, fisheries and forests are important means and pre-requisites for eradicating hunger and poverty. This is widely recognized by *The Voluntary Guidelines on the Responsible Governance of Tenure of Land, Fisheries and Forests in the Context of National Food Security*, which were officially endorsed by the Committee on World Food Security on 11 May 2012 (FAO, 2012). Since then, implementation of these guidelines by national governments has been encouraged by the G20, Rio+ 20 and the United Nations General Assembly.

However, in practice the rights of access, use and tenure of indigenous peoples, pastoralists, family farmers and other small-scale producers continue to be largely ignored and overruled by the imperatives of modernization and economic development. In several ongoing projects designed to modernize the countryside and 'scientifically' manage the land, the rights of access and use of small-scale producers and custodians of biocultural landscapes are actively suppressed as part of their forced displacement (see for example http://ejatlas.org). In many other cases, local communities receive minimal compensation for their loss of tenure rights as part of schemes designed to 'encourage' them to migrate to cities for alternative means of livelihood. According to the UN Council for Human Rights, women are the most harmed by these developments.[23] Moreover, these processes of organized exclusion also undermine the livelihoods of many more rural people because each link in the food chain offers economic niches for millers, butchers, carpenters, tool makers and mechanics, bakers, local milk processors, fishmongers, small shopkeepers and collectors of food losses and waste. With the demise of the local economies, ecologies and food systems that provide livelihoods, employment and socio-cultural meaning, this potentially large share of the active working population also has no choice but to migrate to urban areas.[24]

The *Global Atlas of Environmental Justice* documents some of the socio-environmental conflicts responsible for the forced displacement and outward migration of peasant farmers and other rural people living in forests, farmland, pastures, rangelands, mountains, wetlands, common lands and peri-urban areas (Temper *et al.*, 2015; http://ejatlas.org). This online interactive map also tracks the 'spaces of resistance' which local communities define as 'mobilizations against particular economic activities in which environmental impacts are a key element of their grievances' (Temper *et al.*, 2015). Powerful actors who benefit from these modernizing schemes all rely on knowledge that both drives and legitimizes what Harvey (2004) has called 'accumulation by dispossession'. These might range from land and water grabs for the expansion of industrial agriculture and agrofuel plantations (Borras *et al.*, 2011; Mehta *et al.*, 2012; van der Ploeg *et al.*, 2015); to conservation-induced displacements of indigenous and local communities living in ICCAs and protected areas (Corpuz, 2016; Brockington and Igoe, 2006; Ghimire and Pimbert, 1997b; Vidal, 2016); to foreign investments in carbon markets and REDD+[25] schemes (Lohmann, 2011; Moreno *et al.*, 2015) as payments for ecosystem services (Sullivan, 2011); and to 'green grabbing' (Fairhead *et al.*, 2012) and new forms of commodification of nature (Büscher *et al.*, 2014; Moore, 2016).

In the current crisis of capitalism (Harvey, 2015) and its need to secure access to ever more land, water, energy and mineral resources for its continued expansion, peasant family farmers and other small-scale producers are thus increasingly harmed by the knowledge embedded in the discourse, law, coercion and violence used to deny them of their tenure rights over land and natural resources. In this context, misleading, simplistic and ahistorical perspectives (see Chapters 4, 5 and 7) and policy narratives are widely used by external actors and science-based bureaucracies to blame people for environmental degradation and to justify the imposition of standard environmental management packages which damage their livelihoods (Leach and Mearns, 1996). For example, public policy and interventions designed to modernize nomadic pastoralism and rangeland management in Africa and Asia are based on dominant views of equilibrium ecology that stress the damaging potential of livestock grazing, the threats of degradation and desertification and the need to control livestock numbers and grazing movements (Chapter 4; Scoones, 1994). The loss of forests, soil erosion, desertification, climate change, the mismanagement of water and the destruction of wildlife and fisheries are all problems that are seen to require intervention to prevent further deterioration (Corpuz, 2016; Homewood, 2008; Leach and Mearns, 1996; Molle, 2008; Pimbert and Gujja, 1997; Pimbert and Pretty, 1995). Local misuse of resources and the environment is consistently defined as the principal cause of destruction. All too often, 'by depicting resource users (the local ones) as wild, destructive (or illiterate, uneducated, backward or non-innovative), state resource management agencies think they can justify their use of militaristic environmental protection' (Peluso, 1993).

Much of this 'dominant knowledge' rests on a series of myths that manifest themselves through the neglect of local people – their knowledge, priorities, management systems, local institutions and social organization. It also rests on highly questionable scientific concepts, unproven assumptions and outdated knowledge. For instance, notions of climax ecology together with Malthusian views have sustained a flawed theory that relic forest on the northern margins of Guinea's forest zone in West Africa is degraded and degrading. Vegetation forms that ecologists and policy makers have used to indicate forest loss, such as forest patches in savannah, are – according to historical evidence and the knowledge of local resource users – the result of landscape enrichment by people (Fairhead and Leach, 1996).

In Chapter 4, Sian Sullivan and Katherine Homewood focus on knowledge about pastoralism in African drylands, exploring how problematic the notions of 'non-equilibrium ecology' and 'nomadism' are for scholars and policy makers. They ask why equilibrium concepts have been so strongly naturalized within science and policy communities, to the detriment of the possibilities for self-determination by the peoples who live in these environments. And why are non-equilibrium framings of dryland dynamics apparently so threatening to states and experts? Sullivan and Homewood's remarkable chapter shows how the un-muting of other peoples' knowledge and realities can help transform the nature of knowledge (its concepts, categories and embodied values) to support the goals of self-determination, food sovereignty and biocultural diversity.

However, the policy (or crisis) narratives enabling today's new cycles of capital accumulation and dispossession are robust, hard to challenge and slow to change (Roe, 1991). They play a key role in policy and project-level decision making. They help frame research, structure options, define relevant data and exclude other views within bureaucracies and professional circles. Orthodox views on economic progress and dominant environmental crisis narratives endure across time and space, despite the concerted challenges made to basic concepts and practices (Chapters 4, 5 and 7; Hoben, 1995; Leach and Mearns, 1996; Rist, 2013; Roe, 1991).

Natural resource management bureaucracies in particular do not easily reassess, update or transform the scientific and environmental knowledge which frames and legitimizes their operational procedures and policies for the governance and management of forests, water and land (Chapters 3 and 4). Knowledge about people-environment interactions is a central element of their organizational culture (the combination of the individual beliefs, opinions, shared knowledge, norms and values of the members of an organization). A majority of theorists argue that organizational culture is the most fundamental level at which transformation needs to take place (Goetz, 1997; Michael, 1995; Reed, 2006). No matter how radically processes, structures and systems are reformed within natural resource management bureaucracies, the changes will remain largely superficial and ultimately without effect if environmental knowledge and other aspects of organizational culture are left untouched (Long and Long, 1992; Westley, 1995).

These crisis narratives and disabling organizational cultures are thus major constraints for the implementation of the FAO's guidelines (FAO, 2012) and the proposals contained in the draft *United Nations Declaration on the Rights of Peasants and Other People Working in Rural Areas* (Golay, 2015; La Vía Campesina, 2008b) as well as the *United Nations Declaration on the Rights of Indigenous Peoples* (UNDRIP, 2007).

Transforming the culture and practices of natural resource management bureaucracies and research organizations must therefore go hand in hand with the construction of knowledge for agroecology, biocultural diversity and food sovereignty (Chapter 8).

Experts and ignorance-based management of ecosystems and natural resources

Appropriate transdisciplinary knowledge that embraces the inherent uncertainty, dynamism and complexity of ecosystems is key for designing pathways to sustainability and building resilience to change. Such knowledge can help understand how complex systems adapt and recover from large-scale instabilities such as climate change and global market volatility, as well as from more localized disturbances like floods, pest outbreaks, fires and tornadoes. Theories that explain non-linear dynamics and surprises in the behaviour of complex ecosystems are especially important as bases for adaptive approaches to the management of landscapes and natural resources (Gunderson *et al.*, 2012).

Competing knowledge and theories of ecosystem dynamics – from equilibrium-centered views to co-evolutionary perspectives – generate sharply different management regimes. In Chapter 3, Kristen Blann and Stephen Light maintain that the persistence of the equilibrium-centered view in most science-based natural resource management agencies leads to short-term policies and practices that jeopardize the long-term sustainability and resilience of ecosystems and landscapes. They argue that it is essential that the adaptive management and governance of landscapes and natural resources be underpinned by a co-evolutionary perspective. This views human and natural systems as complex entities that are continually adapting and co-evolving through cycles of change. This co-evolutionary perspective on ecosystem dynamics can guide the transition to sustainability and socio-ecological resilience by emphasizing the need for adaptive social learning and a move away from maximum sustained yield targets.[26] Adaptive ecosystem management is thus key for sustaining the Earth's continued capacity to support contemporary human societies, especially at a time when planetary limits are being exceeded in terms of nitrogen loads, biodiversity loss, global temperature increases and other critical indicators (Smith *et al.*, 2015; Steffen *et al.*, 2015; World Meteorological Organization, 2017).

With the increasing frequency and severity of natural and human-made disasters across the world (Brauch *et al.*, 2011), it has become more urgent to understand the features and qualities of a system that must be maintained or enhanced in order to achieve sustainability and resilience (Walker and Salt, 2012). Current knowledge emphasizes the possibility of building resilience through community self-organization and agency: paying particular attention to people-place connections, values and beliefs, knowledge and social learning and economic diversification (Berkes and Ross, 2013; Gómez-Baggethun *et al.*, 2012). Community-owned solutions through community peer-to-peer exchange are identified as essential for long-term success and sustainability (Holt-Giménez, 2002; Rotarangi and Stephenson, 2014; Tschirhart *et al.*, 2016). This body of knowledge also stresses the importance of subsidiarity, the sharing of power in collaborative management and the vital contributions of polycentric and horizontal networks of local organizations in the local adaptive management and governance of ecosystems and natural resources (Borrini-Feyerabend and ICCA Consortium, 2012; Ostrom, 1990, 2010; Pimbert, 2008; CBD, 2004). Regenerating the material basis of food sovereignty, agroecology and biocultural diversity depends on communities using this transdisciplinary knowledge in the local adaptive management of landscapes (farmlands, forests, rangelands, wetlands, commons etc.) and their ecosystem functions (water purification, pollination, recycling of nutrients, carbon sinks etc.).

For example, family farmers, pastoralists and indigenous agricultural and forest-based communities manage biodiversity at various scales (FAO, 1999; Netting, 1993; Oldfield and Alcorn, 1987; Zimmerer, 2010). Their local organizations co-ordinate the knowledge and collective action that helps create dynamic and heterogeneous landscape mosaics of fields, gardens, orchards, pastures, woodlands, agroforestry and ecosystem patches. Although this agricultural biodiversity and the local knowledge associated with its management are essential for landscape resilience to climate change, their roles and those of local organizations are largely neglected or overlooked by researchers and policy makers around the world (Global Alliance for the Future of Food, 2016; Mijatovicć *et al.*, 2013). Although complexity and resilience thinking are beginning to be widely embraced by academics and professionals who see them as necessary for tackling today's pressing social-ecological challenges (Rogers *et al.*, 2013), in practice there is still little official recognition of the importance of local organizations and their roles in the adaptive management of ecosystems and resources (Bigg and Satterthwaite, 2005).

More generally, outdated equilibrium-centered views of ecosystem dynamics persist and continue to justify short-term management policies that undermine long-term sustainability. At least five mutually reinforcing factors keep many national governments locked into this natural resource management regime: reductionist science; top-down command-and-control approaches; a mismatch between reality and the economic assumptions of policy decisions; market fundamentalism;[27] and engineering-dominated capital-intensive solutions (Chapter 3). Furthermore, non-equilibrium thinking may well generate a subliminal resistance because it radically undermines the superior positioning of 'experts' by emphasizing a fundamental 'unknowability' in predicting the behaviour of complex systems. As Sullivan and Homewood point out for pastoral landscapes, non-equilibrium thinking also creates huge problems for conservationists and land-use planners who wish to clear (purify) landscapes of people and livestock so as to return these environments to a desired (imagined) pristine state (Chapter 4).

Worldwide, the largely sectoral, expert-centered ecosystem-management institutions need to be fundamentally transformed (Biggs *et al.*, 2010; Borrini-Feyerabend *et al.*, 2007; Finger-Stich and Finger, 2002; Pimbert, 2004). To address this epistemological and political challenge, Blann and Light (Chapter 3) reflect on how a science for ecosystem sustainability and socio-ecological resilience could be co-produced by different knowledge holders (scientists, family farmers, pastoralists and other small-scale producers). They emphasize in particular the importance of pedagogical, organizational and policy changes to construct knowledge for eco-logical sustainability and resilience, and embrace 'ignorance-based' management in the face of uncertainty and unpredictability.

Along with Pimbert (Chapter 8), Blann and Light argue that the construction of knowledge(s) for agroecology, biocultural diversity and food sovereignty partly depends on fundamental changes in the 'ways of knowing' and 'ways of working' of research institutions and natural resource management bureaucracies. Reversals in normal professional practice and inclusive citizen participation in decision making are key for transformation.

Epistemic injustice undermines peoples' knowledge and agency

The idea of building on the diversity of peoples' knowledge is at the heart of the practice of food sovereignty and agroecology (Boxes 1.1 and 1.4). Moreover, the inextricable link between biological and cultural diversity is mediated by the knowledge of indigenous and local communities. By valuing and working with peoples' knowledge, advocates of agroecology, biocultural diversity and food sovereignty seek to reverse what Boaventura de Souza Santos describes as 'cognitive injustice' and 'epistemicide' – the failure to recognize the fundamental right of different knowledges and ways of knowing to exist and give meaning to peoples' lives (Santos, 2014).

In Chapter 6, Nina Moeller's ethnographic study of access and benefit sharing (ABS) negotiations by a bioprospecting project in the Ecuadorian Amazon highlights the challenges of realizing epistemic justice in practice. Moeller's account describes the encounter between the Kichwa-speaking Napo Runa peoples and ProBenefit, a large bioprospecting project funded by the government of Germany. At one level, Moeller's careful study of this microcosm highlights some of the attitudes, beliefs and behaviours that continue to marginalize and suppress both peoples' knowledge and their ways of knowing. The Kichwa's account of the subtle – and sometimes overt – prejudice, racism and paternalism of 'scientific' professionals largely echoes the contemporary experience of many wo/men indigenous peoples, pastoralists, artisanal fishers and peasant family farmers. In this regard, descriptions and critiques of development biases against local and gendered knowledge (e.g. Chambers, 1983; Harcourt, 2016; Kabeer, 1994; Richards, 1985; Smith, 1999) continue to be relevant today.

At another level, Moeller's description of Kichwa ways of knowing highlights what are often incommensurable values between scientific and indigenous knowledge systems. For the Kichwa people, knowledge is constructed through 'plants that speak' and the *experience* of their intimate relationship with the forest, which acts like a teacher and helper. Their knowledge thus depends on the continued existence of the plants and the forests that sustain Kichwa livelihoods, culture and spirituality. Similarly, the Quechua communities in the Potato Park in the province of Cusco (Peru), have described the indivisibility of this biocultural heritage as the 'knowledge, innovations and practices of indigenous and local communities that are collectively held and are inextricably linked to traditional resources and territories, local economies, the diversity of genes, species and ecosystems, cultural and spiritual values, and customary laws shaped within the socio-ecological context of communities' (ANDES and IIED, 2005).

These insights into the indivisible nature of indigenous and peasant knowledge systems have major implications for the protection of traditional knowledge and ABS agreements[28] encouraged by the Convention on Biological Diversity,[29] the World Intellectual Property Organisation (WIPO)[30] and the International Treaty on Plant Genetic Resources for Food and Agriculture.[31]

For example, Moeller (Chapter 6) argues that no matter how fair and equitable, bioprospecting and ABS schemes directly contribute to the destruction of the very foundations of traditional knowledge because the value of the latter is calculated solely in market terms and human motivation is reduced to an economic cost-benefit analysis. In situations where the conditions of people's autonomous subsistence are being destroyed, bioprospecting and ABS agreements reduce the protection of traditional knowledge to a hegemonic economic construct compatible with the needs of capital accumulation and expansion of the global market economy (see also Chapters 4 and 7). Assigning a market value to traditional knowledge ultimately facilitates ABS regimes that are extractive, unfair, patent friendly and easily captured by corporations involved in seeds, pharmaceuticals, new natural product development and the life industry (Baumann *et al.*, 1996; ETC, 2011; GRAIN, 2012).

As nature becomes increasingly linked to a tradeable and financialized world, valuable indigenous and peasant knowledge is thus drawn into financial circulation in hitherto unprecedented ways. In this global process of financialization, ontologies and categories of ecology along with the episteme of peoples' knowledge are being replaced by those of 'natural capital' and 'ecosystem goods and services' (Sullivan, 2011). For example, the web portal 'Ecosystem Marketplace' offers information updates and investment and price trend data on carbon, water and biodiversity markets. The website states 'We believe that … markets for ecosystem services will one day become a fundamental part of our economic system, helping give value to environmental services that, for too long, have been taken for granted'. The aim of the information portal is to 'spur the development of new markets' and 'facilitate transactions' (Ecosystem Marketplace, 2016). Ideas and understandings of nature are thus being reconfigured through the production of hegemonic knowledge and this unfolding global process of financialization. These ideas are then institutionalized by powerful organizations like the World Bank, international conservation NGOs and the Convention on Biodiversity (Brockington and Duffy, 2010; Corson and MacDonald, 2012; Moore, 2016; Sullivan, 2011, 2013; IUCN and UNEP, 2012). For example, it is unlikely that 'payments for ecosystem services' (PES)[32] would exist today without the particular framing of global environmental problems by the Millennium Ecosystem Assessment[33] in the early twenty-first century (Kosoy and Corbera, 2010; Fairhead *et al.*, 2012). Conservation and development projects based on PES and ABS schemes are problematic precisely because they introduce market valuations on traditional knowledge protection into areas of life which had previously been oriented by different values. The commodification of ecosystem services denies the multiplicity of culturally specific ways of knowing and valuing nature since it requires that a single exchange-value is adopted for trading (Kosoy and Corbera, 2010).

More generally, policies that aim to protect *only* the intellectual component of knowledge systems are inappropriate for most indigenous and local communities. They fail to take into account more holistic worldviews and practices in which knowledge is closely dependent on and intimately linked with the biological, cultural and landscape components of peoples' knowledge systems (Posey, 1996; UNDRIP, 2007). Traditional and indigenous knowledge systems need to be protected and strengthened as a whole, including all interlinked elements involved in inter-generational transmission (e.g. languages, customary norms and practices, spirituality) as well as traditional territories and resources. Mutually constitutive and ongoing interactions among language, culture, local institutions, landscapes, natural resources and territories are vital for protecting and sustaining traditional knowledge. The holistic nature of indigenous knowledge systems thus calls for approaches that protect bundles of rights over tangible and intangible attributes that are of spiritual, aesthetic, cultural, ecological and economic value to indigenous and local communities (Posey, 1996; Posey and Dutfield, 1996). These alternative approaches stress the importance of engaging in value practices that construct the 'outside' of capitalism as a counter-hegemonic form of traditional knowledge protection which truly safeguards the conditions in which traditional knowledge can flourish (Moeller, in Chapter 6).

Holistic approaches to the protection of indigenous and peasant knowledge are thus essential for the realization of 'cognitive' or 'epistemic justice' (Santos, 2014; Fricker, 2007). Cognitive justice emphasizes the right for different forms of knowledge and their associated practices, livelihoods and socio-ecological contexts to coexist. As Visvanathan argues, cognitive justice is 'the constitutional right of different systems of knowledge to exist as part of a dialogue and debate' (Visvanathan, 2005). This implies the continued existence of 'the ecologies that would let these forms of knowledge survive and thrive not in a preservationist sense but as active practices' (Visvanathan, 2005). But it is not just the continued existence of the ecology of these places that matters. The roles which they play in people's lives and the meaningful relationships which people maintain with these place-based ecologies are more important in this context.

Articulating and claiming this right to cognitive justice *by, for* and *with* hitherto excluded actors is a key challenge for the proponents of food sovereignty, agroecology and biocultural diversity. Two questions stand out here. The first — explored by Rist in Chapter 7 — asks to what extent does the realization of cognitive justice depend on regenerating modern subsistence economies that are not exclusively characterized by market rationalities and values, and largely consist of self-provisioning practices through which the fundamental needs of people are satisfied?[34] The second is explored by Pimbert in Chapter 8, and asks how can ideas of cognitive justice help transform the production of knowledge and research on food, agriculture and human wellbeing?

Decolonizing economics

Social movements for agroecology, biocultural diversity and food sovereignty do not reject trade and economic exchanges *per se*. However, they are highly critical of current policies for growth because they are responsible for the economic genocide of unprecedented numbers of family farmers and rural livelihoods throughout the world (Chapter 5; Bello and Baviera, 2011; Perez-Vitoria, 2005). Women in particular have been more harmed than men by the deepening inequalities, insecure employment and social unrest that have marked the last four decades of neo-liberalism. At the same time, the degradation of living conditions in poorer households nearly everywhere has translated into an increase in levels of violence, particularly domestic and sexual violence, of which women are the main victims (UN Women, 2015).

A 'decolonization of the mind' is needed to construct another economics capable of working for the wellbeing of people and the planet:

> What is needed is a new creation of the imagination that is of unprecedented importance [...] a creation which would put at the centre of human life other meanings than the mere expansion of production and consumption, one which would offer goals in life that are recognized by other human beings as being worthwhile. [...] This is not only necessary to avoid the final destruction of the planet's environment, but it is also and especially needed to rescue fellow human beings from psychological and moral misery.
>
> *(Castoriadis, 1996)*[35]

In Chapter 7, Gilbert Rist draws on a wealth of historical and anthropological evidence to show how mainstream economic science rests on quasi-religious beliefs and assumptions that are deeply committed to commodifying social relations and nature. Rist demonstrates that 'economic science' is highly ethnocentric and reflects parochial views that prevailed in Europe between the late eighteenth and mid-nineteenth century. Mainstream neo-classical economics rests on a set of half-truths and presuppositions that are shown to be either obsolete or just plain wrong — including the idea that all individuals are self-interested and rational calculators with unlimited wants. This construction of a universal *Homo economicus* makes it possible for economic policies to sanctify the market as the device for regulating human interaction (Marglin, 2010; Latouche, 2003). The superiority of 'economic efficiency', the 'commodity economy' and 'financial markets' are celebrated and imposed in discourse, policy and practice to the detriment of the 'care economy', where women traditionally have a predominant responsibility (Carrasco, 1999; Guerin, 2003; Mies and Bennholdt Thomsen, 1999; Praetorius, 2015).

Rist undermines the faith of economics at the deepest level. By inviting us to think outside the box, he awakens our social imagination to the possibility of creating forms of economics that can nurture diverse definitions of wellbeing and begin to heal the 'metabolic rift' (cf. Marx, 1981; Wittman, 2009) between society and nature (Chapter 7; Rist, 2011, 2013). His critical analysis of the history of economic anthropology offers intellectual resources for exiting the dismal 'science of economics' that props up capitalism and modernizing development (Hill, 1986). For example, Rist reminds us that throughout history, economic exchange has been based on radically different principles such as reciprocity, solidarity and gift relations that have all helped to embed economics in society (Mauss, 1966; Polanyi, 1957). In turn, these highly varied forms of economic organization have supported cultural diversity and plural definitions of human wellbeing (Latouche, 1998; Polanyi, 1968; Rahnema, 2003; Rahnema and Bawtree, 1997).

Chapter 7 thus addresses a central question for the future of food security and human wellbeing: what kind of economic knowledge(s) can be constructed for new models of economic exchange that directly support the spread of food sovereignty, the uptake of agroecology and increases in biocultural diversity? A key challenge for activist scholars and social movements is to re-invent 'economics' by building on Rist's radical critique and other traditions of knowledge, including solidarity economics (Utting, 2015), the economics of de-growth (Latouche, 2003; D'Alisa *et al.*, 2014), feminist economics (Waring, 1988; Gibson-Graham, 2016), Gandhian economics (Kumarappa, 1951), participatory economics (Hahnel, 2005, 2016) and anarchist economics (Shannon *et al.*, 2012).

Today this fundamental re-thinking of economics is urgent. Throughout the industrial food system and its related sectors (energy, manufacturing, infrastructure etc.), there is a direct relationship between the increases in productivity made possible through the use of automated technology, bio-science innovations, re-engineering and downsizing on the one hand and the permanent exclusion of growing numbers of people from employment and decent livelihoods on the other. The ecological crisis and this erosion of the link between job creation and wealth creation call for alternative economic paradigms that support for example:

1. local autonomous spaces and opportunities for the generation of use values rather than exchange values (Illich, 1973; Granstedt, 2012);
2. a more equitable distribution of productivity gains through a significant reduction in working hours and a fair sharing of work and free time between men and women (Gollain, 2004; Méda, 1998);

3. the re-localization of plural economies that combine both market-oriented activities with non-monetary forms of economic exchange based on barter, reciprocity, gift relations and solidarity (Laville, 2013; Merlant *et al.*, 2003; Passet, 2012);

4. the rethinking of money and the use of alternative local currencies, time banks, barter and cooperative exchange to turn scarcity into sustainable forms of abundance in re-territorialised economies (Lietaer and Hallsmith, 2011; Lietaer and Dunne, 2013);

5. a guaranteed and unconditional minimum income for all (Murray and Pateman, 2012; Mylondo, 2010);

6. diverse forms of tenure based on co-operative, communal and collective rights of access, use and control over land, water, forests, seeds and knowledge – with clear limits on private property rights that all too often enable the enclosure of the commons and public goods by powerful actors (Almeida *et al.*, 2015; Cabannes, 2014; Dardot and Laval, 2014);

7. a shift from globalized, centralized and increasingly corporate-controlled linear food systems to decentralized and locally controlled circular economy systems that link food and energy production with water and waste management in both rural and urban areas (Bookchin, 1986; Jones *et al.*, 2012; Pimbert, 2012; Webster, 2015);

8. a reduction in carbon and ecological footprints to maintain a good quality of life in rural and urban areas through a controlled process of de-growth in consumption and production based on the '8 Rs': Re-evaluate, Re-conceptualize, Restructure, Redistribute, Re-localize, Reduce, Reuse, Recycle (D'Alisa *et al.*, 2014; Latouche, 2009, 2011; Sinai, 2013);

9. an equitable process of contraction and convergence in carbon footprints and resource consumption (energy, minerals…) which recognizes that whilst fundamental human needs are universal, their satisfiers vary according to culture, region and historical conditions (Max-Neef *et al.*, 1989).

Democratic deficits in the production of knowledge

Whilst much scientific, technological and policy research is done in the name of small-scale producers – and claimed to be for their ultimate benefit – there is usually no meaningful participation of small-scale producers in deciding key research questions with scientists, in co-producing and validating knowledge and in risk and sustainability assessments of research and innovations. And women everywhere are the most excluded from the governance of research – despite the fact that they comprise 43% of the world's agricultural labour force (and up to 70% in some countries) and are vitally important custodians of knowledge on food, farming and land stewardship (FAO, 2016).

Given this democratic deficit, it is not surprising that today's politics of knowledge do not support the construction of alternatives to industrial food and agriculture. Dominant machineries of knowledge production ensure that agri-food systems continue to be geared towards maximizing productivity and yield through processes of homogenization, industrialization, inclusion in global markets, vertical integration and de-territorialization (IPES-Food, 2016). Institutional lock-in situations severely hinder or stop the development of knowledge for transformation – not just in agroecology (Chapter 2), but also in other fields of knowledge important for the expansion of food sovereignty and biocultural diversity. Overall, the current relations of forces between food sovereignty movements and the power of capital are tipped against food providers and food consumers in favour of corporate interests, often with substantial support from both the neo-liberal capitalist and socialist state.

Powerful corporate actors in the private sector increasingly control the directions and outcomes of food and agricultural research in the natural and social sciences. Globally, the private sector accounts for at least 45% of the world's total spending on food and agricultural R&D (Fuglie *et al.*, 2016). In OECD countries the private sector is by far the largest funder of agricultural research, accounting for over 50% of research spending (Alston *et al.*, 1998). For example, in the UK private sector funding amounts to 61% of the total spend on research in agricultural science and technology (HM Government, 2016). Moreover, large corporations have been able to disproportionately influence the directions and outcomes of agricultural research in the public sector too. Substantial funds from philanthropic capitalist foundations such as The Gates Foundation,[36] private-public sector partnerships, patents and other intellectual property rights all ensure that agricultural research selectively favours the production of knowledge and innovations that reflect and reinforce the interests of agri-food corporations and their shareholders (e.g. hybrid seeds, proprietary technologies, neo-liberal food and agricultural policies etc.).

Direct corporate interference in science is also deep and widespread. Extensive monitoring by the Union of Concerned Scientists (UCS) shows how corporations in the USA seek to influence every step of the scientific and policy-making process in order to shape decisions to suit their interests, and avoid regulatory oversight. According to a major study by the UCS, corporations rely on several abusive methods to discipline knowledge and punish disobedient scientists (Union of Concerned Scientists, 2012; see also www.ucsusa.org). These include:

1. Corrupting the science: corporations suppress research, intimidate scientists, manipulate research designs and results, ghostwrite scientific articles and selectively publish results that favour their interests and priorities.
2. Limiting the effectiveness of regulating bodies: companies attack and undermine the science behind policy and risk assessments, hinder the regulatory process, corrupt advisory panels, exploit the 'revolving door' between corporate and government employment and withhold information from the public.
3. Exploiting the law and judicial pathways: corporations have collectively expanded their influence on the judicial system and actively use the courts to undermine science as well as bully and silence scientists.

It is also noteworthy that close-knit groups of high-level scientists and scientific advisors, with deep links to industry and politicians, organize highly effective campaigns to mislead the public and deny well-established scientific knowledge on the dangers of global warming and pesticides. Historians of science have shown how in the USA these 'merchants of doubt' have acted to trivialize, marginalize, vilify and silence scientific evidence established over the last four decades by public-funded research (Oreskes and Conway, 2010). Their discourses not only vehemently deny accepted scientific knowledge; they also provide the rhetoric and stories needed to legitimize an unflinching commitment to growth economics and productivist industrial agriculture.

Towards new ways of knowing

Given this increasingly organized and networked power of business and science (Castells and Cardoso, 2005; Vitali *et al.*, 2011), social movements are faced with the challenge of democratizing research and reclaiming control over the production of knowledge for the public good. This requires a radical shift from the existing top-down and increasingly corporate-controlled research system to an approach which devolves more power and control to food providers and citizen-consumers in the governance of research and production of knowledge. In Chapter 8, Pimbert suggests that inventing more democratic ways of knowing depends on two complementary approaches:

1. democratizing science and technology research, with increased funding for public research and transdisciplinary approaches that include peoples' knowledge; and
2. de-institutionalizing research for autonomous learning and action, with an emphasis on strengthening horizontal networks of grassroots research and innovation as well as citizen oversight over the production of knowledge.

In this book's concluding chapter, Pimbert critically discusses changes in research methodologies, organizations, policies and practices that can facilitate the transformation of knowledge for food sovereignty, agroecology and biocultural diversity.

In this context, actively developing more autonomous and transdisciplinary ways of knowing to produce new and relevant knowledge depends on forms of participation that resonate with visions of a more direct democracy:

> Pursuing civilisation today would therefore mean allowing each potential citizen-subject within society to become real subjects, by offering them [....] a genuine autonomy to exercise their ability to give themselves laws and construct rules with others. [...] More specifically, this implies giving to individuals the means to participate [...] in the daily construction of the rules of living together, and to rethink political, social and economic relationships in

order to civilize them at a deep level, through the permanent exercise of the freedom to participate.

(Méda, 2000)

This understanding of participation is consistent with one of the clearest demands of the food sovereignty movement: the ability of citizens to exercise their fundamental human right to decide their own food and agricultural policies – including framing policies for research and setting priorities for the production of knowledge.

In sum, this book builds on Pierre Bourdieu's idea that 'political subversion presupposes cognitive subversion' (Bourdieu, 1982). Contributing authors offer transformative knowledge in areas that can help realize the mutually supportive principles of food sovereignty, agroecology and biocultural diversity:

- equitable rights of access to territories and a socially just use of resources (Chapters 3, 4, 5 and 7);
- farming and land use in the image of nature, the protection of ecosystems and natural resources (Chapters 2, 3 and 4);
- the assurance of the right to food, water and the benefits of ecosystems as basic human rights rather than commodities to be purchased through trade and global markets (Chapters 5, 6 and 7);
- cognitive justice for the protection of peoples' knowledge and ways of knowing (Chapters 4, 6 and 8); and
- citizens' and peoples' fundamental human rights to contest and construct the knowledge that underpins their food, agricultural, economic, environment and social policies (Chapter 8).

Notes

1 The editor of this volume acknowledges the influence and intellectual contributions of the thousands of wo/men family farmers, indigenous peoples, pastoralists and other small-scale food providers involved in over 17 years of participatory action research on the regeneration of local food systems, ecologies and livelihoods (www.diversefoodsystems. org) and the democratic governance of agricultural research (www.excludedvoices.org).

2 In this volume the term 'governance' refers to the set of political, social, economic and administrative systems, rules and processes (1) which determine the way decisions are taken and implemented by various actors; and (2) through which decision makers are held accountable.

3 A food system includes all the elements (environment, people, inputs, processes, infrastructures, institutions etc.) and activities that relate to the production, processing, distribution, preparation and consumption of food and the outputs of these activities, including socio-economic and environmental outcomes (adapted from a range of other definitions: Ericksen *et al.*, 2010; HLPE, 2014; Tansey and Worsley, 1995).

4 The concept of citizen is at times understood to exclude indigenous peoples and minority ethnic groups who are not considered to be part of the nation state. However, the word citizen is originally derived from the latin *civis* and was in use before the emergence of the nation state. Citizen referred to individuals active in a public body and involved in the management of community affairs. In this volume I use the word citizen in this broad sense to include all people living and working in a given country.

5 La Vía Campesina is an international movement that brings together peasant organizations of small and medium-sized producers, agricultural workers, landless people, women farmers, migrants and indigenous communities from Africa, Asia, the Americas and Europe. It is an autonomous, pluralistic movement, independent of all political, economic or other denominations. La Vía Campesina (LVC) comprises about 164 local and national organizations in 73 countries and represents about 200 million farmers altogether. For more details see: https://viacampesina.org/en.

6 The World Food Summit in 1996 adopted the following as a definition of food security: 'Food security exists when all people at all times have physical and economic access to sufficient, safe and nutritious food to meet their dietary needs and food preferences for an active and healthy life' (FAO, 1996). This definition is based on four dimensions of food security. Food availability: the availability of sufficient quantities of food of appropriate quality, supplied through domestic production or imports. Food access: access by individuals to adequate resources (entitlements) for acquiring appropriate foods for a nutritious diet. Utilization: utilization of food through adequate diet, clean water, sanitation and health care to reach a state of nutritional wellbeing where all physiological needs are met. Stability: to be food secure, a population, household or individual must have access to adequate food at all times.

7 The organizers of the Nyéléni 2007 Forum on Food Sovereignty were: La Vía Campesina, see http://viacampesina.org; ROPPA: Le Réseau des Organisations Paysannes et de Producteurs de l'Afrique de l'Ouest (Network of farmers and producers organizations of West Africa), see www.roppa.info and www.cnop-mali.org; The World March of Women, see www.worldmarchofwomen.org; Friends of the Earth International, see www.foe.co.uk; World Forum of Fish Harvesters and Fishworkers (WFFP), see http://worldfisherforum.org; NGO members of the Food Sovereignty Network, see www.peoplesfoodsovereignty.org/; and IPC – International NGO/CSO Planning Committee for Food Sovereignty, see www.foodsovereignty.org.

8 The right to the city can be understood as the collective right to democratically control the production and use of urban space and urban processes. 'To claim the right to the city … is to claim some kind of shaping power over the processes of urbanization. Over the ways in which our cities are made and remade, and to do so in a fundamental and radical way' (Harvey, 2012).

9 Indigenous peoples are defined by the Special Rapporteur of the UN Economic and Social Council Sub-Commission on Prevention of Discrimination and Protection of Minorities as follows: 'Indigenous communities, peoples and nations are those which, having a historical continuity with pre-invasion and pre-colonial societies that developed on their territories, consider themselves distinct from other sectors of the societies now prevailing on those territories, or parts of them. They form at present non-dominant sectors of society and are determined to preserve, develop and transmit to future generations their ancestral territories, and their ethnic identity, as the basis of their continued existence as peoples, in accordance with their own cultural patterns, social institutions and legal system' (UN ECOSOC, 1986). According to the UN International Labour Office (ILO), indigenous peoples constitute about 5% of the world's population, or nearly 370 million people spread across over 70 countries (www.ilo.org/global/topics/indigenous-tribal/lang–en/index.htm).

10 A policy discourse is an ensemble of norms, rules, views, ideas, concepts and values that govern practice and behaviour, and help interpret social and environmental realities.

11 Pers. comm., Ibrahim Coulibaly, President of the Coordination Nationale des Organisations Paysannes (CNOP), 2015.

12 For reviews of the relevant literature see IAASTD, 2009 and IPES-Food, 2016.

13 See www.ethnobiology.net/what-we-do/core-programs/global-coalition-2/declaration-of-belem.

14 See www.fao.org/giahs/giahs-home/en.

15 Analysis of farmland distribution by country income level shows that in the richest countries farms larger than 20 ha operate 70% of land, while in the poorest countries

70% of land is operated by farms smaller than five ha (Adamopoulos and Restuccia, 2014).

16 While comprehensive, this analysis underestimated the contributions of *all* small-scale producers to food security and agricultural investments because it did not include information on nomadic pastoralists, forest dwellers, fisherfolk or indigenous hunter gatherers, for example. Although limited by the data available, it is estimated that worldwide these different categories of food providers (family farmers, pastoralists, artisanal fishers, indigenous forest dwellers etc.) collectively produce around 70% of the total food – with about 80% of this produce consumed locally or in the country (Pimbert, 2008; FAO, 2014).

17 Thomas Malthus argued that: 'By nature human food increases in a slow arithmetical ratio; man himself increases in a quick geometrical ratio unless want and vice stop him. The increase in numbers is necessarily limited by the means of subsistence. Population invariably increases when the means of subsistence increase, unless prevented by powerful and obvious checks' (Malthus, 1798). The Malthusian paradigm generally advocates population control programmes, to ensure resources for current and future populations. Political and economic elites who feel threatened by the growing numbers of commoners consider birth control as an important means of checking future conflict over their property. Neo-Malthusian views often find favour today with the elites on the issues of overpopulation, food and resource scarcity as well as environmental degradation.

18 Among the multiple terms used to describe small-scale, family-based producers (e.g. smallholders, traditional farmers, subsistence gardeners, petty producers etc.), the term 'peasant' is often laden with negative values and prejudice in many different countries and languages. The idea that 'peasants' symbolize 'backwardness' is being contested by farmers and citizens as they envision new peasant agrarian cultures and food sovereignty (Masioli and Nicholson, 2011).

19 Referring to today's peasantries, van der Ploeg argues that 'Central to the peasant condition, then, is the struggle for autonomy that takes place in a context characterized by dependency relations, marginalization and deprivation. It aims at and materializes as the creation and development of a self-controlled and self-managed resource base, which in turn allows for those forms of co-production of man and living nature that interact with the market, allow for survival and for further prospects and feed back into and strengthen the resource base, improve the process of co-production, enlarge autonomy and, thus reduce dependency [...] Finally, patterns of cooperation are present which regulate and strengthen these interrelations' (van der Ploeg, 2009).

20 Framing is a way of selecting, organizing, interpreting and making sense of a complex reality to provide guideposts for knowing, analyzing, persuading and acting (Schön and Rein, 1994).

21 Over the last 20 years, the share of biotechnologies in the agricultural research budget of the European Framework Programs has increased almost fourfold, amounting to 75% of the total budget in 2013. In contrast, the budget for research on organic agriculture has been stagnant, with 7% of the total in 2013 (Baret *et al.*, 2015).

22 Monsanto is today the world's largest seed company and owns nearly a quarter of the global seed market. For more details see www.monsanto.com.

23 According to the Advisory Committee of the UN Human Rights Council: 'Smallholder farmers, landless people, tenant farmers, agricultural labourers and people living from traditional fishing, hunting and herding activities are among the most discriminated and vulnerable people in many parts of the world. Every year, thousands of peasant farmers are victims of expropriation of land, forced evictions and displacements – a situation that is reaching an unprecedented level owing to the new phenomenon of the global 'land grab'. At the same time, traditional fishing communities are increasingly threatened by the industrialization of fishing activities; people living from hunting activities, by the creation of development projects; and pastoralists, by conflicts with farmers over land and water resources. All together, these people constitute 80% of the world's hungry; peasant

women are particularly affected by hunger and poverty, largely as a result of discrimination in access to and control over productive resources, such as land, water and credit' (UN Human Rights Council, 2012).

24 As numbers vary with country income levels, there are no accurate and comprehensive data on the numbers of people whose livelihoods depend on activities associated with the various links in the food chain and the web of local food systems.

25 Reducing emissions from deforestation and forest degradation (REDD+) is a mechanism developed by Parties to the United Nations Framework Convention on Climate Change (UNFCCC). REDD+ creates a financial value for the carbon stored in forests by offering incentives for developing countries to reduce emissions from forested lands and invest in low-carbon paths to sustainable development (www.unredd.net/about/un-redd-programme.html).

26 A level of crop production or exploitation that is maintained by restricting the quantity harvested to avoid long-term depletion. For example in fishery management, sustainable yield is defined as the number of fish that can be caught without reducing the base of fish stock.

27 Market fundamentalism is a quasi-religious faith that unregulated markets will somehow always produce the best possible results, and a strong belief that free market policies will solve most economic and social problems.

28 The concept of Access and Benefit Sharing (ABS) is derived from the Convention on Biological Diversity which, among other objectives, seeks to ensure the fair and equitable sharing of benefits arising from genetic resources. The Nagoya Protocol is a supplementary international agreement to the Convention on Biological Diversity which provides a legal framework for implementing that objective: *The Nagoya Protocol on Access to Genetic Resources and the Fair and Equitable Sharing of Benefits Arising from their Utilization to the Convention on Biological Diversity.* See https://www.cbd.int/abs/about/default.shtml/

29 The Convention on Biological Diversity (CBD) entered into force on 29 December 1993. The full text is available at https://www.cbd.int/convention/text/default.shtml.

30 The Intergovernmental Committee of the World Intellectual Property Organization (WIPO) is currently facilitating negotiations to develop an international legal instrument on *Sui Generis* Laws on Traditional Cultural Expressions/Expressions of Folklore. See: www.wipo.int/tk/en/indigenous.

31 The International Treaty on Plant Genetic Resources for Food and Agriculture (ITPGRFA) was adopted by the Thirty-First Session of the Conference of the Food and Agriculture Organization of the United Nations on 3 November 2001. The full text of the ITPGRFA can be accessed at www.planttreaty.org/content/texts-treaty-official-versions.

32 Payments for ecosystem services can be defined as a voluntary transaction whereby a well-defined ecosystem service, or a land-use likely to secure that service, is 'bought' by at least one buyer from at least one provider – if, and only if, the provider secures the provision of the service (IUCN and UNEP, 2012 – see http://unep.ch/etb/areas/pdf/IPES_IUCNbrochure.pdf).

33 The Millennium Ecosystem Assessment (2005) is the science-policy assessment that gave birth to the payment for ecosystem services (PES) concept. Initiated in 2001, the objective of the Millennium Ecosystem Assessment was to assess the consequences of ecosystem change for human wellbeing and the scientific basis for action needed to enhance the conservation and sustainable use of those systems and their contribution to human wellbeing (www.millenniumassessment.org/en/Index-2.html).

34 A definition of the 'good life' implies different ways of satisfying *fundamental human needs.* Max-Neef and his colleagues have identified nine fundamental human needs, namely: *subsistence* (for example, health, food, shelter, clothing); *protection* (care, solidarity, work etc.); *affection* (self-esteem, love, care, solidarity and so on); *understanding* (among others: study, learning, analysis); *participation* (responsibilities, sharing of rights and duties); *leisure/idleness* (curiosity, imagination, games, relaxation, fun); *creation* (including intuition,

imagination, work, curiosity); *identity* (sense of belonging, differentiation, self-esteem and so on), *freedom* (autonomy, self-esteem, self-determination, equality). While fundamental human needs are universal, their satisfiers vary according to culture, region and historical conditions (Max-Neef *et al.*, 1989).

35 My translation.
36 Founded by Bill and Melinda Gates, the Gates Foundation (or Bill & Melinda Gates Foundation, abbreviated as BMGF) is the largest private foundation in the world. Agricultural development is one of the largest initiatives of the BMGF. According to its website, the Gates Foundation has committed more than US $2 billion to agricultural development efforts to date, primarily in sub-Saharan Africa and South Asia. The Gates Foundation invests in strategic areas to help address the challenges faced by farmers in the developing world, including: research and development; agricultural policies; access and market systems; advocacy and strategic partnerships with donor countries, multilateral institutions, private foundations and other organizations. See www.gatesfoundation.org.

References

Adamopoulos, T. and D. Restuccia (2014) 'The size distribution of farms and international productivity differences', *The American Economic Review*, 104(6): 1667–1697.

Akram-Lodhi, A. H. (2013) 'How to build food sovereignty', Conference paper presented at the *International Conference on Food Sovereignty: A Critical Dialogue*, Yale University, 14–15 September 2013, Yale University: New Haven, CT.

Agarwal, B. (2014) 'Food sovereignty, food security and democratic choice: Critical contradictions, difficult conciliations', *The Journal of Peasant Studies* 41(6): 1247–1268.

Almeida, F., G. Borrini-Feyerabend, S. Garnett, H. Jonas, A. Kothari, E. Lee, M. Lockwood, F. Nelson and S. Stevens (2015) *Collective Land Tenure and Community Conservation*, Policy Brief of the ICCA Consortium, No. 2, The ICCA Consortium in collaboration with Maliasili Initiatives and Cenesta, Tehran.

Alston, J. M., P. G. Pardey and J. Roseboom (1998) 'Financing agricultural research: International investment patterns and policy perspectives', *World Development* 26(6): 1057–1071.

Altieri, M. A. (1987) *Agroecology: The Scientific Basis of Alternative Agriculture*, Westview Press, Boulder.

Altieri, M. A. (1995) *Agroecology: The Science of Sustainable Agriculture*, Westview Press, Boulder.

Altieri, M. A., C. I. Nicholls, A. Henao and M. A. Lana (2015) 'Agroecology and the design of climate change-resilient farming systems', *Agronomy for Sustainable Development* 35(3): 869–890.

Alvarez, S. E., E. Dagnino and A. Escobar (1998) *Cultures of Politics/Politics of Cultures: Re-visioning Latin American Social Movements*, Westview Press, Boulder.

Amin, S. (1976) *Unequal Development: An Essay on the Social Formations of Peripheral Capitalism*, Monthly Review Press, New York.

Anderson, C., M. P. Pimbert and C. Kiss (2015) *Building, Defending and Strengthening Agroecology. A Global Struggle for Food Sovereignty*, The Centre for Agroecology, Water and Resilience and the Centre for Learning on Sustainable Agriculture, Leusden, The Netherlands.

ANDES and IIED (2005) *Protecting Community Rights over Traditional Knowledge: Implications of Customary Laws and Practices*, Research Planning Workshop, Cusco, Peru, 20–25 May 2005, ANDES and IIED, London.

Andrée, P., J. Ayres, M. J. Bosia and M. J. Massicotte (2014) *Globalisation and Food Sovereignty. Global and Local Change in the New Politics of Food*, University of Toronto Press, Toronto.

Araghi, F. A. (1995) 'Global depeasantization, 1945–1990', *The Sociological Quarterly* 36(2): 337–368.

Bakunin, M. (1982) *On Anarchism*, Edited works by S. Dolgoff, Black Rose Books, Montreal.

Bakunin, M. (1987) *Statism and Anarchy*, Cambridge University Press, Cambridge.

Baret, P., P. Marcq, C. Mayer and S. Padel (2015) *Research for Transition: Research and Organic Farming in Europe*, Université Catholique de Louvain and Organic Research Centre, report commissioned by The Greens/EFA in the European Parliament, Brussels.

Baumann, M., J. Bell, F. Koechlin and M. Pimbert (1996) *The Life Industry. Biodiversity, People and Profits*, ITDG Publications, London.

Bello, W. and M. Baviera (2011) 'Capitalist agriculture, the food price crisis and peasant resistance', In: H. Wittman, A. A. Desmarais and N. Wiebe (Eds.) *Food Sovereignty: Reconnecting Food, Nature and Community*, Food First, Oakland.

Bensin, B. M. (1928) Agroecological characteristics description and classification of the local corn varieties chorotypes. Book. (Publisher unknown.)

Bensin, B. M. (1930) 'Possibilities for international cooperation in agroecological investigations', *Int. Rev. Agr. Mo. Bull. Agr. Sci. Pract.* (Rome) 21, 277–284.

Berger, J. (1978) 'Towards understanding peasant experience', *Race & Class* 19(4): 346–359.

Berkes, F. and H. Ross (2013) 'Community resilience: Toward an integrated approach', *Society & Natural Resources* 26(1): 5–20.

Bernstein, H. (2014) 'Food sovereignty via the "peasant way": A sceptical view', *The Journal of Peasant Studies* 41(6): 1031–1063.

Beuchelt, T. D. and D. Virchow (2012) 'Food sovereignty or the human right to adequate food: Which concept serves better as international development policy for global hunger and poverty reduction?', *Agriculture and Human Values* 9(2): 259–273.

Bhatt, V. V. (1982) 'Development problem, strategy, and technology choice: Sarvodaya and socialist approaches in India', *Economic Development and Cultural Change* 31(1): 85–99.

Bigg, T. and D. Satterthwaite (2005) *How to Make Poverty History: The Central Role of Local Organizations in Meeting the MDGs*, International Institute for Environment and Development, London.

Biggs, R., F. R. Westley and S. R. Carpenter (2010) 'Navigating the back loop: Fostering social innovation and transformation in ecosystem management', *Ecology and Society* 15(2): 9.

Bliss, T. (2011) 'The Urbal fix', *City* 15(1): 105–119.

Bookchin, M. (1986) *Toward an Ecological Society*, Black Rose Books, Montreal.

Bookchin, M. (1989) *Remaking Society. Pathways to a Green Future*, South End Press, Boston, MA.

Bookchin, M. (1995) *From Urbanization to Cities. Towards a New Politics of Citizenship*, Cassell, London.

Bookchin, M. (1998) *The Spanish Anarchists: The Heroic Years, 1868–1936*, AK Press, Edinburgh.

Borras, S. M., R. Hall, I. Scoones, B. White and W. Wolford (2011) 'Towards a better understanding of global land grabbing: An editorial introduction', *The Journal of Peasant Studies* 38(2): 209–216.

Borrini-Feyerabend, G. and ICCA Consortium (2012) *Bio-cultural Diversity Conserved by Indigenous Peoples and Local Communities – Examples and Analysis*, Cenesta, IUCN, UNDP, GEF, SGP and GIZ on behalf of BMZ, Teheran.

Borrini-Feyerabend, G., M. P. Pimbert, M. T. Farvar, A. Kothari and Y. Renard (2007) *Sharing Power. A Global Guide to Collaborative Management of Natural Resources*, Routledge, London.

Bourdieu, P. (1982) *Language and Symbolic Power*, Harvard University Press, Cambridge, MA.

Brauch, H. G., Ú. Oswald Spring, C. Mesjasz, J. Grin, P. Kameri-Mbote, B. Chourou, P. Dunay and J. Birkmann (2011) *Coping with Global Environmental Change, Disasters and Security: Threats, Challenges, Vulnerabilities and Risks*, Springer-Verlag, Berlin Heidelberg.

Brem-Wilson, J. (2015) 'Towards food sovereignty: Interrogating peasant voice in the UN Committee on World Food Security', *The Journal of Peasant Studies* 41(1): 73–95.

Brent, Z. W., C. M. Schiavoni and A. Alonso-Fradejas (2015) 'Contextualising food sovereignty: The politics of convergence among movements in the USA', *Third World Quarterly* 36(3): 618–635.

Brockington, D. and J. Igoe (2006) 'Eviction for conservation: A global overview', *Conservation and Society* 4(3): 424–470.

Brockington, D. and R. Duffy (2010) 'Capitalism and conservation: The production and reproduction of biodiversity conservation', *Antipode* 42(3): 469–484.

Brokensha, D. W., D. M. Warren and O. Werner (1980) *Indigenous Knowledge Systems and Development,* University Press of America, Michigan.

Burnett, K. and S. Murphy (2014) 'What place for international trade in food sovereignty?' *The Journal of Peasant Studies* 41(6): 1065–1084.

Büscher, B., W. Dressler and R. Fletcher (2014) *Nature™ Inc. Environmental Conservation in the Neoliberal Age*, University of Arizona Press, Tucson, AZ.

Cabannes, Y. (2014) 'Les formes coopératives, communautaires et collectives d'occupation du foncier et leur contribution à la fonction sociale du foncier et du logement', In: *La terre est à nous! Pour la fonction sociale du logement et du foncier, résistances et alternatives*, Passerelle, Ritimo/Aitec/Citego.

Campbell, B. M., P. Thornton, R. Zougmoré, P. van Aster and L. Lipper (2014) 'Sustainable intensification: What is its role in climate smart agriculture?' *Current Opinion in Environmental Sustainability* 8: 39–43.

Carrasco, C. (1999) *Mujeres y economia*, Icaria, Barcelona.

Carrol, C. R., J. H. Vandermeer and P. M. Rosset (1990) *Agroecology*, McGraw Hill Publishing Company, New York.

Castells, M. and G. Cardoso (2005) *The Network Society: From Knowledge to Policy,* Johns Hopkins Center for Transatlantic Relations, Washington, DC.

Castoriadis, C. (1996) *La Montée de l'Insignifiance. Les carrefours du labyrinthe* IV, Seuil, Paris.

CBD (2004) *The Ecosystem Approach*, Secretariat of the Convention on Biological Diversity, Montreal.

CFS (2009) *Reform of the Committee on World Food Security*. The Committee on World Food Security (CFS), 35th Session on 14–17 October 2009, Rome. Accessed on 23 December 2016 at ftp://ftp.fao.org/docrep/fao/meeting/018/k7197e.pdf.

Chambers, R. (1983) *Rural Development: Putting the Last First*. Routledge, London.

Chambers, R. (1997) *Whose Reality Counts? Putting the Last First*, Intermediate Technology Development Group, London.

Chambers, R. (2008) *Revolutions in Development Inquiry*, Earthscan, London.

Chambers, R., A. Pacey and L. A. Thrupp (1989) *Farmer First: Farmer Innovation and Agricultural Research*, Intermediate Technology Development Group, London.

Chayanov, A. V. (1989) 'The Peasant Economy', *Collected Works*, Ekonomika, Moscow.

Claeys, P. (2012) 'The creation of new rights by the food sovereignty movement: The challenge of institutionalizing subversion', *Sociology* 46(5): 844–860.

Claeys, P. (2015) *Human Rights and the Food Sovereignty Movement. Reclaiming Control*, Routledge, London.

Clapp, J. and D. Fuchs (2009) *Corporate Power in Global Agrifood Governance*, MIT Press, Cambridge MA.

Colchester, M. (2004) *Salvaging Nature. Indigenous Peoples, Protected Areas and Biodiversity Conservation*, World Rainforest Movement and Forest Peoples Programme, Moreton-in-Marsh.

Collier, G. and E. L. Quaratiello (2004) *Basta! Land and the Zapatista Rebellion in Chiapas*, Food First Books, Berkeley, CA.

Corpuz,V.T. (2016) *Report of the Special Rapporteur of the Human Rights Council on the Rights of Indigenous Peoples*,Victoria Tauli Corpuz, UN General Assembly, New York.

Corson, C. and K. I. MacDonald (2012) 'Enclosing the global commons:The convention on biological diversity and green grabbing', *The Journal of Peasant Studies*, 39(2): 263–283.

CSM (2016) *Connecting Smallholders to Markets. An Analytical Guide*, Civil Society Mechanism (CSM) in the Committee on World Food Security (CFS), Rome.

D'Alisa, G., F. Demaria and G. Kallis (2014) *Degrowth: A Vocabulary for a New Era*, Routledge, London.

Dardot, P. and C. Laval (2014) *Commun. Essai sur la Révolution au XXI Siècle*, La Découverte, Paris.

Delgado Wise, R. and H. Veltmeyer (2016) *Agrarian Change, Migration and Development*, Agrarian Change and Peasant Studies Series, Fernwood Publishing, Nova Scotia.

DeLonge, M. S., A. Miles and L. Carlisle (2016) 'Investing in the transition to sustainable agriculture', *Environmental Science & Policy* 55(1): 266–273.

De Schutter, O. (2010) *Agro-ecology and the Right to Food*, report submitted by the Special Rapporteur on the Right to Food, UN General Assembly, Human Rights Council, New York.

Desmarais, A. A. (2007) *La Vía Campesina: Globalization and the Power of Peasants,* Pluto Press, London.

Desmarais, A. A. and H. Wittman (2014) 'Farmers, foodies and First Nations: Getting to food sovereignty in Canada', *The Journal of Peasant Studies* 41(6): 1153– 1173.

Desmarais, A. A. and P. Nicholson (2013) 'La Vía Campesina. An Historical and Political Analysis', *La Vía Campesina's Open Book: Celebrating 20 Years of Struggle and Hope*, La Vía Campesina, Harare.

Dharampal (1983) *Indian Science and Technology in the Eighteenth Century: Some Contemporary European Accounts* (with a foreword by Dr. D. S. Kothari and introduction by Dr. William A. Blanpeid), Impex India, Delhi, 1971; reprinted by Academy of Gandhian Studies, Hyderabad, 1983.

Dowie, M. (2009) *Conservation Refugees. The Hundred Year Conflict between Global Conservation and Native Peoples*, MIT Press, Cambridge MA.

Durning, A. T. (1993) 'Guardians of the land: Indigenous peoples and the health of the earth', *Worldwatch Paper* 112,Worldwatch Institute,Washington DC.

EAFF, PROPAC and ROPPA (2011) *Agricultural Investment Strengthening Family Farming and Sustainable Food Systems in Africa* (4–5 May 2011, Mfou,Yaoundé, Cameroun), Synthesis Report, EuropAfrica, Rome.

EAFF, PROPAC and ROPPA (2013) *Les agriculteurs familiaux luttent pour des systémes alimentaires durables. Synthése des rapports des réseaux régionaux africains sur les modéles de production, la consommation et les marchés*, EuropAfrica, Rome.

Ecosystem Marketplace (2016) *Ecosystem Marketplace–Making the Priceless Valuable*, www.ecosystemmarketplace.com, accessed on 24 August 2016.

ECVC (2013) 'Proposal of position text on agroecology for the European Coordination Vía Campesina', European Coordination Vía Campesina (ECVC), Brussels, www.eurovia.org.

Edelman, M. (2014) 'Food sovereignty: Forgotten genealogies and future regulatory challenges', *The Journal of Peasant Studies* 41(6): 959–978.

Edelman, M., J. C. Scott, A. Baviskar, S. M. Borras Jr., D. Kandiyoti, E. Holt-Gimenez, T. Weis and W. Wolford (Guest Eds.) (2014) *Global Agrarian Transformations Volume 2: Critical Perspectives on Food Sovereignty, The Journal of Peasant Studies 41(6)*.

Ericksen, P. J., B. Stewart, J. Dixon, D. Barling, P. Loring, M. Anderson and J. Ingram (2010) 'The value of a food system approach', In: J. Ingram, P. Ericksen and D. Liverman (Eds.) *Food Security and Global Environmental Change*, Routledge, London.

Escobar, A. (1996) 'Constructing nature: Elements for a post structuralist political ecology', In: R. Peet and M. Watts (Eds.) *Liberation Ecologies*, Routledge, London.

Esteva, G. and M. S. Prakash (1998) *Grassroots Post-Modernism. Remaking the Soil of Cultures*, Zed Books, London.

ETC (2011) *Who Will Control the Green Economy? Corporate Concentration in the Life Industries*, ETC Group, Ottawa.

ETC (2016) *Software vs. Hardware vs. Nowhere. Year-end Status of the Ag. Mega-Mergers. Deere & Co. is Becoming 'Monsanto in a Box'*, ETC Group, Ottawa.

European Parliament (2016), *The New Alliance for Food Security and Nutrition*, Report of the Committee on Development, A8-0169/2016, European Parliament, Brussels.

EU SCAR (2012) *Agricultural Knowledge and Innovation Systems in Transition: A Reflection Paper*, Standing Committee on Agricultural Research (SCAR) of the European Union, Brussels.

Fairhead, J. and M. Leach (1996) *Misreading the African Landscape: Society and Ecology in a Forest-Savanna Mosaic*, Cambridge University Press, Cambridge.

Fairhead, J., M. Leach and I. Scoones (2012) 'Green grabbing: A new appropriation of nature?', *The Journal of Peasant Studies* 39(2): 237–261.

FAO (1996) *World Food Summit Plan of Action*, UN Food and Agriculture Organization, Rome.

FAO (1999) 'Agricultural biodiversity', *FAO Multifunctional Character of Agriculture and Land: Conference Background Paper* No. 1, Maastricht, September 1999, UN Food and Agriculture Organization, Rome.

FAO (2012) *Voluntary Guidelines on the Responsible Governance of Tenure of Land, Fisheries, and Forests in the Context of National Food Security*, UN Food and Agriculture Organization, Rome.

FAO (2014) *The State of Food and Agriculture. Innovation in Family Farming.* UN Food and Agriculture Organization, Rome.

FAO (2016) *Meeting our Goals. FAO's Programme for Gender Equality in Agriculture and Rural Development,* UN Food and Agriculture Organization, Rome.

Fédération Nature & Progrés (2012) Lettre ouverte à Stéphane Le Foll, Ministre de l'agriculture, de l'agroalimentaire et de la forêt, 17 December 2012, Fédération Nature & Progrés, Alès.

Fedick, S. L and B. A. Morrison (2004) 'Ancient use and manipulation of landscape in the Yalahau region of the northern Maya lowlands', *Agriculture and Human Values* 21(2): 207–219.

Finger-Stich, A. and M. Finger (2002) 'State versus participation: Natural resources management in Europe', *IIED and IDS Institutionalizing Participation Series*, International Institute for Environment and Development (IIED), London.

Ford, A., and R. Nigh (2015) *The Mayan Forest Garden: Eight Millennia of Sustainable Cultivation of Tropical Woodlands*, Left Coast Press, Walnut Creek.

Foucault, M. (1991) *The Foucault Reader: An Introduction to Foucault's Thought*, Penguin, London.

Francis, C., G. Lieblein, S. Gliessman, T. A. Breland, N. Creamer, R. Harwood, L. Salomonsson, J. Helenius, D. Rickerl, R. Salvador, M. Wiedenhoeft, S. Simmons, P. Allen, M. Altieri, C. Flora and R. Poincelot (2003) 'Agroecology: The ecology of food systems'. *Journal of Sustainable Agriculture* 22(3): 99–118.

Fricker, M. (2007) *Epistemic Injustice: Power and the Ethics of Knowing*, Oxford University Press, Oxford.

Friedmann, H. (2005) 'From colonialism to green capitalism: Social movements and emergence of food regimes', In: F. H. Buttel and P. McMichael (Eds.) *New Directions in the Sociology of Global Development*, Elsevier Press, Oxford.

Fuglie, K., P. Heisey and D. Schimmelpfennig (2016) *Private Industry Investing Heavily, and Globally, in Research to Improve Agricultural Productivity*, United States Department of Agriculture Economic Research Service, Washington, DC.

Gahman, L. (2016) 'Food sovereignty in rebellion: Decolonization, autonomy, gender equity, and the Zapatista solution', *The Solutions Journal* 7(4): 77–83.

Garnett, T., M. C. Appleby, A. Balmford, I. J. Bateman, T. G. Benton, P. Bloomer, B. Burlingame, M. Dawkins, L. Dolan, D. Fraser, M. Herrero, I. Hoffmann, P. Smith, P. K. Thornton, C. Toulmin, S. J. Vermeulen and H. C. J. Godfray (2013) 'Sustainable intensification in agriculture: Premises and policies', *Science* 341(6141): 33–34.

Ghimire K. B. and M. P. Pimbert (1997a) 'Social change and conservation: An overview of issues and concepts', In: K. B. Ghimire and M. P. Pimbert (Eds.) *Social Change and Conservation*, Earthscan/Routledge, London.

Ghimire, K. B. and M. P. Pimbert (1997b) *Social Change and Conservation: Environmental Politics and Impacts of National Parks and Protected Areas*, Earthscan/Routledge, London.

Gibson-Graham, J. K. (2016) 'Building community economies: Women and the politics of place', In: W. Harcourt (Ed.) *The Palgrave Handbook of Gender and Development: Critical Engagements in Feminist Theory and Practice*, Palgrave MacMillan, London.

Gliessman, S. R. (1990) *Agroecology: Researching the Ecological Basis for Sustainable Agriculture*, Springer, New York.

Gliessman, S. R. (1991) 'Ecological basis of traditional management of wetlands in tropical Mexico: Learning from agroecosystems', In: M. L. Oldfield and J. B. Alcorn (Eds.), *Biodiversity: Culture, Conservatism, and Ecodevelopment*, Westview Press, Boulder, CO.

Gliessman, S. R. (1998) *Agroecology: Ecological Processes in Sustainable Agriculture*, Sleeping Bear Press, Ann Arbor, MI.

Gliessman, S. R. (2014) *Agroecology: The Ecology of Sustainable Food Systems*, CRC Press, Boca Raton, FL.

Gliessman, S. R. (2015) 'Agroecology. Roots of resistance to industrialised food systems', In: V. Ernesto Méndez, C. Bacon, R. Cohen and S. R. Gliessman (Eds.) *Agroecology. A Transdisciplinary, Participatory and Action Oriented Approach*, CRC Press, Boca Raton, FL.

Global Alliance for the Future of Food (2016) *The Future of Food: Seeds of Resilience. A Compendium of Perspectives on Agricultural Biodiversity from Around the World*, The Global Alliance for the Future of Food.

Goetz, A. M. (1997) *Getting Institutions Right for Women in Development*, Zed Books, London.

Golay, C. (2015) 'Negotiation of a United Nations Declaration on the Rights of Peasants and Other People Working in Rural Areas', *Academy Brief* No. 5, Geneva Academy of International Humanitarian Law and Human Rights, Geneva.

Gollain, F. (2004) *A Critique of Work: Between Ecology and Socialism*, International Institute for Environment and Development (IIED), London.

Gómez-Baggethun, E., V. Reyes-García, P. Olsson and C. Montes (2012) 'Traditional ecological knowledge and community resilience to environmental extremes: A case study in Doñana, SW Spain', *Global Environmental Change* 22(3): 640–650.

Gómez-Pompa, A. and A. Kaus (1992) 'Taming the wilderness myth', *Bioscience*, 42(4): 271–279.

Gonzales de Molina, M. (2010) 'A guide to studying the socio-ecological transition in European agriculture', *Sociedad Espanola de Historia Agraria DT-SEHA* 10 No. 6.

GRAIN (2012) *The Great Food Robbery. How Corporations Control Food, Grab Land and Destroy the Climate*, Pambazuka Press, Nairobi.

Granstedt, I. (2012) *Du chômage à l'autonomie conviviale*, Éditions À plus d'un titre, La Bauche.

Grey, S. and R. Patel (2015) 'Food sovereignty as decolonization: Some contributions from Indigenous movements to food system and development politics', *Agriculture and Human Values* 32(3): 431–444.

Guerin, I. (2003) *Femmes et economie solidaire*, Declée de Brouwer, Paris.

Gunder Franck, A. (1978) *Dependent Accumulation and Underdevelopment*, Monthly Review Press, New York.

Gunderson, L. H., C. R. Allen and C.S. Holling (2012) *Foundations of Ecological Resilience*, Island Press, Washington, DC.

Habermas, J. (1984) *The Theory of Communicative Action*, Beacon Press, Boston.

Hahnel, R. (2005) *Economic Justice and Democracy: From Competition to Cooperation,* Routledge, New York.

Hahnel, R. (2016) *Participatory Economics and the Next System*, The Next System Project, accessed 22 December 2016 at http://thenextsystem.org/wp-content/uploads/2016/03/NewSystems_RobinHahnel.pdf.

Handy, J. (2009) 'Almost idiotic wretchedness: A long history of blaming peasants', *The Journal of Peasant Studies* 36(2): 325–344.

Handy, J. and C. Fehr (2010) 'Drawing forth the force that slumbered in peasants' arms: *The Economist*, high agriculture and selling capitalism', In: H. Wittman, A. A. Desmarais and N. Wiebe (Eds.) *Food Sovereignty: Reconnecting Food, Nature and Community*, Food First, Oakland.

Harcourt, W. (2016) *The Palgrave Handbook of Gender and Development: Critical Engagements in Feminist Theory and Practice*, Palgrave MacMillan, London.

Harvey, D. (2004) 'The "new" imperialism: Accumulation by dispossession', *Socialist Register* 40: 63–87.

Harvey, D. (2012) *Rebel Cities. From the Right to the City to the Urban Revolution*, Verso, London.

Harvey, D. (2015) *Seventeen Contradictions and the End of Capitalism*, Profile Books, London.

Hecht, S. B. (1995) 'The evolution of agroecological thought', In: M. Altieri (Ed.) *Agroecology. The Science of Sustainable Agriculture*, Intermediate Technology Development Group, London.

Henderson, T. P. (2017) 'State–peasant movement relations and the politics of food sovereignty in Mexico and Ecuador', *The Journal of Peasant Studies* 44(1): 33–55.

Herber, L. (1962) *Our Synthetic Environment*, Knopf, New York.

Hernández Xolocotzi, E. (1977) *Agroecosistemas de México: contribuciones a la enseñanza, investigación y divulgación agrícola*, Segunda edición 1981, Colegio de Postgraduados, Chapingo, Mexico.

Hernández Xolocotzi, E. (1985) *Xolocotzia: Obras de Efraím Hernández Xolocotzi*, Tomo 1, Revista de Geografia Agricola, Universidad Autónoma de Chapingo, Texcoco, Mexico.

Hernández Xolocotzi, E. (1987) *Xolocotzia: Obras de Efraím Hernández Xolocotzi*, Tomo 2, Revista de Geografia Agricola, Universidad Autonóma de Chapingo, Texcoco, Mexico.

Herzen, A. (1992) *My Past and Thoughts: The Memoirs of Alexander Herzen*, University of California Press, Oakland.

Hill, P. (1986) *Development Economics on Trial. The Anthropological Case for Prosecution*, Cambridge University Press, Cambridge.

HLPE (2013) *Investing in Smallholder Agriculture for Food Security,* a report by the High Level Panel of Experts on Food Security and Nutrition of the Committee on World Food Security, Rome.

HLPE (2014) *Food Losses and Waste in the Context of Sustainable Food Systems*, a report by the High Level Panel of Experts on Food Security and Nutrition of the Committee on World Food Security, Rome.

HLPE (2016) *Sustainable Agricultural Development for Food Security and Nutrition: What Roles For Livestock?*, a report by the High Level Panel of Experts on Food Security and Nutrition of the Committee on World Food Security, Rome.

HM Government (2016) *Private and Public Sector Funding of Agri-Tech R&D: FY 2012/2013*, Her Majesty's Government, London.

Hoben, A. (1995) 'Paradigms and politics: The cultural construction of environmental policy in Ethiopia', *World Development* 23(6): 1007–1021.

Holt-Giménez, E. (2002) 'Measuring farmers' agroecological resistance to Hurricane Mitch in Central America', *Gatekeeper Series* No. SA102, International Institute for Environment and Development (IIED), London.

Homewood, K. (2008) *Ecology of African Pastoralist Societies*, Ohio University Press, Athens, Ohio.

IAASTD (2009) *Agriculture at a Crossroads: Synthesis Report,* International Assessment of Agricultural Knowledge, Science, Technology for Development. Island Press and UNEP, UNDP, FAO, UNESCO, The World Bank, Global Environment Facility and WHO.

Illich, I. (1973) *Tools for Conviviality,* Marion Boyars Publishers, London.

INRA (2010) *A Ten Year Strategy. An Orientation Document,* INRA, Paris. Accessed on 23 December 2016, http://institut.inra.fr/en/Research-and-results/Strategies.

IPES-Food (2016) *From Uniformity to Diversity: A Paradigm Shift from Industrial Agriculture to Diversified Agroecological Systems,* International Panel of Experts on Sustainable Food Systems, Université Catholique de Louvain, Louvain-la-Neuve, Belgium.

IUCN-CEESP (2008) 'Recognising and supporting indigenous & community conservation – ideas & experiences from the grassroots', *CEESP Briefing Note* 9. IUCN and CEESP, Gland and Tehran.

IUCN and UNEP (2012) *Developing International Payments for Ecosystem Services. Towards a Greener World Economy*, International Union for Conservation of Nature and Natural Resources (IUCN), Gland.

Jones, G. H. (1936) *The Earth Goddess: A Study of Native Farming on the West African Coast*, Longmans, Green and Co., London.

Jones, A., M. P. Pimbert and J. Jiggins (2012) *Virtuous Circles: Values, Systems, Sustainability*, IIED and IUCN CEESP, London.

Juffe-Bignoli, D., N. D. Burgess, H. Bingham, E. M. S. Belle, M. G. de Lima, M. Deguignet, B. Bertzky, A. N. Milam, J. Martinez-Lopez, E. Lewis, A. Eassom, S. Wicander, J. Geldmann, A. van Soesbergen, A. P. Arnell, B. O'Connor, S. Park, Y. N. Shi, F. S. Danks, B. MacSharry and N. Kingston (2014) *Protected Planet Report 2014*, United Nations Environment Programme and World Conservation Monitoring Centre, Cambridge.

Kabeer, N. (1994) *Reversed Realities: Gender Hierarchies in Development Thought*, Zed Books, London.

Kay, L. E. (1997) *The Molecular Vision of Life: Caltech, the Rockefeller Foundation, and the Rise of the New Biology*, Oxford University Press, Oxford.

King, F. H. (1911) *Farmers of Forty Centuries. Permanent Agriculture in China, Korea, and Japan*, originally published by Mrs F. H. King, Madison, Wisconsin (reprinted in 2004 by Dover Publications Inc.).

Klages, K. H. W. (1928) 'Crop ecology and ecological crop geography in the agronomic curriculum', *Journal of the American Society of Agronomy* 20: 336–353.

Kloppenburg, J. (2014) 'Re-purposing the master's tools: The open source seed initiative and the struggle for seed sovereignty', *The Journal of Peasant Studies* 41(6): 1225–1246.

Knorr Cetina, K. D. (1999) *Epistemic Cultures. How the Sciences Make Knowledge*, Harvard University Press, Cambridge.

Koohafkan, P. and M. A. Altieri (2012) *Globally Important Agricultural Heritage Heritage Systems. A Legacy for the Future*, UN Food and Agriculture Organization, Rome.

Kosoy, N. and E. Corbera (2010) 'Payments for ecosystem services as commodity fetishism', *Ecological Economics* 69(6): 1228–1236.

Kothari, A., C. Corrigan, H. Jonas, A. Neumann and H. Shrumm (2012) 'Recognising and supporting territories and areas conserved by indigenous peoples and local communities:

Global overview and national case studies', *Technical Series* no. 64, Secretariat of the Convention on Biological Diversity, ICCA Consortium, Kalpavriksh and Natural Justice, Montreal.

Kothari, A., F. Demaria and A. Acosta (2014) 'Buen Vivir, Degrowth and Ecological Swaraj: Alternatives to sustainable development and the Green Economy', *Development* 57(3–4): 362–375.

Kropotkin, P. (1892) *La Conquête du Pain*, Tresse et Stock éditeurs, Paris.

Kropotkin, P. (1898) *Fields, Factories, and Workshops, or, Industry Combined with Agricultures and Brain Work with Manual Work*, revised 1913, Rpt. Benjamin Blom, New York.

Kuhn, T. S. (2012) *The Structure of Scientific Revolutions*, 4th ed., University of Chicago Press, Chicago.

Kumarappa, J. C. (1951) *Gandhian Economic Thought*, 1st ed., Library of Indian Economics, Vora, Bombay.

Lamphere, J. A and E. A. East (2016) 'Monsanto's biotechnology politics: Discourses of legitimation', *Environmental Communication*, Taylor and Francis online.

Latouche, S. (1998) *L'Autre Afrique. Entre Don et Marché*, Albin Michel, Paris.

Latouche, S. (2003) *Décoloniser l'imaginaire: La Pensée créative contre l'économie de l'absurde*, Parangon, Paris.

Latouche, S. (2009) *Farewell to Growth*, John Wiley Publishers, London.

Latouche, S. (2011) *Vers une société d'abondance frugale: Contresens et controverses sur la décroissance*, Fayard - Mille et une nuits, Paris.

La Vía Campesina (1996) *Food Sovereignty: A Future Without Hunger*, Declaration at the World Food Summit hosted in 1996 by the UN Food and Agriculture Organization, Rome, www.acordinternational.org/silo/files/decfoodsov1996.pdf.

La Vía Campesina (2007) 'Nyéléni Declaration on Food Sovereignty: 27 February 2007, Nyéléni village, Selingue, Mali', *The Journal of Peasant Studies* 36(3), July 2009: 676–763.

La Vía Campesina (2008a) *Declaration of Maputo*, V International Conference of La Vía Campesina, 9–22 October, Maputo, Mozambique.

La Vía Campesina (2008b) *Declaration of Rights of Peasants – Women and Men*, Peasants of the World Need an International Convention on the Rights of Peasants, La Vía Campesina.

Laville, J. L. (2013) *L'économie solidaire: Une perspective internationale*, Fayard/Pluriel, Paris.

Leach, M. and R. Mearns (1996) *The Lie of the Land. Challenging Received Wisdom on the African Environment*, James Curry, Oxford.

Lefébvre, H. (1968) *Le droit à la ville*, Anthopos, Paris.

Leval, G. (1975) *Collectives in the Spanish Revolution*, Freedom Press, London.

Levidow, L., M. P. Pimbert and G. Vanloqueren (2014) 'Agroecological research: Conforming—or transforming the dominant agro-food regime?', *Agroecology and Sustainable Food Systems* 38(10): 1127–1155.

Lietaer, B. and G. Hallsmith (2011) *Creating Wealth*. New Society Publishers, Gabriola Island.

Lietaer, B. and J. Dunne (2013) *Rethinking Money: How New Currencies Turn Scarcity into Prosperity,* Berrett-Koehler Publishers, San Francisco.

Lincoln, Y. S., S. A. Lynham and E. G. Guba (2011) 'Paradigmatic controversies, contradictions and emerging convergences, revisited', In: N. K. Denkin and Y. Lincoln (Eds.) *The SAGE Handbook of Qualitative Research*, Sage Publications Inc., London.

Long, N. and A. Long (1992) *Battlefields of Knowledge: Interlocking of Theory and Practice in Social Research and Development*, Routledge, London.

LOA (2005) *Loi d'Orientation Agricole*, Government of Mali, Bamako.

Lohmann, L. (2011) 'Financialization, commodification and carbon: The contradictions of neoliberal climate policy', In: L. Panitch, G. Albo and V. Chibber (Eds.) *Socialist Register 2012: The Crisis and the Left* 48: 85–107.

Lowder, S. K., B. Carisma and J. Skoet (2015) 'Who invests how much in agriculture in low- and middle-income countries. An empirical review', *The European Journal of Development Research* 27(3): 371–390.

Lowder, S. K., J. Skoet and T. Raney (2016) 'The number, size, and distribution of farms, smallholder farms, and family farms worldwide', *World Development* 87: 16–29.

Lowrance, R., B. R. Stinner and G. J. House (1984) *Agricultural Ecosystems: Unifying Concepts*, Wiley-Interscience, New York.

McKeon, N. (2014) *The New Alliance for Food Security and Nutrition. A Coup for Corporate Capital?*, Transnational Institute (TNI), Amsterdam.

McKeon, N. (2015) *Food Security Governance Empowering Communities, Regulating Corporations*, Routledge, London.

McMichael, P. (2009) 'A food regime genealogy', *The Journal of Peasant Studies* 36(1): 139–169.

McMichael, P. (2014) 'Historicizing food sovereignty', *The Journal of Peasant Studies* 41(6):933–957.

McRae, R. J., S. B. Hill, F. R. Mehuys and J. Henning (1990) 'Farm scale agronomic and economic conversion from conventional to sustainable agriculture', *Advances in Agronomy* 43: 155–198.

Maffi, L. (2001) *On Biocultural Diversity: Linking Language, Knowledge, and the Environment*, Smithsonian Institution Press, New York.

Maffi, L. and E. Woodley (2010) *Biocultural Diversity Conservation. A Global Sourcebook*, Routledge, London.

Malthus T. R. (1798) *An Essay on the Principle of Population*, First Edition, with William Godwin's Essay 'Of Avarice and Profusion', CreateSpace Independent Publishing Platform, London.

Marglin, S. A. (2010) *The Dismal Science: How Thinking Like an Economist Undermines Community*, Harvard University Press, Cambridge MA.

Martinez-Alier, J. (2002) *The Environmentalism of the Poor. A Study of Ecological Conflicts and Valuation*, Edward Elgar, Cheltenham.

Marx, K. (1981) *Capital, vol. III*, Vintage, New York.

Masioli, I. and P. Nicholson (2011) 'Seeing like a peasant. Voices from La Vía Campesina', In: H. Wittman, A. A. Desmarais and N. Wiebe (Eds.) *Food Sovereignty: Reconnecting Food, Nature and Community*, Food First, Oakland.

Masson, D., A. Paulos and E. Beaulieu Bastien (2017) 'Struggling for food sovereignty in the World March of Women', *The Journal of Peasant Studies* 44(1): 56–77,

Mauss, M. (1966) *The Gift: Forms and Functions of Economic Exchange in Archaic Societies*, Cohen & West, London.

Max-Neef, M., A. Elizalde, M. Hopenhayn, F. Herrera, H. Zemelman, J. Jataba and L. Weinstein (1989) 'Human scale development: An option for the future', *Development Dialogue* 1989(1): 5–80.

Méda, D. (1998) *Le Travail, Une Valeur en Disparition*, Flammarion, Paris.

Méda, D. (2000) *Qu'est-ce que la richesse?* Champs Flammarion, Paris.

Mehta, L., J. G. Veldwisch and J. Franco (2012) 'Water grabbing? Focus on the (re) appropriation of finite water resources', *Water Alternatives* 5(2): 193–207.

Méndez, V. E., C. M. Bacon, R. Cohen and S. R. Gliessman (2016) *Agroecology. A Transdisciplinary, Participatory and Action-Oriented Approach*, CRC Press, Boca Raton.

Merlant, P., R. Passet and J. Robin (2003) *Sortir de l'économisme. Une alternative au capitalisme neoliberal*, Editions de l'Atelier, Paris.

Merrill, R. (1976) *Radical Agriculture*, Harper & Row Publishers, New York.

Michael, D. N. (1995) 'Barriers and bridges to learning in a turbulent human ecology', In: L. H. Gunderson, C. S. Holling and Stephen Light (Eds.) *Barriers and Bridges to the Renewal of Ecosystems and Institutions*, Columbia University Press, New York.

Mies, M. and V. Bennholdt Thomsen (1999) *The Subsistence Perspective: Beyond the Globalised Economy*, Zed Books, London.

Mijatović, D., F. Van Oudenhoven, P. Eyzaguirre and T. Hodgkin (2013) 'The role of agricultural biodiversity in strengthening resilience to climate change: Towards an analytical framework', *International Journal of Agricultural Sustainability* 11(2): 95–107.

Millennium Ecosystem Assessment (2005) *Ecosystems and Human Well-being: Synthesis*, Island Press, Washington, DC.

Molle, F. (2008) 'Nirvana concepts, narratives and policy models: Insights from the water sector', *Water Alternatives* 1(1): 131–156.

Monsanto Tribunal, The (2016) 'Human rights violations, crimes against humanity and ecocide', legal case prepared for The Monsanto Tribunal held on 15–16 October 2016, The Hague.

Moore, J. W. (2015) *Capitalism in the Web of Life. Ecology and the Accumulation of Capital*, Verso Books, London.

Moore, J. W. (2016) *Anthropocene or Capitalocene? Nature, History and the Crisis of Capitalism*, PM Press, Oakland.

Moreno, C., D. Speich Chassé and L. Fuhr (2015) 'Carbon metrics – global abstractions and ecological epistemicide', *Ecology* 42, Heinrich Böll Foundation, Berlin.

Mühlhäusler, P. (1996) *Linguistic Ecology: Language Change and Linguistic Imperialism in the Pacific Rim*, Routledge, London.

Murray, M.C. and C. Pateman (2012) *Basic Income Worldwide. Horizons of reform*, Palgrave Macmillan, London.

Mylondo, B. (2010) *Ne Pas Perdre Sa Vie À La Gagner. Pour un Revenu de Citoyenneté*, Editions du Croquant, Bellecombes en Bauge.

Netting, R. M. (1993) *Smallholders, Householders: Farm Households and the Ecology of Intensive Sustainable Agriculture*, Stanford University Press, Stanford.

Nyéléni (2007) *Declaration of the Forum for Food Sovereignty*, Nyéléni Village, Sélingué, Mali, 23–27 February 2007, https://nyeleni.org/IMG/pdf/DeclNyeleni-en.pdf.

Nyéléni (2015) *Declaration of the International Forum for Agroecology*, International Planning Committee for Food Sovereignty website, www.foodsovereignty.org/forum-agroecology-nyeleni-2015/.

Oldfield, M. L. and J. B. Alcorn (1987) 'Conservation of traditional agroecosystems', *BioScience* 37(3):199–208.

Oreskes, N. and E. M. Conway (2010) *Merchants of Doubt. How a Handful of Scientists Obscured the Truth on Issues from Tobacco Smoke to Global Warming*, Bloomsbury Press, London and New York.

Ostrom, E. (1990) *Governing the Commons: The Evolution of Institutions for Collective Action*, Cambridge University Press, Cambridge.

Ostrom, E. (2010) 'Beyond markets and states: Polycentric governance of complex economic systems', *American Economic Review,* 100(3): 641–672.

Partant, F. (1999) *La fin du développement. Naissance d'une alternative?*, Actes Sud, Arles.

Partant, F. (2002) *Que la crise s'aggrave*, Parangon, Paris.

Passet, R. (2012) *La Bioéconomie de la dernière chance*, Editions Les Liens qui Libèrent, Paris.

Patel, R. (2007) *Stuffed and Starved. Markets, Power and the Hidden Battle for the World Food System*, Portobello Books, London.

Peluso, N. L. (1993) 'Coercing conservation? The politics of state resource control', *Global Environmental Change* 3(2): 199–217.

Perez-Vitoria, S. (2005) *Les paysans sont de retour*, Actes Sud, Arles.

Perez-Vitoria, S. (2015) *Manifeste pour un XXI siècle paysan*, Actes Sud, Arles.

Perfecto, I., J. Vandermeer and A. Wright (2009) *Nature's Matrix: Linking Agriculture, Conservation and Food Sovereignty*, Earthscan/Routledge, London.

Perfecto, I. and J. Vandermeer (2017) 'A landscape approach to integrating food production and conservation', In: I. J. Gordon, H. H. T. Prins and G. R. Squire (Eds.) *Food Production and Nature Conservation. Conflicts and Solutions*, Routledge, London.

Pimbert, M. P. (2004) 'Institutionalising participation and people-centered processes in natural resource management', IIED and IDS *Institutionalising Participation Series*, International Institute for Environment and Development, London.

Pimbert, M. P. (2007) 'Transforming knowledge and ways of knowing for food sovereignty', *Reclaiming Diversity and Citizenship Series*, International Institute for Environment and Development, London.

Pimbert, M. P. (2008) 'The role of local organisations in sustaining local food systems, livelihoods and the environment', In: M. P. Pimbert (Ed.) *Towards Food Sovereignty: Reclaiming Autonomous Food Systems*, International Institute for Environment and Development, London and Rachel Carson Centre, Munich.

Pimbert, M. P. (2010) *Towards Food Sovereignty: Reclaiming Autonomous Food Systems*, multimedia e-book, available at www.environmentandsociety.org/mml/pimbertmichel-towards-foodsovereignty reclaiming-autonomous-food-systems, International Institute for Environment and Development, London and Rachel Carson Centre, Munich.

Pimbert, M. P. (2012) 'Fair and sustainable food systems: From vicious cycles to virtuous circles', *IIED Policy Brief*, International Institute for Environment and Development, London.

Pimbert, M. P. (2015a) 'Agroecology as an alternative vision to conventional development and climate-smart agriculture', *Development* 58(2/3): 286–298.

Pimbert, M. P. (2015b) 'Circular food systems', In: K. Albala (Ed.) *The SAGE Encyclopedia of Food Issues*, SAGE Publications, Inc., Thousand Oaks.

Pimbert, M. P. and B. Gujja (1997) 'Village voices challenging wetland management policies: Experiences in participatory rural appraisal from India and Pakistan', *Nature and Resources* 33(1): 34–42.

Pimbert, M. P. and J. N. Pretty (1995) 'Parks, people and professionals. Putting "participation" into protected area management', *UNRISD Discussion Paper* No. 57, United Nations Research Institute for Social Development, Geneva.

Pimbert, M. P. and J. N. Pretty (1998) 'Diversity and sustainability in community-based conservation', In: A. Kothari, R. V. Anuradha, N. Pathak and B. Taneja (Eds.) *Communities and Conservation: Natural Resource Management in South and Central Asia*, UNESCO and Sage Publications, London and New Delhi.

Pimbert, M. P., K. Tran-Thanh, E. Deléage, M. Reinert, C. Trehet and E. Bennett (2006) 'Farmers' views on the future of food and small scale producers', *IIED Reclaiming Diversity and Citizenship Series,* International Institute for Environment and Development, London.

Polanyi, K. (1957) *The Great Transformation*, Beacon Press, Boston, MA.

Polanyi, K. (1968) *Primitive, Archaic, and Modern Economies*, Ed. George Dalton, Anchor Books, New York.

Posey, D. A. (1996) *Traditional Resource Rights: International Instruments for Protection and Compensation for Indigenous Peoples and Local Communities*, International Union for Conservation of Nature (IUCN), Gland.

Posey, D. A. (Ed.) (1999) *Cultural and Spiritual Values of Biodiversity*, UNEP and Practical Action, London.

Posey, D. A. and G. Dutfield (1996) *Beyond Intellectual Property: Toward Traditional Resources Rights for Indigenous Peoples and Local Communities*, International Development Research Centre, Ottawa.

Praetorius, I. (2015) 'The care-centered economy. Rediscovering what has been taken for granted', *Economy + Social Issues* 16, The Heinrich Böll Foundation, Berlin.

Proudhon, P. J. (1979) *The Principle of Federation*, Transl. Vernon Richards, University of Toronto Press, Toronto.

Purcell, M. (2013) 'Possible worlds: Henri Lefébvre and the right to the city', *Journal of Urban Affairs* 36(1): 141–154.

Rabhi, P. (1989) *L'Offrande au crepuscule*, Éditions de Candide, Lavilledieu.

Rabhi, P. and J. Caplat (2015) *L'agroécologie. Une éthique de vie*, Actes Sud and Colibris, Arles.

Rahnema, M. (2003) *Quand la misère chasse la pauvreté*, Fayard/Actes Sud, Paris.

Rahnema, M. and V. Bawtree (1997) *The Post-Development Reader*, Zed Books, London.

Reed, M. (2006) 'Organisational theorising: A historically contested terrain', In: S. R. Clegg, C. Hardy, T. Lawrence and W. R. Nord (Eds.) *Handbook of Organisation Studies*, Sage Publications, London.

Richards, P. (1985) *Indigenous Agricultural Revolution. Ecology and Food Production in West Africa*, Hutchinson Education, London.

Rist, G. (2011) *The Delusions of Economics. The Misguided Certainties of a Hazardous Science*, Zed Books, London.

Rist, G. (2013) *Le développement. Histoire d'une croyance occidentale*, 4 revised edition, Presses de Sciences Po, Paris.

Roe, E. M. (1991) 'Development narratives, or making the best of blueprint development', *World Development* 19(4): 287–300.

Rogers, K. H., R. Luton, H. Biggs, R. Biggs, S. Blignaut, A. G. Choles, C. G. Palmer and P. Tangwe (2013) 'Fostering complexity thinking in action research for change in social-ecological systems', *Ecology and Society* 18(2): 31.

Rosado-May, F. J. (2015) 'The intercultural origin of agroecology. Contributions from Mexico', In: V. Ernesto Méndez, C. Bacon, R. Cohen and S. R. Gliessman (Eds.) *Agroecology. A Transdisciplinary, Participatory and Action Oriented Approach*, CRC Press, Roca Baton.

Ross, E. (1998) *The Malthus Factor: Poverty, Politics and Population in Capitalist Development*, Zed Books, London.

Ross, P. and Y. Cabannes (2014) *21st Century Garden Cities of To-Morrow. A Manifesto*, published online at www.Lulu.com, Lulu Press, Inc.

Rosset, P. M. and M. A. Altieri (1997), 'Agroecology versus input substitution: A fundamental contradiction of sustainable agriculture', *Society and Natural Resources* 10(3): 283–295.

Rosset, P. M, B. Machín Sosa, A. María Roque Jaime and D. Rocio Ávila Lozano (2011) 'The Campesino-to-Campesino agroecology movement of ANAP in Cuba: Social process methodology in the construction of sustainable peasant agriculture and food sovereignty', *The Journal of Peasant Studies* 38(1): 161–191.

Rosset, P. M and M. E. Martínez-Torres (2014) 'Food sovereignty and agroecology in the convergence of rural social movements', In: D. H. Constance, M-C. Renard and M. G. Rivera-Ferre (Eds.) *Research in Rural Sociology and Development Series, Vol. 21: Alternative Agrifood Movements: Patterns of Convergence and Divergence*, Emerald Group Publishing, Bingley, UK.

Rotarangi, S. and J. Stephenson (2014) 'Resilience pivots: Stability and identity in a social-ecological-cultural system', *Ecology and Society,* 19(1): 28 .

Royal Society (2009) *Reaping the Benefits: Science and the Sustainable Intensification of Global Agriculture*, Royal Society, London.

Santos, Boaventura de Souza (2014) *Epistemologies of the South. Justice against Epistemicide*, Paradigm Publishers, Boulder.

Schiavoni, C. M. (2017) 'The contested terrain of food sovereignty construction: Toward a historical, relational and interactive approach', *The Journal of Peasant Studies*, 44(1): 1–32.

Schön, D. A and M. Rein (1994) *Frame Reflection: Toward the Resolution of Intractable Policy Controversies*, Basic Books, New York.

Scoones, I. (1994) *Living with Uncertainty: New Directions for Pastoral Development in Africa*, Intermediate Technology Development Group, London.

Sevilla Guzmán, E. (2011) *Sobre los Origenes de la Agroecologia en el Pensamiento Marxista y Libertaria*, AGRUCO-Plurales Editores, Cochabamba, Bolivia.

Sevilla Guzmán, E. and G. Woodgate (2015) 'Transformative agroecology. foundations in agricultural practice, agrarian social thought, and sociological theory', In: V. Ernesto Méndez, C. Bacon, R. Cohen and S. R. Gliessman (Eds.) *Agroecology. A Transdisciplinary, Participatory and Action-oriented Approach*, CRC Press, Roca Baton.

Shanin, T. (1987) *Peasants and Peasant Societies*, Blackwell Publishers, Oxford.

Shannon, D., A. J. Nocella II and J. Asimakopoulous (Eds.) (2012) *The Accumulation of Freedom. Writings on Anarchist Economics*, AK Press, Edinburgh.

Sinai, A. (2013) *Penser la Décroissance. Politiques de l'Anthropocène*, Les Presses de Sciences Po, Paris.

Smith, L.T. (1999) *Decolonising Methodologies: Research and Indigenous People*, Zed Books, London.

Smith, S. J., J. Edmonds, C. A. Hartin, A. Mundra and K. Calvin (2015) 'Near-term acceleration in the rate of temperature change', *Nature Climate Change* 5(4): 333–336.

Sourisseau, J. M. (2015) *Family Farming and the Worlds to Come*, Springer, Netherlands.

Steffen, W., K. Richardson, J. Rockström, S. E. Cornell, I. Fetzer, E. M. Bennett, R. Biggs, S. R. Carpenter, W. de Vries, C. A. de Wit, C. Folke, D. Gerten, J. Heinke, G. M. Mace, L. M. Persson, V. Ramanathan, B. Reyers and S. Sörlin (2015) 'Planetary boundaries: Guiding human development on a changing planet', *Science* 347(6223): 1–15.

Steiner, R. (1974) *Agriculture. A Course of Eight Lectures*, The Biodyamic Agricultural Association, Rudolf Steiner House, London.

Sullivan, S. (2011) 'Banking nature? The financialisation of environmental conservation', *OAC PRESS Working Papers Series*, No. 8, Open Anthropology Cooperative Press, London.

Sullivan, S. (2013) 'Banking nature? The spectacular financialisation of environmental conservation', *Antipode* 45(1): 198–217.

Tansey, G. and A. Worsley (1995) *The Food System: A Guide*, Routledge, London.

Temper, L., D. del Bene and J. Martinez Allier (2015) 'Mapping the frontiers and front lines of global environmental justice: the EJAtlas', *Journal of Political Ecology* 22: 256–278.

Thompson, E. P. (1991) *The Making of the English Working Class*, Penguin, London.

Tilzey, M. (2017) 'Reintegrating economy, society, and environment for cooperative futures: Polanyi, Marx, and food sovereignty', *Journal of Rural Studies*, http://dx.doi.org/10.1016/j.jrurstud.2016.12.004.

Toledo, V. M. (1988) *La diversidad biologica de Mexico – Ciencia y Desarrollo*, Concyt, Mexico City.

Tornaghi, C. (2016) 'From urban agriculture to urban agroecology: Seeking food justice in the dis-abling city', *Antipode* 49(3): 781–801.

Trauger, A. (2015) *Food Sovereignty in International Context. Discourse, Politics and Practice of Place*, Routledge, London.

Tschirhart, C., J. Mistry, A. Berardi, E. Bignante, M. Simpson, L. Haynes, R. Benjamin, G. Albert, R. Xavier, B. Robertson, O. Davis, C. Verwer, G. De Ville and D. Jafferally (2016) 'Learning from one another: Evaluating the impact of horizontal knowledge exchange for environmental management and governance', *Ecology and Society* 21(2): 41.

UNCTAD (2013) *Trade and Environment Review 2013: Wake Up before It Is Too Late: Make Agriculture Truly Sustainable Now for Food Security in a Changing Climate*, United Nations Conference on Trade and Development (UNCTAD), Geneva.

UNDRIP (2007) *United Nations Declaration on the Rights of Indigenous Peoples*, United Nations, Geneva.

UN ECOSOC (1986) *Study of the Problem of Discrimination against Indigenous Populations*, United Nations Economic and Social Council, New York.

UNESCO (2008) *Links between Biological and Cultural Diversity – Concepts, Methods and Experiences*, report of an international workshop, UNESCO and The Christensen Fund, Paris.

UNESCO (2010) *Atlas of the World's Languages in Danger*, UNESCO Publishing, Paris.

UN Human Rights Council (2012) *Final Study of the Human Rights Council Advisory Committee on the Advancement of the Rights of Peasants and Other People Working in Rural Areas*, Human Rights Council, Nineteenth Session Agenda item 5, 24 February 2012, Document A/HRC/19/75, UN Human Rights Council, Geneva.

UN Women (2015) *Progress of the World's Women 2015–2016: Transforming Economies, Realizing Rights*, UN Women, New York.

Union of Concerned Scientists (2012) *Heads They Win, Tails We Lose: How Corporations Corrupt Science at the Public's Expense*, Union of Concerned Scientists, Cambridge, MA.

Union of Concerned Scientists (2015) *Counting on Agroecology. Why We Should Invest More in the Transition to Sustainable Agriculture*, Union of Concerned Scientists, Cambridge, MA.

Utting, P. (2015) *Social and Solidarity Economy: Beyond the Fringe?* Zed Books, London.

Vandermeer, J. (2010) *The Ecology of Agroecosystems*, Jones and Bartlett Publishers, Burlington, MA.

van der Ploeg, J. D. (2009) *The New Peasantries: New Struggles for Autonomy and Sustainability in an Era of Empand Globalization*, Earthscan/Routledge, London.

van der Ploeg, J. D. (2013) *Peasants and the Art of Farming: A Chayanovian Manifesto*, Fernwood, Halifax.

van der Ploeg, J. D. (2014) 'Peasant-driven agricultural growth and food sovereignty', *The Journal of Peasant Studies* 41(6): 999–1030.

van der Ploeg, J. D., J. C. Franco and S. M. Borras Jr (2015) 'Land concentration and land grabbing in Europe: A preliminary analysis', *Canadian Journal of Development Studies* 36(2):147–162.

Vidal, J. (2016) 'The tribes paying the brutal price of conservation', *The Guardian*, 28 August 2016.

Visvanathan, S. (2005) 'Knowledge, justice and democracy', In: M. Leach, I. Scoones and B. Wynne (Eds.) *Science and Citizens: Globalization and the Challenge of Engagement*, Zed Books, London.

Vitali, S., J. B. Glattfelder and S. Battiston (2011) 'The network of global corporate control', *PLoS ONE,* 6(10): e25995.

Walker, B. and D. Salt (2012) *Resilience Thinking: Sustaining Ecosystems and People in a Changing World*, Island Press, Washington, DC.

Waring, M. (1988) *If Women Counted. A New Feminist Economics*, Harper & Row, New York.

Webster, K. (2015) *The Circular Economy. A Wealth of Flows*, Ellen MacArthur Foundation Publishing, Cowes.

Westley, F. (1995) 'Governing design: The management of social systems and ecosystems management', In: L. H. Gunderson, C. S. Holling and Stephen Light (Eds.) *Barriers and Bridges to the Renewal of Ecosystems and Institutions*, Columbia University Press, New York.

Wezel, A. and V. Soldat (2009) 'A quantitative and qualitative historical analysis of the scientific discipline agroecology', *International Journal of Agricultural Sustainability* 7(1): 3–18.

Wezel, A., S. Bellon, T. Doré, C. Francis, D. Vallod and C. David (2009) 'Agroecology as a science, a movement and a practice. A review', *Agronomy for Sustainable Development* 29(4): 503–515.

Wiebe, N. (2013) 'Women of La Vía Campesina: Creating and occupying our rightful spaces', In: *La Vía Campesina's Open Book: Celebrating 20 Years of Struggle and Hope*, La Vía Campesina.

Windfuhr, M. and J. Jonsén (2005) *Food Sovereignty: Towards Democracy in Localized Food Systems*, ITDG Working Papers, FIAN and Intermediate Technology Development Group, London.

Wittman, H. (2009) 'Reworking the metabolic rift: La Vía Campesina, agrarian citizenship, and food sovereignty', *The Journal of Peasant Studies* 36(4): 805–826.

Wittman, H. (2015) 'From protest to policy: The challenges of institutionalizing food sovereignty', *Canadian Food Studies/La Revue Canadienne des études sur l'Alimentation* 2(2): 174–82.

Wittman, H., A. A. Desmarais and N. Wiebe (2010) *Food Sovereignty. Reconnecting Food, Nature and Community*, Food First, Oakland.

World Meteorological Organization (2017) *WMO Statement on the State of the Global Climate in 2016*, World Meteorological Organisation, Geneva.

Yoxen, E. (1981) 'Life as a productive force: Capitalising the science and technology of molecular biology', In: L. Levidov and R. Young (Eds.) *Science, Technology and the Labour Process*, Marxist Studies Vol. 1, CSE Books, London.

Zimmerer, K. S. (2010) 'Biological diversity in agriculture and global change', *Annual Review of Environment and Resources* 35: 137–166.

2

HOW AGRICULTURAL RESEARCH SYSTEMS SHAPE A TECHNOLOGICAL REGIME THAT DEVELOPS GENETIC ENGINEERING BUT LOCKS OUT AGROECOLOGICAL INNOVATIONS[1]

Gaëtan Vanloqueren and Philippe V. Baret[2]

Introduction

Science and technology are at the core of agricultural change. Fundamental and applied research in biology, chemistry and genetics has resulted in a constant flow of innovations and technical changes that have greatly influenced agricultural systems.

However, the direction of agricultural science and technology (S&T) is now under great scrutiny. International scientific assessments have demonstrated the increasing global footprint of agriculture, including its contribution to climate change (IPCC, 2007; Millennium Ecosystem Assessment, 2005; Steffen *et al.*, 2015), while nongovernment organizations and scientists have long called for radical changes in this field (Union of Concerned Scientists, 1996; Food Ethic Council, 2004; European Science Social Forum Network, 2005; Frison and IPES Food, 2016). Yet now, a radical change has been recommended. The International Assessment of Agricultural Science and Technology for Development has recently and officially called for a reorientation of agricultural science and technology towards more holistic approaches, after a four-year process that involved over 400 international experts (IAASTD, 2008). This panel has already been compared to the Intergovernmental Panel on Climate Change (IPCC), both for the quality of its governance and the importance of its recommendations, which are straightforward: 'Successfully meeting development and sustainability goals and responding to new priorities and changing circumstances would require a fundamental shift in agricultural knowledge, science and technology' (IAASTD, 2008). Furthermore, the IAASTD calls for greater support of agroecological approaches, which it considers of great potential for world agriculture. In contrast, the role of genetic engineering was the sole element of controversy in the final statement, which is weak on this point.

If the IAASTD recommendations, as well as those of the IPCC and Millennium Ecosystem Assessment, are to be taken seriously and implemented, we need to understand why the current agricultural S&T landscape has not sufficiently supported holistic and agroecological approaches, while other agricultural innovations, such as transgenic crops, were able to flourish.

In this chapter, we focus on the development of genetic engineering and agroecology, two important trends within biological and agricultural sciences during the second half of the twentieth century. Both genetic engineering and agroecology were insignificant or non-existent scientific branches before the early 1970s. Scientists and public authorities could theoretically see them as two complementary fields of research with equal potential to improve agricultural systems. Genetic engineering and its vital complementary discipline, molecular biology, have attracted more research funds than agroecology in recent decades (IPES Food, 2016; Watts and Scales, 2015). Agroecology has not acquired such momentum although its influence is also growing (Parrott and Marsden, 2002; Pretty *et al.*, 2003; Union of Concerned Scientists, 2015; Méndez *et al.*, 2013).

It is beyond the scope of this chapter to assess the advantages and drawbacks of the two trends. What drove us to compare them is the necessity to explain their development differential. Is this differential only due to the intrinsic superiority of genetic engineering compared with agroecology, or can it be methodologically explained by other factors? If so, which ones?

The use of concepts from the evolutionary line of thought (evolutionary economics) – such as technological paradigms and trajectories, technological regimes, path dependence and lock-in – is vital in explaining this development differential.

In the first section, we discuss genetic engineering and agroecology as two technological paradigms that make sense and science. Technological paradigms are a concept taken from the study of industrial innovations that has seldom been used to analyze agricultural innovations. The second section explains how a Systems of Innovation (SI) approach can be used to analyze the factors (determinants of innovation) that influence the choice of technological paradigms as well as the development of technological trajectories within agricultural research systems (ARS). The third section offers a systematic description of these determinants, whose combination induces an imbalance between genetic and agroecological engineering. ARS emergent properties such as path dependence and lock-in are analyzed in the fourth section. In the final section we discuss the issues arising from our observations.

Technological paradigms and trajectories, from factories to farmers' crops

Technological paradigms and trajectories

The concepts of 'technological paradigms' and 'technological trajectories' have been suggested by Dosi (1982) to allow research to go beyond the 'demand-pull' and 'technology-push' theories of technological change. While Dosi initially introduced his concepts in the field of technological change within industrial structures, it was later argued that they could be extended to agriculture (Possas *et al.*, 1996).

Dosi defined a technological paradigm as a 'model and a pattern of solution of selected technological problems, based on selected principles derived from natural sciences and on selected material technologies' (Dosi, 1982). A technological paradigm defines an idea of 'progress' by embodying prescriptions on which directions of technological change to pursue and which to neglect. This is a broad analogy with the Kuhnian definition of a scientific paradigm, which determines the field of enquiry, the problems, the procedures and the tasks (Kuhn, 1962). A technological trajectory is the 'pattern of normal problem solving activity (i.e. of progress) on the ground of a technological paradigm' or, in other words, the improvement pattern of concrete solutions based on a paradigm.

Applications of these concepts in agriculture vary widely. Parayil (2003) described the Green Revolution and the Gene Revolution as two technological trajectories.[3] Biotechnology, including agricultural biotechnologies, was soon presented as a new technological paradigm (Russel, 1999) and several authors have analyzed particular technological trajectories in agrochemical and agro-biotech industries (Joly and Lemari, 2004; Chataway *et al.*, 2004; Dibden *et al.*, 2013).

Genetic engineering and agroecology, or 'agroecological engineering'

So far, genetic engineering and agroecology had not been compared as two technological paradigms that rely on two different scientific paradigms, pursue different objectives and are composed of different sub-trajectories (Table 2.1).

Genetic engineering is the deliberate modification of the characteristics of an organism by the manipulation of its genetic material. The main technology upon which this process is based is transgenesis, following the discovery of the recombinant DNA technique in 1973. The best known applications of genetic engineering in agriculture are transgenic herbicide-tolerant plants: soybean or insect-resistant Bt maize in the USA. The fundamental strategy in genetic engineering is either to modify the plants to allow them to be productive in adverse conditions, such as those caused by pests, pathogens, drought, saline environments and unfertile soils; or to design plants for new objectives, such as altered nutritional content. This goal fits the scientific paradigm that underlies genetic engineering, i.e. reductionism. Genetic engineering has been described as a new technological paradigm (Orsenigo, 1989; Feindt, 2012), although this conceptualization has not yet been much explored in the literature.

Eleven years after their introduction transgenic crops were grown on 114 million hectares in 23 countries (James, 2007).[4] The progress of genetic engineering has been relatively fast. In the US, the number of field trial permits issued rose from zero in 1986 to 107 in 1991, to more than 1,000 every year since 1998. In 2005 the number of field trials permits had already reached 12 000 (Information Systems for Biotechnology, 2006).

Agroecology emerged from the convergence of ecology and agronomy (Dalgaard *et al.*, 2003). It is the application of ecological science to the study, design and management of sustainable agroecosystems (Altieri, 1995). We use the term 'agroecological engineering' in this chapter to put the two technological paradigms on an equal footing.[5] 'Agroecological engineering' refers to the fact that agricultural systems can be 'engineered' by applying agroecological principles, just as plants are 'engineered' by transgenesis in 'genetic engineering'. The term 'agroecological engineering' has seldom been used, except occasionally in China (Yan and Zhang, 1993).[6]

Agroecological engineering is an umbrella concept for different agricultural practices and innovations such as biological control, cultivar mixtures, agroforestry systems, habitat management techniques (for instance, strip management or beetle banks around wheat fields), or natural systems agriculture aiming at perennial food-grain-producing systems. Crop rotations, soil fertility improvement practices, mixed crop and livestock management and intercropping are also included. Some applications involve cutting-edge technologies, while others are old practices (for instance, traditional systems that provide significant insights into agroecology). Globally, hundreds of agricultural systems are based on agroecological principles – from rice paddies in China to mechanized wheat systems in the US, although data are not as accurate as for the acreage of transgenic crops (Parrott and Marsden, 2002; Pretty *et al.*, 2003; Wezel *et al.*, 2014).

The scientific paradigm on which agroecological engineering relies is ecology (and holism). The objective is to design productive agricultural systems that require as few agrochemicals and energy inputs as possible, and instead rely on ecological interactions and synergisms between biological components to produce the mechanisms that will enable the systems to boost their own soil fertility, productivity and crop protection (Altieri, 1995; Kremen and Miles, 2012). Some aspects of agroecological engineering may be related to biomimicry (Benyus, 1997; Kirschenmann, 2007). While the objective of genetic engineering is to improve only a single element of the agroecosystem (modifying existing plants or designing new plants), the objective of agroecological engineering is to improve the structure of the agricultural system and 'to make every part of the structure work well' (Liang, 1998; see also Kremen *et al.*, 2012; Altieri *et al.*, 2017).

In a dynamic perspective, three conceptual levels may be discerned: (1) the two technological paradigms: genetic engineering and agroecological engineering; (2) the technological trajectories (the progress within these two paradigms); and (3) the various sub-trajectories (the concrete implementations of each paradigm, meaning Bt-resistant and herbicide-resistant plants for genetic engineering; agroforestry and habitat management strategies for agroecological engineering – see Table 2.1).

TABLE 2.1 Genetic engineering and agroecological engineering are two different technological paradigms

Technological paradigms	Genetic engineering	Agroecological engineering
Basic definition	Deliberate modification of the characteristics of an organism by the manipulation of its genetic material	Application of ecological science to the study, design and management of sustainable agroecosystems
Implicit objective	Engineering plants: Modify plants to our best advantage by making them productive in adverse conditions or by designing them to fit new objectives	Engineering systems: Improve the structure of an agricultural system to make every part work well; rely on ecological interactions and synergisms for soil fertility, productivity and crop protection
Scientific paradigm underlying the technological paradigm	Reductionism	Ecology and holism
Examples of sub-trajectories progressing along the technological paradigm	Bt insect resistant plants, herbicide-tolerant plants, virus-resistant plants etc.	Biological control, cultivar mixtures, agroforestry, habitat management techniques etc.

Paradigms and the real world

The comparison of two broad and archetypal paradigms may seem too caricatured or simplistic to be useful. Yet the dual opposition between genetic engineering and agroecology already exists in the real world, both in science and in society. Proponents of both paradigms claim that their paradigm is the only one able to feed the world and solve environmental issues, and that the other paradigm puts the world at great risk. Paradigms consequently influence science and technology choices. This fact justifies using these concepts in a comparative framework.

Several authors have used paradigms to analyze the models at stake in the agrifood sector. Lang and Heasman (2004) have convincingly put forward the concept of 'food paradigms'. They have argued that the Life Sciences Integrated paradigm and the Ecologically Integrated paradigm were competing to replace the Productionist paradigm in food systems. Allaire and Wolf (2004), who focus on food innovations, similarly picture three 'innovation paradigms': an old one (the mass-production and consumption Fordist model) and two new ones (the first is represented by the segmentation of products within supermarkets, and the second by products with strong identities such as those available at farmers' markets). The competition between rival 'agrifood paradigms' has also been put forward by Marsden and Sonnino (2005) and Morgan *et al.* (2006).[7] This point of view has already been used to analyze the debate about the possibility of coexistence between genetic engineering and alternative agricultures (Devos *et al.*, 2009; Levidow and Boschert, 2008). Such analyses avoid a fake black and white vision of the agrifood sector. In reality, they enable researchers to analyze the trends and choices at stake in the agrifood sector, all of which are vital for democratic S&T choices.

Our analysis focuses on agricultural innovations. This scope is therefore much more limited than the agrifood paradigm perspectives chosen by the authors mentioned above.

Three remarks must be made as the real world is obviously not as clear-cut as theoretical concepts:

- Hybrid situations exist. Systems biology, for instance, focuses on interactions between components of biological systems, such as the enzymes and metabolites in a metabolic pathway. It thus combines a focus on ever-smaller levels of the living systems (from molecular biology and reductionism) with an interest in interactions (from the systems approach).
- Within trajectories, there is a wide spectrum of diversity. For instance, biological control of insects can result in innovations such as the mass release of predator insects, which is an efficient intervention but has no impact on the practice of monoculture, an important cause of insect problems. If designed in the agroecological paradigm, biological control can lead to habitat management solutions (landscape ecology) such as beetle banks and strip management, which have a structural effect on disease control (e.g. Ekroos *et al.*, 2014; Levie *et al.*, 2005). Some agroecological approaches may also be used in conventional systems. In practice, agricultural innovations are used in agricultural systems with various degrees of closeness to agroecological principles. In fact, farmers combine various types of innovations that stem from different trajectories.
- Agroecological engineering is not to be confused with organic farming. Organic farming has many principles in common with agroecology. Organic farmers have implemented many agroecological innovations in their crops, although they may in certain cases also replicate the productivist approach that goes against agroecological principles (Dupuis, 2000; Guthman, 2000, 2014; IPES Food, 2016).

How agricultural research systems shape innovation choices

One of the main questions behind Dosi's concepts of technological paradigms was 'How does a paradigm emerge in the first place and how was it "preferred" to other possible ones?' (Dosi, 1982). Dosi's hypothesis was that economic forces, together with institutional and social factors, operate as a 'selective device' (selection environment) by influencing criteria such as feasibility and profitability at each level, from research to development.

Dosi's selective device has been overshadowed by a similar yet stronger concept of technological regime. Technological regimes are the (sets of) rules of the game that guide the direction of technological innovation and use (Possas *et al.*, 1996).[8] Different approaches have analyzed the various factors shaping the technological regime: the relative price of resources (Hayami and Ruttan, 1985), the factors of technology adoption by farmers (Sunding and Zilberman, 2001) and the public policies and market-related factors (Bailey *et al.*, 2010; Tait *et al.*, 2001; Bijman and Tait, 2002). Concerning genetic engineering in particular, Parayil (2003) has demonstrated that the key factors in the emergence of the Gene Revolution (compared with the Green Revolution) were not only the advances of cellular and molecular biology, but also the revolution in information technologies and global economic forces, such as the new rules of global finance and free trade, or consolidations and strategic alliances in the agricultural input industry. Russel (1999) and Wield *et al.* (2010) focus on the aspects of international political economy that encouraged biotechnologies, specifically the structural power of the US government and American companies.

We maintain however that the analysis is not yet complete for genetic engineering, and almost non-existent where agroecological engineering is concerned. Moreover, the advantages of a systematic comparison have not been exploited.

Theoretical concepts: analyzing the current technological regime through an SI approach

Our approach is in the realm of systems of innovation (SI) approaches (Edquist, 1997; Lundvall, 1992; Nelson, 1993; Soete *et al.*, 2010). The SI approach analyzes the components of systems of innovation, their functions and the relationships between components on a national, sectoral or regional scale. We focus on agricultural research systems (ARS), which are a large part of the SI that shapes agricultural S&T. ARS are composed of organizations (such as scientists, universities, private companies); and of institutions, such as R&D policies (Borrás and Edquist, 2013; Marrocu *et al.*, 2013). A simplified representation of ARS and SI is shown in Figure 2.1.

Within ARS, we have identified the factors that influence the choice of technological paradigms and the development of technological trajectories. These factors are 'determinants of innovation': social, cultural, economic and/or political factors that act positively or negatively on the development of technological trajectories (Edquist, 2001). It is the addition and combination of these determinants that collectively form the technological regime.

We could have conducted a similar analysis for other concepts. Pestre's concept of 'knowledge production regimes' (2003) encompasses institutions, beliefs and practices as well as political and economic regulations that define the place and role of the sciences. Friedmann (2005) has suggested the concepts of 'food regimes' to encompass broader historical patterns of agricultural production and food consumption (see also McMichael, 2016). Our specific focus on ARS narrows

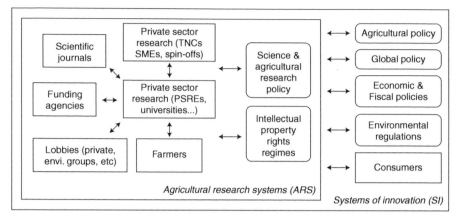

FIGURE 2.1 Agricultural research systems (ARS) are a part of systems of innovation (SI)

Notes: Rectangles represent actors. Rectangles with rounded angles represent policies and regulations. Not all interactions are represented. TNC, transnational corporations; SME, small and medium enterprises; PSREs, public-sector research establishments.

our scope. Besides, the use of evolutionary economics concepts is coherent with the relevance of this line of thought for agriculture (Marechal *et al.*, 2008).

Sources and methodology

The sources for our analysis of determinants of innovation are manifold:

- Interviews with scientists and stakeholders in five agrifood chains (wheat, apple, sugar beet, maize, soybean) (20–30 interviews each) in two countries (Belgium and Argentina).
- Participant observation of public forums on agriculture, science and innovation in Belgium, France and the UK, as well as in Brussels for European Union (EU) institutions.
- An analysis of key policy documents, such as White Papers from public authorities (from the US National Research Council, to the European Commission, to the Food and Agriculture Organization, FAO).
- A multidisciplinary literature review.

The analysis of the determinants of innovation uses (1) evidence and a few illustrative quotes from our surveys of stakeholders in ARS; (2) logical reasoning using results and conclusions from published research; and (3) specific illustrative cases of transgenic plants and/or agroecological innovations.

Our approach was not carried out on a national scale, unlike many SI approaches. The international division of research has already taken place (Edquist, 2011; Pardey and Beintema, 2001). Russel (1999) emphasizes the need to apply the idea of territory loosely and include transnational aspects, even if national differences exist in ARS and SI (Tait *et al.*, 2001; Tödtling *et al.*, 2009). This cross-country, multi-source approach is considered useful and valid for agricultural research in both developed and developing countries.

Two assumptions are made. First, 'agricultural research' comprises agricultural as well as biological sciences. Secondly, genetic engineering is closely associated with molecular biology, the basic science on which it rests, even if molecular biology has other goals and is also related to agroecology.

Special attention is intentionally paid to the public sector and to the influence of determinants on agroecological engineering, a field much less explored than genetic engineering. Cultivar mixtures and agroforestry systems are used as examples of agroecological innovations that have already lived up to their reputation, while biological control has an intermediate status.[9]

The results of our analysis sum up all the ARS determinants that together structure the current technological regime. We assume these determinants to be predominant in the choice of technological paradigms and the development of technological trajectories. This does not mean that each determinant is valid in all cases (i.e. present in all research institutions and true for every scientist). Consequently, each determinant must be understood as part of a whole. In fact, the systems approach (Checkland, 1981) assumes that the overall performance of a complex of elements depends not only on the characteristics of each element, but also on the interactions between these elements. It is thus the aggregation of the various determinants that matters.

Determinants of innovation shape a technological regime that induces an imbalance between genetic and agroecological engineering

The determinants of innovation fall into three main categories: agricultural science policies, private sector research and public sector research (Table 2.2). The following sections explore how each determinant affects genetic and agroecological engineering.

Agricultural science policies

Policies influence technological paradigms in four different ways: the choice of research orientations, the relationships between public and private sectors, the power of lobbies and the role of media and lobbies.

Research orientations: focus on growth, competitiveness and biotechnologies

Science policies are explicitly and increasingly oriented towards growth and national competitiveness. These goals are clearly stated in key policy documents,

TABLE 2.2 Determinants of innovation in agricultural research systems that induce an imbalance between genetic and agroecological engineering

Categories	Subcategories	Determinants of innovation
(1) Agricultural science policies	Research orientations	Focus on growth, competitiveness and biotechnologies
	Relationships between public and private sectors	Public-private partnerships
		Public-private division of innovative labour
	Influence of lobbies	Imbalance in the power of lobbies
	Media	The media channel public opinion towards a single paradigm
(2) Private sector	Research orientations	Focus on biotechnologies and importance of patents
(3) Public sector	Cultural and cognitive routines (values and world views of scientists)	Assumptions on current and future agricultural systems
		Assumptions on past agricultural systems
		Assumptions on the nature and value of innovations
	Organization within research systems (rules of the game)	Views of complexity and framing of agricultural research
		Assessment of the performance of agricultural innovations
		Specialisation vs. interdisciplinarity
		'Publish or perish'
		Technology transfer mission: patents, spin-offs and extension

including the EU's 2007-2013 research and development (R&D) framework programme (European Commission, 2005a, 2005b), the Europe 2020 growth strategy (European Commission, 2013) and the US National Innovation Act (Congress of the US, 2005). Since the early 1980s, biotechnologies have been intimately linked with these objectives (European Council, 1981; National Research Council, 1987). Most countries then implemented specific policies on agricultural biotechnologies such as transgenic plants. These policies are still strongly supported in the United States (NRC, 1998) as well as in the European Union, despite the 1999–2004 de facto moratorium on transgenic crops (European Commission, 2002, 2004). International organizations have also supported genetic engineering, though calling for caution and asking for a specific investment in pro-poor technologies, programmes and policies (UNDP, 2001; FAO, 2004).

Genetic engineering has benefited from the creation of a broad, favourable environment, which included funds, specific infrastructures (such as the European Molecular Biology Laboratory) and a workforce trained in molecular techniques – a request expressed in early policy documents (National Research Council, 1987).

During our surveys, scientists mentioned the fact that molecular biology continued to be important in all EU programmes even in the 2000s: 'You had to have a molecular biologist in your research project for it to be accepted'. The increased importance of molecular biology has affected scientific institutions themselves. An analysis of the recruitment of scientists at the French *Institut National de la Recherche Agronomique* (INRA) demonstrates that the share of molecular biologists grew from less than 10% in the 1970s to more than 20% of total job opportunities between 1988 and 1997 (Mignot and Poncet, 2001). In Europe, strong consumer opposition to transgenic plants and the 1999–2004 de facto moratorium on the commercialization of transgenic plants have had a strong negative impact on the development of genetic engineering, with multinationals pulling out of R&D in Europe. EU-supported research on transgenic plants was also partly redirected towards the life sciences linked to human health. Nevertheless, research in genetic engineering continued and the number of field tests rose after the end of the moratorium.

In contrast, agroecological engineering has not been linked to growth and competitiveness goals. Sustainable agriculture only featured more noticeably on research agendas from the late 1990s onwards. 'Sustainable agriculture research and education' programmes in the US, 'agrienvironmental schemes' in the EU and organic farming research programmes facilitated the development and the adoption of agroecological innovations. Some agroecological sub-trajectories even benefited from the greater interest in molecular biology (biocontrol for the identification of useful biocontrol agents). However, research at the agroecosystem level has not developed as intensely as research at the molecular level. Some research institutions even lost some agronomists and soil microbiologists. Between 1982 and 1988, the substantial increase in funds, faculty and students dedicated to biotechnology in US land-grant universities was concomitant with a decline in the numbers of plant and animal breeders (Hess, 1991).

Relationships between the public and the private sectors

Two trends in the relationships between public and private research have influenced the technological paradigm: the increased influence of industry through public-private partnerships, and the division of the innovative labour between public and private entities.

Public–private partnerships

The promotion of public-private partnerships (PPPs) is now explicitly part of the missions given to public-sector research establishments, as a mean to transfer technology and knowledge (Muscio *et al.*, 2013; Piesse and Thirtle, 2010; Tait *et al.*, 2001). Examples of PPPs in the realm of biological and agricultural sciences include the well-known alliance between Novartis and the University of California to support basic agricultural genomics research (US$25 million over five years) or plant genomics platforms such as the French initiative Genoplante.

Genetic engineering has benefited more from PPPs than agroecological engineering, because PPPs were only launched on technological trajectories in which private firms had an interest. (Note that firms have invested more in modern agricultural biotechnologies than in agroecological innovations in the last three decades, see the section on 'Private sector research'.) Moreover, PPPs have had an indirect but more profound impact: a change in the culture of science. A key finding of an external evaluation of one of these large PPPs – the University of California-Novartis agreement – found that administrators and university scientists who participated in the partnership tended to define the public good as research that leads to the creation of commercialized products, narrowing the definition of the public good towards private goods (Busch *et al.*, 2004). As Levidow *et al.* (2004) put it, 'even a small proportion of industry funding can influence overall research priorities: the tail can wag the dog'. This trend is favourable to transgenic plants but unfavourable to agroecological innovations with a public good characteristic. In the end, PPPs could induce a redirection of public funds towards the areas of research leading to these partnerships if they are considered likely to have a positive effect on economic growth (Pew Initiative on Food and Biotechnology, 2003; Food Ethic Council, 2004).

Another trend in public research-industry links is privatization. Direct privatization of research infrastructure and resources has only been an important feature of reorganization in the UK. Indirect privatization has happened nevertheless in many countries by giving private research institutes access to public funds, or through the 'industry capture of research programmes', e.g. the increased presence of industrial representatives on the committees establishing research priorities (Alston *et al.*, 1998b, 2001; King *et al.*, 2012). This indirect privatization, like PPPs, favours the innovations that attract the private sector, at the expense of innovations of a public good nature (whose benefits are not exclusively appropriated by the farmer, but are of wider public value as they produce large externalities). These trends are in line with the analysis of industry's increased influence on public science (Slaughter and Leslie, 1997; Naseem *et al.*, 2010).

Public–private division of innovative labour

There is a 'division of innovative labour' (Arora and Gambardella, 1994) among the various public and private research institutions in the agricultural and biological sciences. Public-sector research focuses on basic research while the private sector focuses on applied R&D.

Genetic engineering has benefited greatly from this division of innovative labour, as research on this technological paradigm occurs at all levels (basic, applied and development). An analysis of US patents issued between 1975 and 1998 in the field of biological sciences applied to plant agriculture demonstrates that universities undertake the initial research that contributes to the evolution of technological trajectories and yields the most original and most general work, while start-up companies specialize in turning basic research into applied innovations and large corporations concentrate on later developments (Graff *et al.*, 2003; Graff, 2004;

King *et al.*, 2012). More than 70% of US publications cited in agricultural bio-technology patents are authored by US university researchers,[10] a good measure of the importance of public science (Xia and Buccola, 2005). In other words, bio-tech industries depend on public science much more heavily than other industries (Breschi and Catalini, 2010; McMillan *et al.*, 2000). Contrariwise, the division of innovative labour is not a positive factor if all research stages are not shared out, for instance if the private sector does not invest in applied research and development, which is the case for many agroecological sub-trajectories (see later).

Imbalance in the power of the lobbies

The analysis of the influence of lobbies (providers of agricultural inputs, con-sumer groups, environmental conservation groups) is an integral part of the SI approach, since they influence strategic choices and thus, technological paradigms (Edquist, 1997).

Genetic engineering has received the backing of strong industrial platforms such as Bio in the US or Europabio in the EU. Their lobbying has considerably influenced public policies, such as intellectual property rights (IPR) regimes in the framework of the World Trade Organization, as well as research framework programmes at the European Commission (Balanya *et al.*, 2003; Parayil, 2003). However, they did not manage to stop the 1999–2004 de facto EU moratorium on transgenic crops.

The activity of green lobbies on agroecological engineering is not as straight-forward. Environmental NGOs such as Greenpeace and the Soil Association have put more energy into banning transgenic crops or securing strong regulations than into promoting a research agenda for alternative technological paradigms such as agroecological innovations. However, slowing down one trajectory does not auto-matically result in support for another. Remember that the few scientific organiza-tions that back a stronger research agenda on agroecology (Union of Concerned Scientists, 1996, 2015; European Science Social Forum Network, 2005) have sig-nificantly less clout than mainstream scientific organizations that support genetic engineering (Royal Society *et al.*, 2000).

The media channel public opinion towards a single paradigm

The simplified approach characterizing the mainstream media favours a binary approach concentrating on the benefits and risks of genetic engineering, for better or worse. The stress on potential risks is a drawback, yet the coverage of ambitious possible outcomes has maintained trust in the technology's potential. Media have not adopted thinking on technological choices that would have discussed the comparative advantages of transgenic crops and their alternative options. The archives of *The New York Times* between 1981 and 2008 contain, for instance, 2,696 references to 'genetic engineering' compared to three to 'agroecology', seven to 'agroforestry' and none to 'cultivar mixtures' (*The New*

York Times, 2008).[11] Moreover, agroecological innovations, when considered, are usually presented as innovations for organic agriculture, not as possible agricultural practices in the future. The media's stand is of great importance, given the power they wield over public opinion. As communication theorist Bernard Cohen observed in what became a widely accepted communication theory: 'the press is significantly more than a purveyor of information and opinion. It may not be successful much of the time in telling people what to think, but it is stunningly successful in telling its readers what to think about' (Cohen, 1963). The public's attention is thus drawn to the risks and benefits of genetic engineering, not to the alternatives, such as agroecological engineering.

Private sector research

The private sector is an increasingly important actor in agricultural research, accounting for roughly one-third of global agricultural research spending (Pardey and Beintema, 2001; see also Wright, 2012). This share rises to 50% in OECD countries, where the growth of private R&D is three times that of public research (Alston *et al.*, 1998a). In capitalist market economies, innovation is a tool to generate higher revenues and secure competitiveness, a matter of survival for most private companies. However, private companies do not invest equally in all technological trajectories. R&D strategies rely on the possibility of securing sufficient future revenues from R&D spending. Consequently, private companies focus on innovations that can be protected by patents or other forms of IPR regimes.

A key event gave transnational companies the green light for huge investments in genetic engineering. In 1980, the United States Supreme Court decision in Diamond v. Chakrabarty allowed patenting on microorganisms, and this was later extended to plants (National Research Council, 2002). Companies such as Monsanto or Novartis (now Syngenta) then made strategic decisions to orient their R&D activities towards genetic engineering in the 1980s, and acquire the appropriate companies throughout the 1990s. Between 1976 and 2000, firms invested more in modern agricultural biotechnologies than in other patentable biological innovations, such as biocontrol of pests and diseases (Heisey *et al.*, 2005). Consequently, today three out of four US agricultural biotechnology patents are in the private sector (Graff *et al.*, 2003; Fuglie *et al.*, 2012a, 2012b).

Agroecological innovations have not benefited from this new regime of intellectual property rights. Only a few patentable agroecological innovations have attracted private actors, such as biological control (which leads to patents on methods for rearing biocontrol agents). The private incentives for agroecological research are actually limited as private companies are unable to capture all the benefits resulting from these innovations (Sunding and Zilberman, 2001). For instance, innovations in agroforestry systems can hardly be patented; they are hard to promote as their benefits are very long-term (wood is a long-term product) and their benefits are to a large extent public goods (with positive environmental externalities such as carbon sequestration or biodiversity). Consequently, agroecological innovations such as agroforestry or cultivar mixtures have mainly relied on the public sector for their development.

Public sector research

The internal organization of the public agricultural research sector (universities, national and independent not-for-profit research institutes), as well as cultural and cognitive routines, are also part of the technological regime.

Cultural and cognitive routines (values and world views of scientists)

Cognitive and cultural rules or routines are the assumptions scientists and experts frequently make. They make them look in particular directions and not in others (Dosi, 1982; Nelson and Winter, 1982). It has been acknowledged for a long time that values and world views interfere with science as well as with risk assessment, expertise and public policies (Flores and Sarandón, 2004; Jasanoff, 1990; Lacey, 1999; Stirling, 1999). Assumptions about current, future and past agricultural systems, and assumptions about the nature of innovation, generate an imbalance between the two technological paradigms.

Assumptions about current and future agricultural systems

A common assumption made by scientists about current modern agricultural systems is that they only require small adaptations. Problems such as pesticide risks are acknowledged, but the validity of the model in itself – monoculture, reliance on a high level of external inputs such as fossil fuels – is rarely questioned. Thinking on agriculture remains close to the industrial approach that has characterized agricultural sciences for more than a century (Bawden, 1991), complemented when possible by some soft ecological concepts such as integrated pest management (IPM). As for the future, scientists mainly think in terms of the *most probable* future agricultural systems, not the *most desirable* future systems, i.e. they seem to forecast future agricultural systems by integrating the most probable economic and political trends. These trends are the globalization and liberalization of agricultural commodity markets, two trends that push all regional agricultural systems into global competition (Bernstein, 2014; Cerny, 1997), and the strengthening of the strategies of the dominant actors in agrifood transformation and retailing (Goodman and Watts, 1997; Lee *et al.*, 2012). As these trends exacerbate economic pressures on farmers, the pursuit of input-intensive approaches is thought to be the most probable evolution. Many scientists frame their research around these constraints and behave as if global warming and the rising cost of energy did not demand major policy shifts (Kirschenmann, 2007) or as if there was no alternative to the mainstream economic trends (Patel, 2007).

Genetic engineering fits into these expected trends: it does not entail many changes in current farming systems, such as monoculture. It only uses different types of seed, inputs (herbicides and insecticides) and management schemes and is thus seen as 'potentially transferable' to farmers.

Innovations and systems closest to the principles of agroecology face the opposite situation as they challenge the fundamentals of the current agricultural system, such as monoculture and crop protection, that rely mainly on external interventions. Many scientists do not explore these agroecological innovations because 'it goes against the flow', as a scientist explicitly stated during an interview when asked why cultivar mixtures were not being researched to create systems resistant to fungal diseases. Scientists and stakeholders refer to current social and economic barriers impeding the use of some possible innovations by farmers today to justify the research deficit. Current barriers are seen as permanent immovable obstacles. As a result, some agroecological innovations are considered to be 'theoretically valid' but 'not feasible' in modern agricultural systems, as they 'go against the flow'. The attitude towards genetic engineering is different: the current opposition of consumers in Europe is not seen as an immovable obstacle.

Assumptions about past agricultural systems

Past agricultural systems are rarely seen as sources of insights for innovation in mainstream agricultural science, where modernization remains an important *leitmotiv*. This is a small issue for genetic engineering, which has little need of insight from past agricultural systems. On the contrary, agroecology values past systems as a source of insight for the improvement of current systems. Examples of 'rediscovered' systems are the subtle combinations of rice terraces and agroforestry systems in Madagascar, rice-fish systems in East Asia, Andean *waru-waru* ridge fields that control drought and frost and Mesoamerican *milpa*-solar cropping systems in Mexico (Esquinas-Alcázar, 2006).[12] Such systems are seen as a 'return to old times', worthy of curiosity, but not of real academic interest. Consequently, indigenous knowledge and traditional systems have until recently remained a weak area of agricultural research (Gómez-Baggethun *et al.*, 2010; IAASTD, 2008).

Assumptions about the nature of innovation (biotechnological and agroecological innovations)

Scientists make two important assumptions about the nature and value of innovations that generate an imbalance between technological paradigms. The first difference relates to the nature of innovations. Molecular biology and genetic engineering are seen as 'breakthrough' scientific discoveries, which lead to fundamental or radical innovations. On the other hand, agroecological innovations are seen as 'incremental' innovations, despite agroecology's record of success stories, such as the restoration of traditional Incan terracing systems that increased productivity by as much as 150% (Parrott and Marsden, 2002); wheat-poplar agroforestry systems that produce as much 'grain+wood' output on 1 hectare (ha) as 1.3 hectares of separate monocultures (SAFE, 2005); and barley cultivar mixtures that reduced the incidence of powdery mildew, and cut fungicide use by 80% on 350,000 ha in East Germany between 1984 and 1990 (Vallavieille-Pope, 2004).

Genetic engineering is also seen as a provider of 'total' solutions, while agro-ecological engineering is seen to only offer 'partial' solutions that must be completed by other strategies. The reality is far from this simplistic assumption: plants genetically engineered for resistance to diseases or drought are expected to have an improved, but not complete resistance. Rather than their true potential to solve problems, the 'low-technology' aspect of agroecological innovations is a possible cause of the scant interest they receive in ARS. As Wolfe, a prominent scientist working on cultivar mixtures, puts it: 'Is it just too simple, not making enough use of high technology?' (Wolfe, 2000).

The second difference is the widely shared belief that genetic engineering is of universal value, a belief that does not exist in agroecology (Lacey, 2002). According to this belief, while agroecology would only be of value for some problems and in some regions, genetic engineering would be able to solve all problems in all places. This major assumption in favour of genetic engineering is supported by three major arguments. First, many scientists consider agroecological innovations to be innovations 'for organic agriculture' because agroecology and organic agriculture share common agroecological principles. Research funding is low because organic agriculture is considered a niche market innovation and because of the mainstream view that organic agriculture is unable to feed the world – an opinion disproved by recent research (Badgley *et al.*, 2007; FAO, 2007; IPES Food, 2016).

Second, many scientists reject agroecological innovations as being neither true research nor 'development'. Nothing explains this better than the actual words of a scientific adviser for a public authority responsible for agricultural research funding interviewed as part of this research:

> It is very difficult to finance a research that is not anymore a 'real one', i.e. when the scientists have already put into evidence all the scientific laws they could put into evidence, even if that research project needs a large-scale validation. These projects should systematically go to the Development department but it's hard and it rarely happens: it seems too 'research' for the Development department.

Third, some agroecological innovations are dismissed because their current record of results for commercial real-scale applications is considered too low. This is clear in the case of elicitors of induced resistance, a new possible way to protect crops by inducing plant defence mechanisms. While the early fundamental research on this sub-trajectory dates back to the 1940s (Kuc, 2001), research applied to commercial crops has been much more recent. For instance, the screening of the thousands of possible molecules effective for apple diseases truly started in the 2000s, yet the absence of rapid positive results led many stakeholders to conclude that elicitors were not a solution, while they should actually be considered as a fairly new trajectory, just like transgenic disease-resistant apple trees (Vanloqueren and Baret, 2004).

The fact that the value of agroecology has not been universally acknowledged may also arise from the absence of a strong integrated prospective vision of agroecological engineering, which would take into account the possible synergies between the different agroecological sub-trajectories. Such a vision could lead to breeding wheat cultivars designed to be most productive in cultivar mixtures grown in agroforestry systems which would themselves include beetle banks, and finally be protected by the mass release of aphids for pest biocontrol.

We may conclude from this analysis of cultural routines that scientists have a biased approach to the two technological paradigms analyzed in this chapter. Genetic engineering is recognized as a technological paradigm and trajectory, while this is not the case for agroecological engineering. The scientist's perception of genetic engineering is dynamic: genetic engineering has produced results in the past, it does so today and it has potential for the future: it is a technological paradigm and trajectory. The scientist's view is static when it comes to agroecological engineering: scientists acknowledge that agroecology exists, but they do not consider its innovative possibilities in the same light as those of genetic engineering. This 'variable geometry' approach is in total contradiction to sound science, which should have a balanced vision of the two paradigms, as both make sense and make science (as shown in the first section of this chapter).

Organization of research systems

Five organizational aspects influence technological paradigms and trajectories: the different views of complexity and the framing of agricultural research, the assessment of the performance of agricultural innovations, the specialisation of research, publication pressure and the technology transfer mission.

Views of complexity and the framing of agricultural research

Science deals with complexity by nature. Both molecular biologists and agroecologists agree that living beings and ecosystems are complex systems. Yet the two paradigms deal with different types of complexity. Molecular biology and genetic engineering are about complexity at the cell and the gene levels. The technical and technological developments of the three last decades have greatly helped scientists to deal with this type of complexity. Computers process more and more data while DNA sequencers can sequence longer strands of DNA more quickly. These developments have allowed a 'Taylorization' of research. Scientists in this field now compare their institutions in terms of thousands of Mb (mega base pairs) per day.

Agroecological engineering, on the other hand, is about complexity up to the ecosystem level. The main approach is a systems approach, which does not fit the laboratory realms as well as a reductionist approach. A good example of this complexity is agroecological research into the improvement of coffee groves under high-canopy trees in Central America; this involved the identification of the optimal shade conditions that could minimize the entire pest complex, maximize the beneficial microflora and fauna while maximizing yield and coffee quality (Staver et al., 2001; Jha and Dick, 2008; Perfecto et al., 2014).[13] New software and tools also helped these analyses, but such highly context-dependent research is not open to the standardization processes that were so useful to the development of genetic engineering. Thus, agroecological innovations are thought to be too complex to be dealt with, which could seem paradoxical given the complex technologies used in genetic engineering.

Time and size requirements for research on each paradigm also differ widely. The transposition of a transgene into a host plant can be detected by easy means in the lab within days, and lead to scientific discoveries that are published in renowned scientific journals. In contrast, sound research on a number of agroecological sub-trajectories requires large-scale and long-term on-farm experiments. Proving the positive effect of rice cultivar mixtures on the prevalence of an important disease required more than 3,000 ha of Chinese rice fields, as scale affected results (Zhu et al., 2000). The productivity assessment of wheat-walnut agroforestry systems takes years from planting to publishable results, a requirement that poorly matches the short timeframe of research grants (Auclair and Dupraz, 1999).

Assessment of the performance of agricultural innovations

Scientific and methodological reductionisms also involve greater focus on the assessment of direct, local and short-term impacts, along with the underestimation or neglect of the indirect, global or systemic and long-term impacts of agricultural systems and innovations. Stress is easily laid on measurable variables such as gross yield, rather than those variables that are much more complex to measure, such as sustainability and externalities. This encourages scientists and innovators to focus on yield rather than economic optimum, on monocultures rather than multiple cropping systems. This influence can be traced to the econometric methods of calculating rates of returns on investments in agricultural research, which only take into account one objective (total net benefits, or growth), instead of taking into account externalities and multiple socioeconomic and environmental objectives (Alston et al., 1995; Vanloqueren and Baret, 2008; Garnett et al., 2015).[14]

Classic agricultural performance assessments are favourable to genetic engineering. The benefits of transgenic plants, usually grown in monocultures, are local and direct, and are consequently taken into account. Yet, classic performance measurements hinder agroecological engineering, particularly the sub-trajectories with strong positive environmental or economic externalities. For instance, agroforestry systems are also carbon sinks, they help to improve soil fertility and biodiversity, while also bringing new revenues to farmers.

'Publish or perish' constraints and the organization of scientific publications

Scientists in academic institutions aim to improve scientific knowledge and share discoveries through publications, which are non-market incentives to value priority in scientific discovery. Yet the different technological paradigms lead to dissimilar publication trends and impacts. The difference in academic prestige between the two technological paradigms may be grasped by a simple bibliometric analysis of some of the most appraised scientific journals: genetic engineering features roughly a hundred times more than agroecological engineering in *Nature*, *Science* and *The Proceedings of the National Academy of Sciences*.[15] In general, research representative of genetic engineering is published in scientific journals with impact factors (IFs) as high as 29.3, while agroecological research is published in journals with IFs ranging from 0.4 to 4.5.[16]

Genetic engineering and molecular biology are in fact perfectly adapted to the current publishing constraints. Firstly, the most appraised scientific journals focus on the smallest levels of organization in life (the plant, the cell and the molecular level). Second, the Taylorization of research favours the division of research into a handful of publishable results, from the identification of a particular gene, to the method of transposition into a host plant, to the assessment of its activity in the host plant. Agroecologists also publish parts of their work separately, even though the exact goal of agroecology is to encompass an agroecosystem as a whole. Innovations in the field of agroforestry take years before producing any publishable results. Scientists working on such agroecological trajectories collectively publish fewer papers.

With the growing importance of international rankings and formal research assessment procedures (as in the UK), the difference in 'publication productivity' may become an incentive to hire more molecular biologists in the future, as they contribute more than agroecologists to the global competition for highest rankings. This will, in turn, influence the career choices made by young scientists.

Specialisation versus interdisciplinarity

Genetic engineering thrives with the growing specialisation of science and the Taylorization of research in biotechnology laboratories. It calls for interdisciplinarity, while remaining inside the realm of a restricted number of natural sciences. The scientists involved share common cultures, languages, methods and techniques. In contrast, agroecological engineering requires the greater integration of agronomical, ecological, social and economic dimensions (Altieri, 1989, 1995). Academic barriers to interdisciplinarity are therefore obstacles to the development of agroecological trajectories (Dalgaard *et al.*, 2003; O'Brien *et al.*, 2013).[17] Moreover, the low value given to social sciences in ARS is also an impediment, even though they could help identify and create institutional innovations that improve knowledge-sharing processes, which are vital to the development of agroecology (Uphoff, 2002; Pretty and Bharucha, 2015).

Technology transfer mission: patents, spin-offs and extension

Another mission of agricultural-related public-sector research establishments (PSREs) is to transfer knowledge and technology from basic to applied research to the private sector. To do so, PSREs are expected to file patents on their exploitable results and launch spin-off companies (Tait *et al.*, 2001). However, technological trajectories are not equally suited to generating patentable results. The possibilities for creating spin-offs are also unequal. So universities that rely increasingly on non-public money are encouraged to engage in sub-trajectories that lead to patents and spin-offs.

Extension or technology transfer to farmers is another mission of PSREs and other dedicated centres. The explicit objective is to improve the situation of farmers and help them face new challenges, such as increased international competition and environmental issues. While this may seem a positive factor for both paradigms, these organizations often concentrate on technologies that can be of direct use to farmers (new cultivars, choice of best fungicide mixtures or optimal timing of spraying), with an ensuing improvement and strengthening of the dominant agricultural system (modern input-intensive monoculture). This is mainly a positive factor for genetic engineering, as transgenic crops suit this system. Agroecological innovations on the other hand do not become a priority, as they do not fit into the existing agricultural system and require structural changes (such as breaking away from large-scale monocultures).

To summarize, the overall organization of research systems is broadly more in favour of genetic engineering than agroecological engineering.

Emergent properties: path dependence and lock-in

Current research orientations are not only influenced by one or several of the determinants of innovations analyzed above; they are also shaped by two other processes that can be described as systemic emergent properties of agricultural research systems: path dependence and lock-in.

Path dependence has been suggested to explain the stability of sociotechnical systems. Among several technologies that perform similar functions and compete for adoption by economic agents, one technology may become dominant, even though it may have an inferior long-run potential (David and Arthur, 1985; Arthur, 1989). This process is 'path dependant' as the initial conditions may greatly influence the success of the dominant technology, particularly when increasing returns occur.[18] This process is also self-reinforcing and may lead to a technological lock-in situation in which the dominant technology excludes competing and possibly superior technologies (Liebowitz and Margolis, 1995).

The existence of path dependence and lock-in processes has been observed in agriculture, in such sectors as pest control strategies and breeding (Cowan and Gunby, 1996; Wilson and Tisdell, 2001; Jacobsen et al., 2013). While the concepts of path dependence and lock-in are generally used to analyze the adoption of competing innovations by end users, they are used here to help understand the adoption of competing technological paradigms by scientists and by agricultural research systems.[19]

Plant breeding is one of the rare sciences where the importance of past research efforts is well understood. It has for instance been shown that wheat varieties launched in the US in the early 1990s relied on varieties developed or discovered as long ago as 1873, with over 36% of the varieties incorporated existing before 1940 (Pardey and Beintema, 2001). The concept of 'knowledge stocks' enables more precise quantification of the importance of past research efforts. Knowledge stocks are money measures of the stocks of scientific knowledge (Adams, 1990). US scientists calculated that the accumulated stock of agricultural knowledge in the US in 1995 (from 1850 onwards) was 11 times larger than the amount of agricultural output produced during the same year (Pardey and Beintema, 2001). This means that 'for every $100 of agricultural output, there existed a $1,100 stock of knowledge to draw upon'. These observations are of the utmost importance as they demonstrate that modern agricultural systems rely on a wide scientific base, not only on public subsidies, as generally acknowledged.[20]

Past technological paradigms and their associated trajectories have thus profound and lasting effects on ARS, since current innovations have their roots in past strategic decisions and research efforts. Past science policies were shaped by productivist objectives that were and still are more favourable to genetic engineering than to agroecological engineering. The accumulation in time and the continuous interactions among all the determinants of innovation shape the current technological regime, but have also created a technological and institutional lock-in situation that severely hinders or stops the development of one of the technological paradigms, in this case agroecological engineering, though both paradigms make sense and make science, as seen in the first section of this chapter.

Genetic engineering, while a breakthrough innovation, was not locked-out. It fitted the main scientific approach (positivism and reductionism), as well as the technological regime shaping agricultural systems for decades (current transgenic plants have a 'technological coherence' with the development of pesticides) and finally, the larger political and economic trends that have reshaped the global economic system over the three last decades (Castree, 2010; Parayil, 2003; Patel, 2007). Agroecology, however, has stayed on the margins of the agricultural sciences, as it is distant from the main scientific approach as well as from the technological regime and the larger economic and political dominant trends. Its development has long been too limited to lead to significant increasing returns ('learning by doing', 'network externalities').[21]

Discussion: breaking out of the lock-in situation in agricultural S&T

The existence of a lock-in situation in agricultural research systems is not only of theoretical importance: it has consequences for public action. Lock-in situations justify public intervention if science is understood as a public good. As put forward by Callon and Bowker (1994), science is a public good which must be preserved at all costs because it is a source of variety and of new global developments ('states of the worlds'), and because the market would lead to irreversible situations without it. The sources of irreversibility are numerous because a change of scientific trajectory implies high switch-over costs (Geels, 2004; Horlings and Marsden, 2011). Scientists have been educated in a particular way and have acquired specific competencies that enable them to be best in some domains and not in others. The cost of moving from one research theme to another is too high (knowledge, reputation, networks, access to research grants). Research centres have also invested in infrastructure and machines that need to be paid off, and give them a comparative advantage in one or several very specific scientific areas. These switch-over costs favour incremental progress along an established technological trajectory rather than a change of paradigm and trajectory.

The issue is thus how to break out of this lock-in situation, as incremental progress is just not enough. Agroecological innovations hindered by the lock-in have been analyzed as crucial for our societies, especially in the context of climate change and the need for sustainable agriculture (IAASTD, 2008; IPES Food, 2016).

The practical ways to systematically reduce the imbalance between the two paradigms are beyond the scope of the chapter. We have briefly discussed three aspects that are the key to the necessary shift requested by the IAASTD recommendations: 'fair' forecasting, the importance of 'niches' in innovation policies, and issues of complementarity between technological paradigms. Firstly, fair forecasting exercises need to be performed in order to explore the potential contribution of the two paradigms to solving current and future challenges. Very few analyses have involved forecasting where our agricultural systems could be in 10 or 20 years if S&T policy and agricultural policies massively promoted agroecological innovations; yet forecasting on agricultural biotechnologies has been plentiful (Strategy Unit of the Cabinet Office, 2003; Reiss and Strobel, 2003).

Second, innovation policies must take into account the importance of niches and the true value of agroecological innovations in order to face the challenges of global climate change. Innovation niches are locations where it is possible to deviate from the rules of the existing technological regime (Geels, 2002, 2004). These niches have a crucial role in the stimulation of radical innovations to counterbalance the consequences of path dependence and lock-in. Some innovations, wind turbines for example, may have an initial low performance, but their development in a niche brings about their improvement through learning processes, technical developments, and/or adapted public policies. Agroecological innovations such as cultivar mixtures or agroforestry systems are precisely radical innovations that have both emerged in protected spaces (subsidized schemes, research programmes and the like). Niches are also crucial because they serve as laboratories that prepare us for the wider changes that are occurring or will occur. Today, climate change and the rising cost of energy are key elements that prove that the current technological regime is unfit for current and future needs (Gilbert, 2012; Kirschenmann, 2007). These challenges justify direct support to innovation niches, not to mention fundamental changes in the dominant technological regime.

Third, the issues of complementarity and competition between the two paradigms must be faced. Innovations from both paradigms are supposedly complementary (Conway, 1999). It is expected for instance that drought-tolerant transgenic plants could be used within agroecological systems designed to maximize resilience to climate extremes. However, if technological trajectories are to be used together in the future, their complementarity needs to be widely recognized and collectively thought out. This is not the case today. Proponents of genetic engineering strongly deny the potential of agroecology for feeding future generations. Many agroecologists, on their side, object to genetic engineering, follow a rigorous precautionary approach and argue that classic and marker assisted breeding are sufficient. These scientists postulate that the right model for agriculture is ecology.[22]

Another aspect of this expected complementarity is the uncertainty about the alleged universal value of genetic engineering. Twenty years after the first commercialization of transgenic crops, the second generation of transgenic plants has not materialized: herbicide-resistant and Bt insect-tolerant transgenic crops still make up 99% of the transgenic crops acreage. Moreover, there is great uncertainty about the possibility that 'sustainable' transgenic plants will be developed in the future because of scientific obstacles and structural aspects of the biotech industry (Hubbell and Welsh, 1998).

Coherent complementarity, let alone the acknowledgement of the existence of several innovation pathways, would require clarification on the likely developments in both trajectories, an identification of long-term risks associated with genetic engineering, a shared definition of agricultural sustainability, and a shared vision of future agrifood systems as well as political economy choices.[23]

Conclusions

The concepts of technological paradigms and technological trajectories are useful for explaining and analysing important trends in agricultural science and technology (S&T) at a time when fundamental shifts in agricultural S&T are increasingly recommended. Two of these trends – genetic and agroecological engineering – can be analyzed and compared with these concepts.

The process by which one paradigm is favoured over the other is the result of the interactions among many factors, and not a deliberate and planned movement. The system of innovation (SI) approach is powerful for demonstrating how agricultural research systems are a selection device that influences S&T choices. It leads to an in-depth analysis of all the determinants of innovation (factors influencing S&T choices) ranging from orientation of science policies to scientists' cultural and cognitive routines. The interactions among these determinants shape a technological regime. Genetic engineering – a technological paradigm that is well suited to scientific reductionism – is more successful in this technological regime than agroecological engineering – a paradigm that questions mainstream approaches within agricultural research. The development of agroecological innovations is clearly impeded, while their importance for sustainable agriculture and climate change has been clearly established in recent international reports, such as the International Assessment of Agricultural Science and Technology for Development (IAASTD).

Our analysis contributes to strengthening the relevance of the evolutionary line of thought (evolutionary economics) against the neo-classical approach for agriculture-related issues. The existence of path dependence and lock-in situations in agricultural research legitimizes public intervention. In other words, a global environment favourable to agroecology must be created if the recommendations of the IAASTD are to be implemented. This means not only a more balanced allocation of resources in agricultural research, but also attention to the larger framework that influences S&T choices.

Notes

1   Reprinted from *Research Policy*, 38(6): 971–983. Vanloqueren G. and Baret P.V. (2009) *How agricultural research systems shape a technological regime that develops genetic engineering but locks out agroecological innovations* © 2009, with permission from Elsevier. While the original text has been left untouched, the list of references has been updated. It should be noted that since 2008, the concepts of lock-in and path dependence have gained more traction both in scientific publications and in civil society networks seeking to promote more sustainable food systems. Recent publications making use of these concepts include Stassart and Jamar (2008), Duru and Therond (2015) and Magrini *et al.* (2016). Civil society networks have also used these concepts, including in Belgium (Servigne and Stevens, 2014) and Quebec (Audet *et al.*, 2015). The concepts have been integrated in publications aiming to influence policy-makers both in Europe (Standing Committee on Agricultural Research, 2011; Servigne, 2013) and at the international level (IPES Food, 2016). These concepts are appropriated in various ways by different actors. IPES-Food, for instance, highlights several lock-ins (eight) instead of one systematic lock-in made of several factors.

2   The first author is now affiliated with ICHEC-Brussels School Management, but prepared this paper when working at the University of Louvain. The authors are grateful to Marco Bertaglia, Gauthier Chapelle, Felice Dassetto, Marc Mormont and Frederic Varone for helpful comments on earlier versions of this paper. The analysis and comments made here remain however our sole responsibility. We would also like to thank the editors and three anonymous reviewers for providing insights that helped to shape this article in its present form. Mélanie Braibant, Jasmina Fiasse and Séverine Goret respectively managed interviews in the sugar beet, maize and soybean agrifood chains. This research was conducted with the financial support of the Belgian National Fund for Scientific Research (FNRSFRIA).

3   Some authors refer mainly to the concept of technological trajectories while others use technological paradigms, but the logic is the same as trajectories signify progress along a paradigm.

4 By 2014, GM crops were grown on 181.5 million hectares in 28 countries (James, 2014).

5 In subsequent papers, we have chosen to refer to 'agroecology' and not to 'agroecological engineering', as we did in this paper. This choice is driven by the need to consider agroecology not solely from an innovation-centred prism. We have indeed contributed with others to clarify the social and economic dimensions of agroecology. See in particular Stassart *et al.* (2012), Levidow *et al.* (2014) and Dumont *et al.* (2016), as well as De Schutter and Vanloqueren (2011).

6 A recent paper by Lescourret *et al.* (2015) also uses the term 'agroecological engineering'.

7 See also IPES-Food, 2016; McMichael, 2012; Levidow *et al.*, 2014; and Pimbert, 2015.

8 According to Possas *et al.* (1996), the current technological regime of modern agriculture is the evolutionary result of the intersection of different trajectories that have reached a growing technological coherence over the last 150 years. These developments involve industries (pesticides, fertilizers, seeds and machinery), public research and educational institutions, producers and producer organizations.

9 Cultivar mixtures are an application of the concept of crop heterogeneity (increasing the genetic diversity in a cultivated field to increase crop resistance to biotic and abiotic stresses). Agroforestry embraces land-use systems in which trees are deliberately integrated with crops and/or animals on the same land, usually producing ecological and economic interactions. Biological control is a method of controlling pests and diseases that relies on conservation and/or the enhancement of natural predators and consequently fits the agroecological paradigm. The fact that it is sometimes defined as belonging to agricultural biotechnology should not cause confusion. Biocontrol has an intermediate status in this chapter as is often the case: the impact of the determinants of innovation on biocontrol are in between those affecting genetic engineering and those affecting agroecological engineering.

10 According to King *et al.* (2012) a 'study, using citations data from scientific publications, found that total citations from private research to university publications increased, and that firms citing more university research had a higher likelihood of producing patentable technologies'.

11 *The New York Times* has published more papers on agroecology in the last ten years than it did from 1981 to 2008. However, the figures remain very low and confirm the analysis made in 2008. The interested reader can easily check these statistics on https://www.nytimes.com/

12 These past agricultural systems have today been recognized as Globally Important Agricultural Heritage Systems (GIAHS).

13 Variables include light intensity and rate of humidity, pest complex, diversity of trees and coffee cultivars, shade management, altitude, climate and soil.

14 Advances in environmental economics and ecological economics are yielding new and better-adapted methods, which are nevertheless not yet widely known in ARS and extension services.

15 A simple keyword search for genetic engineering found 358 papers, while a similar keyword search for agroecological engineering found only one paper, two papers for crop mixtures and two for agroforestry. ISI Web of Science, ISI Web of Knowledge, 25 October 2006. Databases = SCI-EXPANDED, SSCI, A&HCI; Timespan = 1987–2006.

16 Journal Citation Reports and ISI Web of Science, ISI Web of Knowledge, 15 November 2006. Technological trajectories were first defined by keyword lists, then the scientific journals with highest publication records for these keyword lists were selected and their IFs checked. Illustrative examples are *Nature Biotechnology* (22.7); *Plant Physiology* (6.1), *Agriculture, Ecosystems & Environment* (1.5), *Agroforestry Systems* (0.7). Another way to look at the same thing is to analyze IFs of the 20 most influential journals (those with highest impact factor) in Institute for Scientific Information (ISI) categories representative of each trajectory. Similar results are found. Publications most representative of genetic engineering are published in the following ISI categories of scientific journals: biotechnology and applied microbiology, plant sciences, biochemistry and molecular biology. The 20 most influential scientific journals in these categories have an IF rating of

between 2.7 and 33.4. Publications most representative of agroecological engineering are agronomy, agriculture/multidisciplinary, ecology and environmental sciences. With the exception of the category 'ecology' (IF between 3.3 and 14.9), most influential journals have an IF of between 0.3 and 5.3.

17 These barriers are cultural and organizational: securing research grants, going on exchange programmes, publishing, gaining recognition, securing a job, or being promoted (Bauer, 1990; Nissani, 1997).

18 Increasing returns are summarized by Callon and Bowker (1994): the more a technology is produced and offered, the more it becomes worthwhile for the supplier to produce it and for the user to consume it. Increasing returns to adoption may be of three types: scale economies, learning economies ('learning by doing', performance is improved as specialized skills and knowledge accumulate through experience) and adaptive expectations (increasing adoption reduces uncertainty among producers and users) (Unruh, 2000).

19 This application of the concepts can be best understood by reading the preceding paragraph and replacing 'technologies' by 'technological paradigms'.

20 Public support to farmers represents 29% of farm receipts in OECD countries (OECD, 2006).

21 On increasing returns, see Callon and Bowker (1994), pp. 407–408.

22 As Weiner (2003) puts it, 'Ecology is a relatively young science that cannot yet deliver answers to many of the questions agricultural researchers are asking. But this does not mean that the answers can be found elsewhere. One cannot solve traffic problems through the engineering of automobiles alone. One needs to use traffic engineering, even if traffic engineering is not as highly developed as automobile engineering'.

23 The two technological paradigms have very different visions of the desirable socio-economic development for the future. A large share of agroecologists challenge economic globalization, agricultural trade liberalization and the current view of what a productive and sustainable agricultural system is (The International Commission on the Future of Food and Agriculture, 2003). Agroecologists privilege alternative food systems operating at a regional scale or based on closer farmer-consumer relationships, or product networks that mobilize localized resources and have strong identities (Chapter 1 this volume; Goodman and Watts, 1997; Whatmore and Stassart, 2003; Allaire and Wolf, 2004).

References

Adams, J. D. (1990) 'Fundamental stocks of knowledge and productivity growth', *The Journal of Political Economy* 98(4): 673–702.

Allaire, G. and S. A. Wolf (2004) 'Cognitive representations and institutional hybridity in agrofood innovation', *Science, Technology & Human Values* 29(4): 431–458.

Alston, J. M., G. W. Norton and P. G. Pardey (1995) *Science under Scarcity: Principles and Practice for Agricultural Research Evaluation and Priority Setting*, Cornell University Press, Ithaca.

Alston, J. M., P. G. Pardey and J. Roseboom (1998a) 'Financing agricultural research: International investment patterns and policy perspectives', *World Development* 26(6): 1057–1071.

Alston, J. M., P. G. Pardey and M. J. Taylor (2001) *Agricultural Science Policy. Changing Global Agendas*, John Hopkins University Press, Baltimore and London.

Alston, J. M., P. G. Pardey and V. H. Smith (1998b) 'Financing agricultural R&D in rich countries: What's happening and why', *The Australian Journal of Agricultural and Resource Economics* 42(1): 51–82.

Altieri, M. A. (1989) 'Agroecology: A new research and development paradigm for world agriculture', *Agriculture, Ecosystems & Environment* 27: 37–46.

Altieri, M. A. (1995) *Agroecology: The Science of Sustainable Agriculture*, 2nd ed. Westview Press, Boulder, Colorado.

Altieri M. A., C. I. Nicholls and R. Montalba (2017) 'Technological approaches to sustainable agriculture at a crossroads: An agroecological perspective', *Sustainability* 9(3): 349.

Arora, A. and A. Gambardella (1994) 'The changing technology of technological change: General and abstract knowledge and the division of innovative labour', *Research Policy* 23(5): 523–532.

Arthur, W. B. (1989) 'Competing technologies, increasing returns, and lock-in by historical events', *Economic Journal* 99(394): 116–131.

Auclair, D. and C. Dupraz (Eds.) (1999) *Agroforestry for Sustainable Land-Use Fundamental Research and Modeling with Emphasis on Temperate and Mediterranean Applications*, Kluwer Academic Publishers, Dordrecht.

Audet, R., S. Lefèvre and M. El-Jed (2015) 'La mise en marché alternative de l'alimentation à Montréal et la transition socio-écologique du système agroalimentaire', *Les cahiers de recherche* OSE 01–2015.

Badgley, C., J. Moghtader, E. Quintero, E. Zakem, M. Jahi Chappell, K. Avilés-Vázquez, A. Samulon and I. Perfecto (2007) 'Organic agriculture and the global food supply', *Renewable Agriculture and Food Systems* 22(2): 86–108.

Bailey, K. L., S. M. Boyetchko and T. Längle (2010) 'Social and economic drivers shaping the future of biological control: A Canadian perspective on the factors affecting the development and use of microbial biopesticides', *Biological Control* 52(3): 221–29.

Balanya, B., A. Doherty, O. Hoedeman, A. Ma'anit and E. Wesselius (2003) *Europe Inc. Regional & Global Restructuring and the Rise of Corporate Power*, Pluto Press, London.

Bauer, H. (1990) 'Barriers against interdisciplinarity: Implications for studies of Science Technology, and Society (STS)', *Science, Technology & Human Values* 15(1): 105–119.

Bawden, R. J. (1991) 'Systems thinking and practice in agriculture', *Journal of Dairy Science* 74(7): 2362–2373.

Benyus, J. (1997) *Biomimicry: Innovation Inspired by Nature*, Harper Perennial, New York.

Bernstein, H. (2014) 'Food sovereignty via the "peasant way": A sceptical view', *The Journal of Peasant Studies* 41(6): 1031–1063.

Bijman, J. and J. Tait. (2002) 'Public policies influencing innovation in the agrochemical biotechnology and seed industries', *Science and Public Policy* 29(4): 245–251.

Borrás, S. and C. Edquist (2013) 'The choice of innovation policy instruments', *Technological Forecasting and Social Change* 80(8): 1513–1522.

Breschi, S. and C. Catalini (2010) 'Tracing the links between science and technology: An exploratory analysis of scientists' and inventors' networks', *Research Policy* 39(1): 14–26.

Busch, L., R. Allison, C. Harris, A. Rudy, B. T. Shaw, T. T. Eyck, D. Coppin, J. Konefal and C. Oliver, with J. Fairweather (2004) *External Review of the Collaborative Research Agreement between Novartis and the University of California*, Institute for Food and Agricultural Standards, Michigan State University.

Callon, M. and G. Bowker (1994) 'Is science a public good?' *Science, Technology and Human Values* 19(4): 395–424.

Castree, N. (2010) 'Neoliberalism and the biophysical environment: A synthesis and evaluation of the research', *Environment and Society: Advances in Research* 1(1): 5–45.

Cerny, P. G. (1997) 'Paradoxes of the competition state: The dynamics of political globalization', *Government and Opposition* 31(2): 251–274.

Chataway, J., J. Tait and D. Wield (2004) 'Understanding company R&D strategies in agrobiotechnology: Trajectories and blind spots', *Research Policy* 33(6–7): 1041–1057.

Checkland, P. B. (1981) *Systems Thinking, Systems Practice*, John Wiley, New York.

Cohen, B. C. (1963) *The Press and Foreign Policy*, Princeton University Press, Princeton.

Congress of the US (2005) *National Innovation Act*. In: 109th Congress of the United States, 1st Session, S. 2109, Washington, DC.

Conway, G. R. (1999) *The Doubly Green Revolution: Food for All in the Twenty-First Century*, Cornell University Press, Ithaca.

Cowan, R. and P. Gunby (1996) 'Sprayed to death: Path dependence, lock-in and pest control strategies', *Economic Journal* 106(436): 521–542.

Dalgaard, T., N. J. Hutchings and J. R. Porter (2003) 'Agroecology, scaling and interdisciplinarity', *Agriculture Ecosystems & Environment* 100(1): 39–51.

David, P. A. and B. Arthur (1985) 'Clio and the economics of QWERTY', *American Economic Review* 75(2): 337.

De Schutter, O. and G. Vanloqueren (2011) 'The new green revolution: How twenty-first-century science can feed the world', *Solutions Journal* 2(4): 33–44.

Devos, Y., M. Demont, K. Dillen, D. Reheul, M. Kaiser and O. Sanvido (2009) 'Coexistence of genetically modified (GM) and non-GM crops in the European Union. A review', *Agronomy for Sustainable Development* 29(1): 11–30.

Dibden, J., D. Gibbs and C. Cocklin (2013) 'Framing GM crops as a food security solution', *Journal of Rural Studies* 29: 59–70.

Dosi, G. (1982) 'Technological paradigms and technological trajectories: A suggested interpretation of the determinants and directions of technical change', *Research Policy* 11(3): 147–162.

Dumont, A., G. Vanloqueren, P. M. Stassart and P. V. Baret (2016) 'Clarifying the socioeconomic dimensions of agroecology: Between principles and practices', *Agroecology and Sustainable Food Systems* 40(1): 24–47.

Dupuis, E. M. (2000) 'Not in my body: rBGH and the rise of organic milk', *Agriculture and Human Values* 17(3): 285–295.

Duru, M. and O. Therond (2015) 'Designing agroecological transitions; A review', *Agronomy for Sustainable Development* 35(4): 1237–1257.

Edquist, C. (1997) *Systems of Innovation: Technologies, Institutions and Organizations*, Pinter/Cassel, London.

Edquist, C. (2001) 'The systems of innovation approach and innovation policy: An account of the state of the art,' Lead paper presented at the DRUID Conference, Aalborg, 12–15 June.

Edquist, C. (2011) 'Systems of innovation: Perspectives and challenges', *African Journal of Science, Technology, Innovation and Development* 2(3): 14–43.

Ekroos, J., O. Olsson, M. Rundlöf, F. Wätzold and H. G. Smith (2014) 'Optimizing agri-environment schemes for biodiversity, ecosystem services or both?', *Biological Conservation* 172: 65–71.

Esquinas-Alcázar, J. (2006) 'Protecting crop genetic diversity for food security: Political, ethical and technical challenges', *Nature Reviews Genetics* 6: 946–953.

European Commission (2002) *Life Sciences and Biotechnology: A Strategy for Europe*, Communication from the Commission to the Council, the European Parliament, the Economic and Social Committee and the Committee of Regions, EC, Brussels.

European Commission (2004) *Plants for the Future: A 2025 Vision for European Plant Biotechnology*, Directorate-General for Research Food Quality and Safety (EUR 21359 EN), EC, Brussels.

European Commission (2005a) *Building the ERA of Knowledge for Growth*, Communication from the Commission, EC, Brussels.

European Commission (2005b) *Proposal for a Decision of the European Parliament and of the Council Concerning the Seventh Framework Programme of the European Community for Research, Technological Development and Demonstration Activities (2007 to 2013)*, EC, Brussels.

European Commission (2013) *Europe 2020 – Europe's Growth Strategy*, EC, Brussels, http://ec.europa.eu/europe2020/pdf/europe_2020_explained.pdf.

European Council (1981) '81/1032/EEC: Council Decision of 7 December 1981 adopting a multi-annual research and training programme for the European Economic Community in the field of biomolecular engineering', *Official Journal* L375(12): 0001–0004.

European Science Social Forum Network (2005) 'Framework programme 7: Towards a real partnership with society?', http://sciencescitoyennes.org/towards-a-real-partnership-with-society-ngo-alliance-on-framework-programme-7-of-the-ec.

FAO (2004) 'The state of food and agriculture: Biotechnology: Meeting the needs of the poor?', *FAO Agriculture Series* No. 35, UN Food and Agriculture Organization, Rome.

FAO (2007) *International Conference on Organic Agriculture and Food Security*, UN Food and Agriculture Organization, Rome, 3–5 May 2007.

Feindt, P. H. (2012) 'The politics of biopatents in food and agriculture, 1950–2010: Value conflict, competing paradigms and contested institutionalisation in multi-level governance', *Policy and Society* 31(4): 281–93.

Flores, C. C. and S. J. Sarandón (2004) 'Limitations of neoclassical economics for evaluating sustainability of agricultural systems: Comparing organic and conventional systems', *Journal of Sustainable Agriculture* 24(2): 77–91.

Food Ethics Council (2004) *Just Knowledge? Governing Research on Food and Farming*, Food Ethics Council, London.

Friedmann, H. (2005) 'From colonialism to green capitalism: Social movements and emergence of food regimes, new directions in the sociology of global development' *Research in Rural Sociology and Development* 11: 229–267.

Fuglie, K., P. Heisey and J. King (2012a) 'The contribution of private industry to agricultural innovation', *Science* 338(6110): 1031–1032.

Fuglie, K., P. Heisey, J. King and D. Schimmelpfennig (2012b) 'Private industry investing heavily, and globally, in research to improve agricultural productivity', *Oregon Wheat* 64(4): 10–13.

Garnett, T., E. Röös and D. Little (2015) *Lean, Green, Mean, Obscene…? What Is Efficiency? And Is It Sustainable?*, Food Climate Research Network, University of Oxford.

Geels, F. W. (2002) 'Technological transitions as evolutionary reconfiguration processes: A multi-level perspective and a case study', *Research Policy* 31(8–9): 1257–1274.

Geels, F. W. (2004) 'From sectoral systems of innovation to socio-technical systems: Insights about dynamics and change from sociology and institutional theory', *Research Policy* 33(6–7): 897–920.

Gilbert, N. (2012) 'One-third of our greenhouse gas emissions come from agriculture', *Nature News*, 31 October 2012, www.nature.com/news/one-third-of-our-greenhouse-gas-emissions-come-from-agriculture-1.11708.

Gómez-Baggethun, E., S. Mingorría, V. Reyes-García, L. Calvet and C. Montes. (2010) 'Traditional ecological knowledge trends in the transition to a market economy: Empirical study in the Doñana natural areas', *Conservation Biology: The Journal of the Society for Conservation Biology* 24(3): 721–729.

Goodman, D. and M. Watts (1997) *Globalising Food: Agrarian Questions and Global Restructuring*, Routledge, London.

Graff, G. D. (2004) 'The division of innovative labor among universities, entrepreneurs, and corporations in generating the technological trajectories of agricultural biotechnology', paper prepared for *Center for Intellectual Property Symposium 2004*, Sweden, 1–3 June 2004.

Graff, G. D., S. E. Cullen, K. J. Bradford, D. Zilberman and A. B. Bennett (2003) 'The public–private structure of intellectual property ownership in agricultural biotechnology', *Nature Biotechnology* 21(9): 989–995.

Guthman, J. (2000) 'Raising organic: An agro-ecological assessment of grower practices in California', *Agriculture and Human Values* 17(3): 257–266.

Guthman, J. (2014) 'Organic farming', In: Guthman, J. (ed.) *Agrarian Dreams: The Paradox of Organic Farming in California*, 2nd ed., University of California Press, Oakland.

Hayami, Y. and V. W. Ruttan. (1985) *Agricultural Development: An International Perspective*, Johns Hopkins University Press, Baltimore.

Heisey, P. W., J. L. King and K. D. Rubenstein (2005) 'Patterns of public-sector and private-sector patenting in agricultural biotechnology', *AgBioForum* 8(2–3): 73–82.

Hess, C. E. (1991) 'Resource allocation to state agbiotech research: 1982–1988', *Bio/Technology* 9(1): 29–31.

Horlings, L. G. and T. K. Marsden (2011) 'Towards the real green revolution? Exploring the conceptual dimensions of a new ecological modernisation of agriculture that could "feed the world"', *Global Environmental Change* 21(2): 441–452.

Hubbell, B. J. and R. Welsh (1998) 'Transgenic crops: Engineering a more sustainable agriculture?', *Agriculture and Human Values* 15(1): 43–56.

IAASTD (2008) *Agriculture at a Crossroads: Executive Summary of the Synthesis Report of the International Assessment of Agricultural Knowledge, Science and Technology for Development (IAASTD)*, Island Press, Washington, DC.

Information Systems for Biotechnology (2006) 'Field test releases in the US' (Database), see www.isb.vt.edu.

IPCC (2007) *IPCC Fourth Assessment Report: Climate Change 2007*, Intergovernmental Panel on Climate Change, Cambridge University Press, Cambridge.

IPES Food (2016) *From Uniformity to Diversity: A Paradigm Shift from Industrial Agriculture to Diversified Agroecological Systems*, International Panel of Experts on Sustainable Food Systems, www.ipes-food.org/images/Reports/UniformityToDiversity_FullReport.pdf.

International Commission on the Future of Food and Agriculture (2003) *Manifesto on the Future of Food*, The International Commission on the Future of Food and Agriculture, San Rossore, Italy.

Jacobsen, S. E., M. Sørensen, S. M. Pedersen and J. Weiner (2013) 'Feeding the world: Genetically modified crops versus agricultural biodiversity', *Agronomy for Sustainable Development* 33(4): 641–662.

James, C. (2007) 'Global status of commercialized biotech/GM crops: 2007', *ISAAA Briefs* No. 37, International Service for Acquisition of Agri-Biotech Applications, Ithaca, NY.

James, C. (2014) 'Global status of commercialized biotech/GM crops: 2014', *ISAAA Brief* 49, International Service for Acquisition of Agri-Biotech Applications, Ithaca, NY.

Jasanoff, S. (1990) *The Fifth Branch: Science Advisers as Policymakers*, Harvard University Press, Cambridge, MA.

Jha, S. and C. W. Dick (2008) 'Shade coffee farms promote genetic diversity of native trees', *Current Biology* 18(24): R1126–R1128.

Joly, P. B. and S. Lemari (2004) 'The technological trajectories of the agrochemical industry: Change and continuity', *Science and Public Policy* 29(4): 259–266.

King, J. L., A. A. Toole and K. O. Fuglie (2012) 'The complementary roles of the public and private sectors in US agricultural research and development', *Economic Brief* 19, United States Department of Agriculture Economic Research Service, Washington, DC.

Kirschenmann, F. (2007) 'Potential for a new generation of biodiversity in agroecosystems of the future', *Agronomy Journal* 99(2): 373–376.

Kremen, C. and A. Miles (2012) 'Ecosystem services in biologically diversified versus conventional farming systems: Benefits, externalities, and trade-offs', *Ecology and Society* 17(4): 40.

Kremen, C., I. Alastair and C. Bacon (2012) 'Diversified farming systems: An agroecological, systems-based alternative to modern industrial agriculture', *Ecology and Society* 17(4): 44.

Kuc, J. (2001) 'Concepts and direction of induced systemic resistance in plants and its application', *European Journal of Plant Pathology* 107(1): 7–12.

Kuhn, T. S. (1962) *The Structure of Scientific Revolutions*, Chicago University Press, Chicago.

Lacey, H. (1999) *Is Science Value Free? Values and Scientific Understanding*, Routledge, London.

Lacey, H. (2002) 'Assessing the value of transgenic crops', *Science and Engineering Ethics* 8(4): 497–511.

Lang, T. and M. Heasman (2004) *Food Wars: Public Health and the Battle for Mouths Minds and Markets*, Earthscan, London.

Lee, J., G. Gereffi and J. Beauvais (2012) 'Global value chains and agrifood standards: Challenges and possibilities for smallholders in developing countries', *Proceedings of the National Academy of Sciences* 109(31): 12326–12331.

Lescourret, F., T. Dutoit and F. Rey (2015) 'Agroecological engineering', *Agronomy for Sustainable Development* 35(4): 1191–1198.

Levidow, L., V. Søgaard and S. Carr (2004) 'Agricultural public-sector research establishments in Western Europe: Research priorities in conflict', *Science and Public Policy* 29(4): 287–295.

Levidow, L. and K. Boschert (2008) 'Coexistence or contradiction? GM crops versus alternative agricultures in Europe', *Geoforum* 39(1): 174–190.

Levidow, L., M. P. Pimbert and G. Vanloqueren (2014) 'Agroecological research: Conforming— or transforming the dominant agro-food regime?', *Agroecology and Sustainable Food Systems* 38(10): 1127–1155.

Levie, A., M. A. Legrand, P. Dogot, C. Pels, P. V. Baret and T. Hance (2005) 'Mass releases of *Aphidius rhopalosiphi* and strip management to control of wheat aphids', *Agriculture, Ecosystems & Environment* 105(1–2): 17–21.

Liang, W. (1998) 'Farming systems as an approach to agro-ecological engineering', *Ecological Engineering* 11(1–4): 27–35.

Liebowitz, S. J. and S. E. Margolis (1995) 'Path dependence, lock-in, and history', *Journal of Law, Economics and Organization* 11(1): 205–226.

Lundvall, B. A. (1992) *National Systems of Innovation: Towards a Theory of Innovation and Interactive Learning*, Pinter, London.

McMichael, P. (2012) 'Food regime crisis and revaluing the agrarian question', In: Almås, R. and Campbell, H. (eds.) *Rethinking Agricultural Policy Regimes: Food Security, Climate Change and the Future Resilience of Global Agriculture*, Vol. 18, *Research in Rural Sociology* and *Development Book Series*, Emerald Group Publishing Limited, Somerville, MA.

McMichael, P. (2016) 'Commentary: Food regime for thought', *The Journal of Peasant Studies* 43(3): 648–670.

McMillan, G. S., F. Narin and D. L. Deeds (2000) 'An analysis of the critical role of public science in innovation: The case of biotechnology', *Research Policy* 29(1): 1–8.

Magrini, M. B., M. Anton, C. Cholez, G. Corre-Hellou, G. Duc, M. H. Jeuffroy and S. Walrand (2016) 'Why are grain-legumes rarely present in cropping systems despite their environmental and nutritional benefits? Analyzing lock-in in the French agrifood system', *Ecological Economics* 126(C): 152–162.

Marechal, K., H. Aubaret-Joachain and J.-P. Ledant (2008) 'The influence of economics on agricultural systems: An evolutionary and ecological perspective', *Centre Emile Bernheim (Solvay Business School) Working Paper* (08/028).

Marrocu, E., R. Paci and S. Usai (2013) 'Proximity, networking and knowledge production in Europe: What lessons for innovation policy?', *Technological Forecasting and Social Change* 80(8): 1484–1498.

Marsden, T. and R. Sonnino. (2005) 'Rural food and agri-food governance in Europe: Tracing the development of alternatives', In: Higgins, V. and Lawrence, G. (eds.) *Agricultural Governance: Globalization and the New Politics of Regulation*, Routledge, London.

Méndez, V. E., C. M. Bacon and R. Cohen (2013) 'Agroecology as a transdisciplinary, participatory, and action-oriented approach', *Agroecology and Sustainable Food Systems* 37(1): 3–18.

Mignot, J.-P. and C. Poncet (2001) 'The industrialization of knowledge in life sciences: Convergence between public research policies and industrial strategies', *Cahiers de recherche du Creden*, 20 January 2001.

Millennium Ecosystem Assessment (2005) *Ecosystems and Human Well-Being: Global Assessment Reports*, Island Press, Washington, DC.

Morgan, K., T. Marsden and J. Murdoch (2006) *Worlds of Food, Place, Power, and Provenance in the Food Chain*, Oxford University Press, Oxford.

Muscio, A., D. Quaglione and G. Vallanti (2013) 'Does government funding complement or substitute private research funding to universities?' *Research Policy* 42(1): 63–75.

Naseem, A., D. J. Spielman and S. W. Omamo (2010) 'Private-sector investment in R&D: A review of policy options to promote its growth in developing-country agriculture', *Agribusiness* 26(1): 143–173.

National Research Council (1987) *Agricultural Biotechnology: Strategies for National Competitiveness*, Report of the Committee on a National Strategy for Biotechnology in Agriculture, The National Academy Press, Washington, DC.

National Research Council (1998) *Designing an Agricultural Genome Program*, report of the Board on Biology and Board on Agriculture, The National Academy Press, Washington, DC.

National Research Council (2002) *Publicly Funded Agricultural Research and the Changing Structure of US Agriculture*, Report, Board on Agriculture and Natural Resources, The National Academies Press, Washington, DC.

Nelson, R. R. (1993) *National Systems of Innovation: A Comparative Study*, Oxford University Press, Oxford.

Nelson, R. R. and S. G. Winter (1982) *An Evolutionary Theory of Economic Change*, Harvard University Press, Cambridge, MA.

New York Times (2008) Online archives, available at www.nytimes.com, accessed 12 August 2008.

Nissani, M. (1997) 'Ten cheers for interdisciplinarity: The case for interdisciplinary knowledge and research', *The Social Science Journal* 34(2): 201–216.

O'Brien, K., J. Reams, A. Caspari, A. Dugmore, M. Faghihimani, I. Fazey and V. Winiwarter (2013) 'You say you want a revolution? Transforming education and capacity building in response to global change', *Environmental Science and Policy* 28: 48–59.

OECD (2006) *Agricultural Policies in OECD Countries at a Glance*, OECD, Paris.

Orsenigo, L. (1989) *The Emergence of Biotechnology*, Pinter, London.

Parayil, G. (2003) 'Mapping technological trajectories of the Green Revolution and the Gene Revolution from modernization to globalization', *Research Policy* 32(6): 971–990.

Pardey, P. G. and N. M. Beintema (2001) *Slow Magic: Agricultural R&D a Century After Mendel*, International Food Policy Research Institute, Washington, DC.

Parrott, N. and T. Marsden (2002) *The Real Green Revolution: Organic and Agroecological Farming in the South*, Greenpeace Environmental Trust, London.

Patel, R. (2007) *Stuffed and Starved. Markets, Power and the Hidden Battle for the World Food System*, Portobello Books Ltd., London.

Perfecto, I., J. Vandermeer and S. M. Philpott (2014) 'Complex ecological interactions in the coffee agroecosystem', *Annual Review of Ecology, Evolution, and Systematics* 45(1): 137–158.

Pestre, D. (2003) 'Regime of knowledge production in society:Towards a more political and social reading', *Minerva* 41(3): 245–261.

Pew Initiative on Food and Biotechnology (2003) 'University-industry relationships: Framing the issues for academic research in agricultural biotechnology', *Agbio- Forum* 8(2–3), 19–20 November 2002.

Piesse, J. and C.Thirtle (2010) 'Agricultural R&D, technology and productivity', *Philosophical Transactions of the Royal Society of London. Series B, Biological Sciences* 365(1554): 3035–3047.

Pimbert, M. P. (2015) 'Agroecology as an alternative vision to conventional development and climate-smart agriculture', *Development* 58(2–3): 286–298.

Possas, M. L., S. Salles and J. M. de Silveira (1996) 'An evolutionary approach to technological innovation in agriculture: Some preliminary remarks', *Research Policy* 25(6): 933–945.

Pretty, J. N., J. I. L. Morison and R. E. Hine (2003) 'Reducing food poverty by increasing agricultural sustainability in developing countries', *Agriculture, Ecosystems & Environment* 95(1): 217–234.

Pretty, J. and Z. P. Bharucha (2015) 'Integrated pest management for sustainable intensification of agriculture in Asia and Africa', *Insects* 6(1): 152–182.

Reiss, T. and O. Strobel (2003) *State of the Art Report Life Sciences (Annex 3), Integrating Technological and Social Aspects of Foresight in Europe (ITSAFE)*.

Royal Society, US National Academy of Sciences, Brazilian Academy of Sciences, Chinese Academy of Sciences, Indian National Science Academy, Mexican Academy of Sciences, Third World Academy of Sciences (2000) *Transgenic Plants and World Agriculture*, National Academy Press,Washington, DC.

Russel, A. (1999) 'Biotechnology as a technological paradigm in the global knowledge structure', *Technology Analysis & Strategic Management* 11(2): 235–254.

SAFE (2005) *Silvoarable Agroforestry For Europe (SAFE)*, project final progress report, INRA, Montpellier, www.montpellier.inra.fr/safe.

Servigne, P. (2013) *Nourrir l'Europe en temps de crise. Vers des systèmes alimentaires résilients.* Report commissioned by Yves Cochet, Groupe des Verts/ALE at the European Parlement.

Servigne, P. and R. Stevens (2014) *Alors, ça vient ? Pourquoi la transition se fait attendre*, Barricade, Brussels.

Slaughter, S. and L. L. Leslie (1997) *Academic Capitalism: Politics, Policies and the Entrepreneurial University*, John Hopkins University Press, Baltimore.

Soete, L., B.Verspagen and B. ter Weel (2010) 'Chapter 27 – Systems of Innovation', In: Hall, B. H. and Rosenberg, N. (eds.), *Handbook of the Economics of Innovation*, Vol. 2: 1159–1180, Elsevier.

Standing Committee on Agricultural Research (2011) *Sustainable Food Consumption and Production in a Resource-constrained World*, Third Foresight Exercise of the European Commission – Standing Committee on Agricultural Research, Brussels.

Stassart, P. and D. Jamar (2008) 'Steak up to the horns! The conventionalization of organic stock farming: Knowledge lock-in in the agrifood chain', *GeoJournal* 73(1): 31–44.

Stassart, P. M., P. Baret, J.-C. Grégoire, T. Hance, M. Mormont, D. Reheul, G.Vanloqueren and M.Visser (2012) 'Trajectoire et potentiel de l'agroécologie, pour une transition vers des systèmes alimentaires durables'. In:Van Dam, D., Streith, M., Nizet, J. and Stassart P. M. (eds.) *Agroécologie. Entre pratiques et sciences sociales.* Educagri éditions, Paris.

Staver, C., F. Guharay, D. Monterroso and R. G. Muschler (2001) 'Designing pest suppressive multistrata perennial crop systems: Shade-grown coffee in Central America', *Agroforestry Systems* 53(2): 151–170.

Steffen, W., K. Richardson, J. Rockström, S. E. Cornell, I. Fetzer, E. M. Bennett, R. Biggs, S. R. Carpenter, W. De Vries, C. A. De Wit, C. Folke, D. Gerten, J. Heinke, G. M. Mace,

L. M. Persson, V. Ramanathan, B. Reyers and S. Sörlin (2015) 'Planetary boundaries: Guiding human development on a changing planet', *Science* 347(6223): 1259855.

Stirling, A. (1999) 'Risk at a turning point?' *Journal of Environmental Medicine* 1(3): 119–126.

Strategy Unit of the Cabinet Office. (2003) *Weighing Up the Costs and Benefits of GM Crops*, The Cabinet Office, London.

Sunding, D. and D. Zilberman (2001) 'The agricultural innovation process: Research and technology adoption in a changing agricultural sector', In: Gardner, B. L. and Rausser, G. C. (eds.), *Handbooks of Agricultural Economics*, Vol. 1A, Elsevier.

Tait, J., J. Chataway and D. Wield (2001) *PITA Project: Policy Influences on Technology for Agriculture: Chemicals, Biotechnology and Seeds*, Final Report, European Commission, Brussels.

Tödtling, F., P. Lehner and A. Kaufmann (2009) 'Do different types of innovation rely on specific kinds of knowledge interactions?', *Technovation* 29(1): 59–71.

UNDP (2001) *Human Development Report 2001: Making New Technologies Work for Human Development*, United Nations Development Programme, New York.

Union of Concerned Scientists (1996) *On Research for Sustainable Agriculture*, testimony before the Subcommittee on Resource Conservation, Research and Forestry, US House of Representatives Committee on Agriculture.

Union of Concerned Scientists (2015) 'Scientist and expert statement of support for public investment in agroecological research', www.ucsusa.org/sites/default/files/legacy/ assets/ documents/food_and_agriculture/scientist-statement-agroecology-7-2-2014.pdf.

Unruh, G. C. (2000) 'Understanding carbon lock-in', *Energy Policy* 28(12): 817–830.

Uphoff, N. (2002) *Agroecological Innovations: Increasing Food Production with Participatory Development*, Earthscan Publications, London.

Vallavieille-Pope, C. (2004) 'Management of disease resistance diversity of cultivars of a species in single fields: Controlling epidemics', *Comptes Rendus Biologies* 327(7): 611–620.

Vanloqueren, G. and P.V. Baret (2004) 'Les pommiers transgéniques résistants à la tavelure - Analyse systémique d'une plante transgénique de 'seconde génération' (Transgenic scab-resistant apple trees: A systems approach to a 'second generation' transgenic plant), *Le Courrier de l'Environnement de l'INRA* 52: 5–20.

Vanloqueren, G. and P.V. Baret (2008) 'Why are ecological, low-input, multi-resistant wheat cultivars slow to develop commercially? A Belgian agricultural 'lock-in' case study', *Ecological Economics* 66(2–3): 436–446.

Watts, N. and I. R. Scales (2015) 'Seeds, agricultural systems and socio-natures: Towards an actor-network theory informed political ecology of agriculture', *Geography Compass* 9(5): 225–36.

Weiner, J. (2003) 'Ecology—the science of agriculture in the 21st century', *Journal of Agricultural Science* 141(3–4): 371–377.

Wezel, A., M. Casagrande, F. Celette, J.Vian, A. Ferrer and J. Peigne (2014) 'Agroecological practices for sustainable agriculture. A review' *Agronomy for Sustainable Development* 34(1): 1–20.

Whatmore, S. and P. Stassart. (2003) 'Guest editorial: What's alternative about alternative food networks?' *Environment and Planning* A 35: 389–391.

Wield, D., J. Chataway and M. Bolo. (2010) 'Issues in the political economy of agricultural biotechnology', *Journal of Agrarian Change* 10(3): 342–366.

Wolfe, M. S. (2000) 'Crop strength through diversity', *Nature* 406(6797): 681–682.

Wilson, C. and C. Tisdell (2001) 'Why farmers continue to use pesticides despite environmental, health and sustainability costs', *Ecological Economics* 39(3): 449–462.

Wright, B. D. (2012) 'Grand missions of agricultural innovation', *Research Policy* 41(10): 1716–1728.

Xia, Y. and S. Buccola (2005) 'University life science programs and agricultural biotechnology', *American Journal of Agricultural Economics* 87(1): 287–293.

Yan, J. S. and J. S. Zhang (1993) 'Advances of ecological engineering in China', *Ecological Engineering* 2(3): 193–215.

Zhu, Y. Y., H. R. Chen, J. H. Fan, Y. Y. Wang, Y. Li, J. B. Chen, J. X. Fan, S. S. Yang, L. P. Hu, H. Leung, T. W. Mew, P. S. Teng, Z. H. Wang and C. C. Mundt. (2000) 'Genetic diversity and disease control in rice', *Nature* 406(6797): 718–722.

3

SUSTAINABILITY SCIENCE AND 'IGNORANCE-BASED' MANAGEMENT FOR A RESILIENT FUTURE

Kristen Blann and Stephen S. Light

Western civilization has reached a watershed in its history of co-evolution with nature. For the past 300 years humans have employed science as a means to subdue and exploit nature. But our unwillingness to embrace error and adapt our societies is eroding our life support systems irreversibly.

Thus science has given us industrial agricultural monocultures that ignore the buffering and stabilizing functions of biodiversity and that are over-dependent on artificial inputs and chemicals; epidemics of wildlife disease; overuse of antibiotics in medicine and food supplies; bio-accumulating toxins; and exotic species invasions, to name but a few. The values, actions and structures that drive science and society are perpetuating such mistakes, from the persistence in practices of soil and water degradation, to ignoring the reality of climate change (see Rees, 2003; Collier and Webb, 2002). The words of Aldo Leopold, in many respects the father of ecological thinking in resource management, summarize the situation: 'We have the sad spectacle of one obsolete idea chasing another around a closed circle, while opportunity goes begging' (cited in Meine, 1988).

Since the Earth Summit in 1992, the global community has realized that a massive transformation is the only way to deal with the unprecedented ascent of regional and global problems. The Earth Summit established a new goal and ethos – sustainability. Pragmatically, sustainability impels a search for new options, technologies and ways of structuring natural and social relationships. The ethos of sustainability focuses on relationships, mutuality and cooperation and the search for symbiotic social and ecological connections.

This chapter outlines how emerging theories of ecosystem dynamics and their implications are essential for the transition to sustainability. First we reflect on how evolving theories of ecosystem behaviour have shaped our approach to resource management, often with grave consequences. We then outline how a co-evolutionary approach to ecosystems might be the way forward, and discuss the pedagogical, organizational and policy obstacles to, and implications of, such an approach.

Theories of ecosystem dynamics

Theories of ecosystem behaviour and management are in transition. They are moving away from reductionist, positive, deterministic notions of optimization, stability and maximum sustained yield towards managing for the resilience and evolutionary adaptive capacity needed to sustain systems' ability to produce resources and ecological services.

One can frame the evolving ecological perspectives around three fundamental categories:

1. Equilibrium-centred view
2. Boundary-centred view
3. Co-evolutionary view.

Particular configurations of technology and social organization tend to generate performance stability domains. We are currently facing a bifurcation point in which we must reorganize around ecological principles to meet the challenge of sustainability, or face regional and even global ecological-economic collapse (modified based on Raskin *et al.*, 2002).

Equilibrium-centred view

Most science-based resource agency policy and management in the US and elsewhere is still anchored in an equilibrium-centred view. This assumes that biophysical systems are deterministic and mechanistic, possessing some equilibrium state to which they will return after short-lived disturbances. This equilibrium is assumed to bound a stable set of ecological processes and functions that is viewed as the 'preferred' incarnation of the ecosystem. In this view, the universe is totally knowable; with time, science will reveal all of nature's mysteries and general principles. Single cause and effect relationships are assumed. Forest plans, biological opinions, engineering designs and harvesting recommendations have been largely predicated on notions of maximizing yield, on the assumption that ecosystems can be managed to maximize yield of a few commodities with stability and predictability. This view tends to adopt the myth that nature, in the end, is infinitely forgiving. This perspective underlies the 'technological optimist' notion that humans will be able to endlessly engineer fixes for ecological problems generated by our own management solutions (e.g. genetic engineering of crops to tolerate herbicide use, saline soils, or drought). It assumes that human ingenuity is a perfect substitute for natural capital.

Yet history shows that the equilibrium-centred view has generated fundamental contradictions in outcomes. A focus on single management targets and reductionist models of ecosystems (see later) creates solutions that might succeed in the short term but eventually backfire in the long haul. When simplistic solutions encounter complex ecological networks of relationships, unanticipated consequences result (see Box 3.1).

BOX 3.1 SIMPLE SOLUTIONS, UNANTICIPATED CONSEQUENCES

Conventional management has typically pursued technological solutions based on the assumption that a 'problem' is a closed system of single cause and effect, independent of other cause and effect relationships. The assumption – particularly once implementation begins – is that we know what the answer is and can design the technology to implement it, and that perfect knowledge and control of the system are attainable. An overreliance on 'final' technological solutions creates vulnerability because partial fixes tend to backfire, creating problems more challenging or expensive than those that the initial solution was created to solve. The end result is an increasingly constrained, over-engineered system that generates new problems and is ultimately more vulnerable to collapse. For example, in the Florida Everglades, society has approached water management as if each problem could be solved independently of other system functions. To solve the 'problem' of seasonal flooding and extensive wetlands, an immense drainage system was constructed through the Central & South Florida Project (CSFP). This drainage – implemented between 1905 and 1927 – caused problems associated with exposing peat land: oxidation, peat fires and water level management, including flooding. The Corps Project attempted to 're-water' the Everglades through a set of pumps, levees and canals, but failed to recognize the park's need for overland flow. Once recognized, a quota system of minimum monthly allocations was established that bore little resemblance to natural flows. The northern Everglades were partitioned into multi-purpose conservation areas. Water level management in the CSFP's Water Conservation Area 2A began converting sawgrass prairie into lakebed, leading to cascading ecological changes threatening the survival of species and entire ecological systems (e.g. tree islands, wood storks, Everglades kites, etc.).

The persistence of the equilibrium–centred view in many resource management agencies means that many ecosystems suffer from short-term management policies that sacrifice long-term sustainability (Ludwig et al., 1993). Ludwig and colleagues explain how 'magic' theories justify short-term or narrow management strategies. The 'magic' in these theories is that in predicting maximum sustainable yields, they simplify uncertainty, dynamism and complexity in ecosystems and ignore the influence of social systems (Ludwig et al., 1993; see also Wu, 2013). In their rush to make things appear tractable for citizens and policy makers, resource managers often side-step reality by ignoring uncertainty and complexity in natural systems. The trade-off of reducing complexity is loss of resiliency. In extreme cases, 'the ecological system loses resilience, the industries become dependent and inflexible, the management agencies become rigid and myopic, and the public loses trust in governance' (Holling, 1995; see also Eakin *et al.*, 2011; Stern and Coleman, 2015).

The boundary-centred view

The boundary-centred view expands on the notion of equilibrium by recognizing the possibility of multiple stability regimes. Ecosystem function is maintained through fluctuations and is able to persist in the face of disturbance and adversity (in other words, it is resilient). Holling (1986) describes the creative destruction and renewal functions of ecosystems; ecosystems do not reach stasis where the conservation of matter and nutrients remains stable. Instead, through events like floods, fire, disease and windthrow, ecosystems go through minor or quantum episodes of change and unpredictability where instabilities can flip a system into another regime of behaviour (Holling, 1995).

Perturbations can move a system from one to another stability domain. For example, lakes move from oligotrophic to eutrophic conditions in response to anthropogenic changes in nutrient inputs, which shift the structure of plant and animal communities. Such ecosystem behaviour is characterized from differing vantage points as being alternative stable states, requiring significant perturbations or interventions to recover once a threshold is reached (Scheffer *et al.*, 2001; Scheffer and Carpenter, 2003).

Some ecologists have suggested that when systems lose their resilience, collapsing into simpler systems, they can be both more stable and more resistant to change. This can reduce options, resulting in irreversible losses of ecosystem services and actual or potential benefit streams. A management approach that strives for resilience would emphasize the need to keep options open, to view events in a larger-than-local context and to emphasize heterogeneity. The fundamental insight is that while patterns may be discerned within given temporal and spatial scales, future events will be unexpected, not predictable. Approaches based on adaptive management and resilience do not require the precise capacity to predict the future, but the qualitative capacity to devise adaptive systems that can absorb and accommodate future shocks in whatever unexpected form they may take.

More recently, many ecologists have begun to question the whole notion of stable states. They contend that ecosystems are entirely historically contingent; they are constellations of independent species interacting and being acted on by forces at multiple scales, but those forces are random, complex, chaotic and in flux. Potential states are nearly infinite and not very stable (Dietz and Stern, 1998). Ecosystem trajectories are as unpredictable and historically contingent as human history, and as socially constructed as our notions of human nature (Lawhon and Murphy, 2012; Sagoff, 2000). Indeed, the very intensity and rapidity of environmental change over the past 100 years suggest that the notion of ecosystem stability is problematic. In the twentieth century, ecosystems have experienced such profound and significant transformations – habitat fragmentation, species invasions, major and minor alteration of nutrient cycles and the atmosphere, species invasions – that even were all human impact on ecosystems to cease immediately, the evolutionary and ecological implications of these changes are likely to play out almost indefinitely before anything resembling stable states re-emerge in ecosystems. Thus, while ecosystem stability as a theoretical construct may become increasingly obsolete, humans do seek to manage natural resource systems for desirable configurations that are stable and resilient over relevant, practically defined temporal and spatial scales. Even as the notions, definitions and interrelationships of resilience, complexity and stability continue to be debated (McCann, 2000; Walker and Abel, 2002), most ecologists agree that management strategies that seek to sustain the resilience of desirable system configurations should hedge their bets: conserve critical functions, processes, components and adaptive capacity; acknowledge complexity and uncertainty; avoid irreversible decisions; and keep future options open (Ludwig et al., 1993; Walker and Salt, 2012).

Co-evolutionary view

A more 'co-evolutionary' perspective is now adding to the notion of natural systems as contingent, discontinuous, interdependent, dynamically interacting, irreducible and unpredictable. An emerging set of works views ecosystems as living systems which respond to environmental influences with structural changes, and these changes in turn alter the future behaviour of those systems (Berkes and Ross, 2013; Maturana and Varela, 1992; Capra, 1996; Kauffman, 2000; Röling and Jiggins, 2001). These structurally coupled living systems are essentially 'learning' systems – complex adaptive systems (Röling and Jiggins, 2001).

Identifying and understanding the dynamics and cross-scale interactions of complex systems that produce 'unexpected' consequences are key to anticipating and responding to future system behaviour. Cross-scale interactions cause 'ecological and economic systems [to be] non-linear and adaptive, [and exhibit] complex and far-from equilibrium dynamics' (Levins, 1998). Positive feedbacks between scales in complex systems are responsible for producing discontinuities and multiple equilibria that will not be predicted by linear models operating at discrete scales (Munn, 1987). Once a system undergoes a transformation to a new equilibrium state, it is difficult if not impossible to return to the previous state: 'gradually bend a stick and suddenly it breaks. Simply releasing the pressure does not cause the stick to become intact again' (Kahn and O'Neill, 1999). The implications for natural resource management are that it is difficult to derive a general theory about system component interactions using assumptions about system parameters such as carrying capacity and maximum population size that are specific to a particular equilibrium state (Pastor et al., 1998).

The challenge is to understand the new system trajectories and guide them towards the goal of a healthy and self-sustaining ecosystem. Novel ways of thinking and acting are needed to create a future where humans can understand enough about ecological dynamics to channel both nature and human behaviour in ways that do not foreclose options and result in irreversible losses of natural capital and evolutionary potential. The fundamental challenge for twenty-first century resource scientists and managers is to discover through ingenuity and hard-won experience how to guide ecosystem trajectories to resilient configurations that can sustain production of integrated ecological and social benefits.

The management challenge in a context of complexity

How can we foster integrated management of social and ecological systems so that they are as a whole more flexible and adaptive than their constituent parts? And what are the implications for science and society under these conditions of irreducible uncertainty? Instead of society and culture being brought under the sway of science, we need instead to be prepared, as both Sun Tzu (Sawyer, 1994) and von Clausewitz (1976) counselled, for 'the totally unexpected ways of battle' (cited in Kaufmann, 2000). For this we need intuition and command genius that require '*konnen*' not '*wissen*'; know-how, instead of know-that (Kauffman, 2000).

A certain amount of complexity is necessary to retain resilience in an ecosystem, yet a certain amount of predictability is desirable in the management of ecosystems. Natural resource managers, driven by political, economic and social goals, often fall into the trap of managing systems for a fixed state. However, the dynamic nature of ecosystems means that such a state is not a long-term possibility (Fischer *et al.*, 2015; Kay and Schneider, 1994). Mismanagement may stem from identification of wrong cause(s), or from the assumption that causation is simple or that a few causes can be summed together in additive fashion. This leads to several predictable mistakes: (1) the significance of the single cause under test can be masked by noise contributed by the unsuspected and uncontrolled factors; (2) the problem mis-diagnosis appears only when two or more causes interact; or (3) the problem appears when there are present any number of sufficient causes which are not mutually exclusive (Hilborn and Stearns, 1982). Although these phrases sound hackneyed and overplayed, such simplification continues to underlie expensive and largely ineffectual efforts (see Box 3.2).

BOX 3.2 CHESAPEAKE BAY: AN ECOSYSTEM, NOT JUST A SINK

As recently as 20 years ago, the Chesapeake Bay in the eastern United States was viewed primarily as a sink – a dumping ground for nutrients and toxics – rather than a self-regulating ecosystem. Consequently, the bay's problems included nutrient enrichment, algae growth, loss of sea grasses and the impact of toxic substances. Early assessment suggested that nutrients coming into the bay from upstream were the source of the collapse of the famous bay fisheries.

Efforts started in 1983 to clean up the bay, but the focus was largely on nutrient reduction (for example, the goal in 1987 was for a 40% reduction in nutrients by 2000; Costanza and Geer, 1995). However, nutrient cleanup over 20 years has not improved the bay's condition (Chesapeake Bay Program Office, 1999). According to the Chesapeake Bay Foundation President, '[In] three and a half years ... the states and federal government have yet to implement any decisive actions' (Huslin, 2003). Five years later, the *Washington Post* reported the failure of oyster-saving efforts in Chesapeake, with fewer bivalves in the bay after a US$58 million campaign (Fahrenthold, 2008).

A National Academy of Sciences report states that changes in nutrients, sea grass beds and the abundance of many species of plants and animals serve not only as indicators of change but also as barriers to restoration (NAS, 2003). Restoration will require a multifaceted approach that looks at re-establishing ecological functions (NAS, 2003). For example, the indigenous oyster (*Crossostrea virginica*) has been reduced to 1% of its historic abundance largely due to fishing pressure over the past 150 years (Costanza and Geer, 1995). Oysters are living water filters, feeding on phytoplankton. Some researchers estimate that the once-abundant oyster beds of the bay could filter a volume of water equal to the entire bay in less than a week (Newell, 1988). Newell was subsequently quoted as saying 'the bay's keepers [ought] to preserve oysters "not for their economic value but for their ecological value"'. However, we should avoid the mistake of thinking that any one management action is going to reclaim the bay; simplistic solutions invariably backfire.

New ecological knowledge systems need to work with the complexity of ecosystems in a constructivist approach to science so that innovation and learning become embedded in management (Röling and Jiggins, 1998). Widening the framework of consideration in natural resource management to include social and economic systems introduces even more complexity, yet also increases the scope of tools available for management (Box 3.3).

Röling and Jiggins contend that given the complexity and inter-relatedness of the new class of resource management problems and dilemmas, humans have displayed very little 'effective knowledge' thus far in responding to the modern global ecological challenge (Röling and Jiggins, 1998). They call for 'ecological rationality',[1] the purposeful and collective redesign of human interactions to maintain the ecological services that we have thus far taken for granted. This is opposed to the economic rationality of the individual that underlies virtually all contemporary policy design and whose cumulative outcome is pathological in the modern context:

> Humans as a major force of nature are a collective cognitive agency. Regenerating the biosphere and building opportunity means purposefully and collectively redesigning human interactions. This appreciation – being

part of nature not separate from it – could make change and instability seem a more 'natural' human condition because humans would not see themselves as insulated from nature, as we do when we act as if we were outside of nature – that is, conquering, or overcoming, or breaking through it. ... If we do indeed comprise a duality with our environment, if we are inescapably part of the complex web of life... then human survival depends on our ability to maintain the ecological services that we have so far taken for granted. Ecological rationality demands that [we] develop the institutions, cognitions, norms, platforms for collective decision-making, cosmos-visions and other social elements that allow us to remain structurally coupled to context.

(Röling and Jiggins, 2001)

BOX 3.3 CO-EVOLUTION IN THE EVERGLADES

The trend in the Everglades since the early 1980s has been to look at ways to optimize the system though operational changes, iterative testing and experiments that have led to design changes without the real adoption of an adaptive management culture. Brought about by the piecemeal approach described in Box 3.1, a series of crises – drought, endangered species, exotic species and general ecosystem decline – have seen demand emerge for more integrated solutions to the problems of water supply and Everglades restoration. This has led to the innovative and unprecedented US$8 billion Comprehensive Everglades Restoration Project (CERP), which centres on an update of the Central & Southern Florida (C&SF) Project described in Box 3.1. It explicitly calls for adaptive management throughout its design, implementation and operation. Under CERP, a team has been assembled of managers and scientists with deep operational experience in Everglades water management. Their aim was to figure out how adaptive management and collaboration could be fully integrated into what had previously been a 'bricks and mortar' approach to management. The new approach recognized a need not just for a re-conceptualization of the system, but for a fundamental reorganization of science, management and policy institutional structures, approaches and paradigms. But even in this case, lead scientists and managers have struggled to reform conventional modes of design, construction and operation that are mired in outdated compartmentalized, standard procedures. Many of the proposed adaptive management experiments have never been fully implemented, and some of the initial political will and enthusiasm has dissipated over time and personnel turnover. The challenge of embracing holistic, resilience-based management remains.

Kauffman (2000) echoes this understanding of duality based on his investigation 'into what it means and is to know and make our world together ... Make no mistake, we autonomous agents mutually construct our biosphere, even as we coevolve with it'. Yet the conventional assumption of rational self-interest maximization is so deeply embedded in powerful social institutions that new understandings of cognition, learning and development of norms are still not being acknowledged. If we are to generate coherence and correspondence between our knowledge systems and institutions, in order to design adaptive resource management systems that work with change, we have to become open to learning more rapidly about complexity and uncertainty.

Evolving understanding and integration of science, ecosystem dynamics, human nature, social systems and the pathologies and barriers to learning inherent in conventional institutional and organizational cultures have radical implications for the way knowledge, economic functions, social relationships and power should be organized to address the sustainability challenge of the coming century. Social, ecological and economic sustainability requires new pedagogical and institutional designs and strategies, and a transformation of the role and structures for knowledge and science in society. New tools for understanding complex ecosystems (such as dynamic simulation modelling, mapping, decision analysis and Bayesian statistics) acknowledge uncertainty as an opportunity for learning and offer the promise of fewer unpleasant surprises.

But the challenges are many. Enlightenment Age theories of nature and instrumentality have fostered policy and institutions that are now trapped in five powerful and interwoven dogmas, which taken together thwart society's future responsiveness to ecological challenges. Together these tenets are highly resistant to recognizing and adapting to new kinds of feedback and other ways of generating effective shared knowledge about the world. Not only do they suppress recognition of the dynamism and uncertainty inherent in management of complex adaptive systems, but they also carry with them considerable defensive inertia so as to preserve and extend the life of their values, techniques and practices.

We briefly look at each of these tenets, and then describe how they might be tackled:

Reductionist science

Reductionist science is the science of parts, as opposed to the science of the integration of parts (Holling, 1986). While reductionist science in molecular biology is extremely helpful in such problems as analysing the impact of endocrine disrupters in alligators, reductionist approaches to solving ecosystem problems tend to lead to narrowly focused solutions to the exclusion of other modes of inquiry and lines of evidence. Rather than looking at ecological problems through increasingly narrow lenses, scientific methods need to be developed that are transdisciplinary and more holistic in approach. In the Kissimmee River Restoration, for example, biologists, hydrologists and engineers worked together to come up with hydrologic criteria (e.g. stage duration, timing, distribution) designed to reproduce the landscape mosaic needed to support colonial wading bird rookeries (Cummins and Dahm, 1995; Toth *et al.*, 1997).

In defining and focusing on problems that readily yield to reductionist analysis, while selectively ignoring those that do not, science continues to propagate the myth that complex problems (e.g. feeding the world, solving drought and soil erosion) can be divided into independent constituent components to which there are simple and independent technological solutions (e.g. genetically modified crop species). But in fact today's problems and their purported solutions deeply interconnect society and nature (Cooke *et al.*, 2016; Levins and Lewontin, 1985; Folke *et al.*, 2011). Unfortunately, corporations, governments and financial institutions are using reductionist science to perpetuate and advance ideologically based policies (e.g. maximum production, technological quick fixes, exploitation of nature) that have delivered short-sighted financial and technological benefits at the expense of future generations. Scientists often either turn a blind eye to how their research is being used politically or they drop the report on the desk of the policy makers without revealing all the unknowns associated with their findings relative to the policy decision at hand (Wesselink *et al.*, 2013; Wilson, 2002). These fixes often do more harm than good, unwittingly increasing the impact, scale and complexity of ecological problems to which society must subsequently adapt (Regier and Baskerville, 1986). Yet on the whole, science remains wilfully ignorant of or completely uninterested in understanding or coping with the messy secondary and tertiary order impacts of ecological and social responses.

Mismatch between reality and the economic assumptions that underlie policy decisions

Micro-economic theory, when employed in policy (e.g. cost-benefit analysis), continues to assume rational maximizing behaviour based on incremental utility analysis (Stakhiv *et al.*, 2003; TEEB, 2010). This assumes perfect, logical and deductive thinking and has similarly reductionist notions of how ecosystems generate benefits and services. While providing short-term economic gain to innumerable forestry, fisheries and range management systems, the theory breaks down when faced with biophysical complexity (Ostrom *et al.*, 2002). The abstract assumptions of economics-based resource management evolved in a relatively 'empty world' context in which actions of policy makers and their consequences for ecosystems and ecological economies were isolated in time and space (Daly, 1991; Gómez-Baggethun, 2015). The benefits of scientific management could be privatized and channelled to support more such development, while secondary and tertiary costs could be externalized to the ecosystem and the commons. There was little political economy to require that externalities be acknowledged either in theory or in practice; theory ignored the question. The impacts on marginalized indigenous and local people, who were often the first to observe and suffer from ecosystem change, were easily disregarded or rationalized as the price of development. But as evidence of ecosystem change and degradation has become more and more apparent, this paradigm has found itself increasingly challenged by new theories in physics, ecology, psychology and mathematics (Light, 2004; Gunderson *et al.*, 1995). It has also found itself increasingly out of sync with social reality, as those who have often paid the price for development and scientific progress (indigenous and local people, domestic workers and minorities) have begun to develop an empowered political voice. Converging lines of argument and theory are challenging the notion that the aggregation of benefit streams exceeds the cost to system resilience and performance as a whole. Many economists still hold dear the concept of rational economic behaviour that assumes complete knowledge (cf. Chapter 7, this volume), yet bounded rationality and complexity theory recognize how complicated 'economic' problems can be, necessitating inductive reasoning and approaches that emphasize the whole rather than the sum of individual parts (Arthur, 1994b; Holland, 1998; Gasparatos *et al.*, 2009). New economic models are needed to support adaptive decision-making processes that do not foreclose options, neglect emergent and holistic impacts, or assume away uncertainty. Yet economic concepts of marginal utility analysis still dominate contemporary policy-making processes.

Top-down command and control

Bureaucratic structures that rely on top-down, command-and-control approaches to management are more responsive to feedback and direction from special interests than to deliberative social democracy. When a management approach does not work, the response is often to make slight adjustments in conventional approaches or institutions that merely increase the control costs while failing to address the underlying cause of the performance decline. For example, incremental and reductionist approaches to resource management in the United States' Pacific Northwest have combined to precipitate larger and larger cascades of detrimental ecological change in salmon runs (National Research Council, 1996). Pacific salmon have disappeared from 40% of their historical range. Hatcheries, designed to compensate for the negative impacts of hydropower, irrigated agriculture and commercial fishing (all economic development solutions), have led to competition between less genetically robust hatchery fish and wild salmon, and the introduction of disease. As a result, the salmon take by native people in the Columbia Basin, which despite its spatial and temporal variability was once a predictable annual phenomenon, has become so unpredictable that it threatens to extinguish both the salmon and a way of life.

Market fundamentalism

Political structures are based on the myth that markets are self-regulating and transparent, that those operating in them are rational utility maximizers with full and symmetric knowledge and equal access, and that unregulated markets therefore lead to the most efficient allocation of resources and outcomes. The inconsistencies and negative social and environmental consequences of political, financial and economic decisions are poorly understood, and minimized in the minds of policy makers as regrettable but forgettable externalities of market dynamics – the cost of progress (Gómez-Baggethun, 2013; Klein, 2015; Stiglitz, 2002). International financial institutions, governments and corporations cling doggedly to their belief in the self-perpetuating and self-serving infallibility of the market; contending that markets are objective, policy-neutral and detached from other social values. But in reality, markets evolve from particular constellations of world views, concepts, values, techniques and practices that are shared with and arise from social communities, and are used by those communities to define legitimate problems and solutions. The preoccupation with market-based solutions is produced and disseminated within powerful social, legal and political institutions which have a tendency towards speculation, monopolistic distortions (Sullivan, 2013) and 'exaltation of markets over legislative criteria, including local democratic priorities' (Phillips, 2002). Individual economic choice is assumed to outperform, and allowed to trump, reasonable collective choice.

Engineering-dominated solutions

With the rapid industrialization of the Northern countries over the past century, increasing attention has been paid to engineering solutions that place nature in straightjackets of consistency. Instead of developing temporary structures to test various solutions to resource problems associated with flood control, water supply, water quality and environmental mitigation, society has invested in optimal, large-scale, irreversible solutions that take advantage of cheap labour, excess engineering capacity and availability of capital (Zimmerer, 2011). As population growth, land and water dynamics, ecological needs and climate change continue to place new demands on society, engineering solutions will continue to be called upon. But the challenge of the future will not be to build monuments to increasing obsolescence, but to develop new methods and approaches that restore ecosystems and reconcile new engineering designs with constantly evolving social and economic needs.

Bringing about change

In an era of no easy answers, there are no simple, technological, or ideological solutions to the ecological problems we face. New standards of human behaviour are required, and nothing short of that will work. Faced with futures that cannot be forecast, fundamentally new responses and policies are required from science-based agencies.

Tackling knowledge, power and privilege

Anything that stands in the way of such resilience [adaptation required in a new context], be it elites, institutions, escapist pathologies, inability to learn, impaired or distorted perceptions of contextual change, or inflexibility of investment, is bound to have grave consequences.

(Röling, 2002)

It is critical to understand the role of entrenched interests and social relationships involved in decision-making processes and outcomes (Scoones and Thompson, 1994). Power is inextricably linked to who participates, whose knowledge counts and who is excluded from decision making (Baker *et al.*, 2013; Hulme, 2010; Robbins, 2011). Thus, power imbalances in society tend to systematically suppress ideas that may have in reality been present for a long time. Many modern environmental predicaments have emerged directly as a consequence of the disproportionate influence given to the conventional reductionist scientific paradigms described earlier. Not all environmental 'surprises' – such as the environmental and human health effects of DDT, pesticides and other biologically and hormonally active agents; climate change; the ozone hole; and the recent discovery that genetically modified corn has reached remote corners of Mexico – were entirely unanticipated. But those who carried that knowledge, the Aldo Leopolds or Rachel Carsons, for example, did not have the power to influence the debate or decisions at the time.

Focusing on the role of power thus reveals that many of what are commonly presented as 'new scientific ideas' reflect instead shifts in the power relations of competing discourses. With hindsight, we recognize that throughout human history, respected visionaries provided warnings, unheeded advice and antidotes to the folly of dominant paradigms. Increased vulnerability in the ecosystem has often been noted by small groups of scientists or local citizens. The critiques of industrial commodity monocultures, technological quick fixes and expert science coming from feminist, environmental, local and indigenous perspectives have been with us for over 50 years. The focus on relationship, function, dynamics, interdependence and process in nature is so often portrayed as 'emerging ideas in ecology' (Botkin, 1990). Yet, none of this is really 'new' science. Rachel Carson, Aldo Leopold, Alfred Whitehead and many others were all advocating, before 1950, for the recognition of interdependence and adoption of more integrative and holistic approaches to science and management (Worster, 1977). The recent focus on integrative science thus reflects the nature of the problems that have begun to attract political attention, as the marginal benefits of advances in conventional science and technology have begun to appear less significant relative to the growing scale and complexity of the problems, and given the pattern of their emergence directly out of technological solutions to past problems (Horgan, 2015).

Modern conventional government decision-making processes are slow and largely impenetrable to citizens. Social learning is limited to those privileged by the process. Power is positioned to maximize its access and maintain its privileged status. The privileging of scientific and technological voices and ways of knowing – often over the objections of local, public and/or minority voices – has a role in generating inequitable or unsustainable outcomes that must be acknowledged.

Finding sustainable solutions in agriculture and natural resource management is dependent on involving citizens, farmers, indigenous and rural people in the debate (Adams *et al.*, 2014; Kloppenburg, 1991; Risvoll *et al.*, 2016). Such perspectives serve to widen discussion, embolden other minority voices and help develop robust responses that avoid catastrophic 'surprise'.

Redesigning science for sustainability

Given the shortcomings of conventional models of reality, knowledge, ecosystems and human nature that still govern natural resource management decision making, new roles and forms of science in decision making are needed. Civic science, adaptive management and sustainability science all advocate expanding the boundaries of conventional science to integrate alternative approaches, leading to a radically transformed science whose contours are only now being mapped (Kloppenburg, 1991; Lee, 1993; Funtowicz and Ravetz, 1993; Gunderson *et al.*, 1995; Irwin, 1995; Toomey, 2016).

Funtowicz and Ravetz's 'post-normal' science, Irwin's 'citizen science' and Lee's 'civic science' suggest the transformation of science by recognizing the role of citizens' knowledge to extend the peer community to those directly affected by an environmental problem or issue (Funtowicz and Ravetz, 1993). The goal is not to undermine science as a method of generating and vetting knowledge, nor to lessen its role or significance in 'science-based' decision-making processes, but to integrate scientific expertise with other assessments, problem definitions and various expertise and to appreciate interconnectedness. Reductionist, deterministic inquiry must be balanced by open acknowledgement of the limitations, uncertainties and risks of science and technology.

Opening up science to a wider community and greater sources of inquiry should enhance the development of effective knowledge for biodiversity conservation and can produce more innovative, creative and robust solutions. The existence of a diversity of paradigms, problem definitions and world views means that flaws in the methods, conclusions and, especially, assumptions in the decision-making processes, are more likely to be challenged at some point during the process. This can help to predict certain types of surprises and unintended consequences that would go unchallenged in traditional decision-making contexts. The result is more robust decision making.

Science for sustainability differs qualitatively from conventional science (National Research Council, 1999). It acknowledges and embraces uncertainty – about the environment, the future and in human understanding and social relationships. It requires the creation and enhancement of options, generating ecological resilience and social trust. It requires new tools such as adaptive management, scenario planning, citizen science, collaborative planning and decision making and conflict resolution. Acknowledging uncertainty implies a shift from 'knowledge-based' to 'ignorance-based' management, predicated on precaution, adaptive management and learning.

Sustainability science requires not just interdisciplinary expertise, but also transdisciplinary synthesis. Ecologists and managers are challenged to transform the intellectual and epistemological assumptions that frame their conservation activities (Kloppenburg, 1991). Such broadening includes questioning the objectivity of science, expanding the umbrella of 'officially sanctioned knowledge', opening the practice of science to more inclusive methodologies and ending the separation between research and management (Irwin, 1995; Funtowicz and Ravetz, 1993; Holling et al., 1995; Toomey, 2016). Multiple realities do exist and different environmental knowledge must be recognized and valued (Kloppenburg, 1991; Turnbull, 1997; Ruiz-Mallén and Corbera, 2013). Valuing traditional, local and indigenous knowledge as sources of understanding and novelty is a vital part of this process. This includes involving grassroots, local movements in monitoring, researching and repairing environmental problems as well as in resource decision making.

However, conventional natural resource management, reductionist sciences and neoclassical economics still dominate disciplinary scientific management, policy and decision-making contexts, where equilibrium and boundary-centred views of ecological systems still dominate. This is despite the emergence of these new scientific paradigms and the growing influence of integrative and interdisciplinary sciences (e.g. ecology, conservation biology, complexity, management and participation sciences, environmental and sustainability sciences), all of which are gradually embracing more co-evolutionary, adaptive, systems perspectives (Holling, 1986; Walker and Salt, 2012).

Reforming governance of natural resources

Recognizing the significant role of power and conventional paradigms as obstacles to adaptive learning suggests that transforming social-ecological systems will require radical reorganization and reform of natural resource management institutions. 'Top-down' systems of governance must be recast to become more collaborative and participative. Resource management institutions and agencies must be restructured and reoriented to embrace the coordination of public-private partnerships, broad public input and participatory models of research and decision making. Conflict resolution, deliberative negotiation of risk and values and social learning processes must all be fully integrated into decision-making processes; they are required to ensure that democratically and socially desired outcomes are negotiated from within a coherent framework that is consistent with ecosystems' natural limits and dynamics.

The contradictions and paradoxes that arise through attempts to negotiate multiple, often conflicting values, as well as to operationalize participation, collaboration, sustainability and ecosystem management, create conflicts within the individual and group that must be resolved through learning. This facilitates 'double loop learning' (see Figure 3.1), in which learning goes beyond the simple acquisition of new data or information to lead to a critical reassessment and re-conceptualization of the assumptions underlying that knowledge, and may possibly be accompanied by a

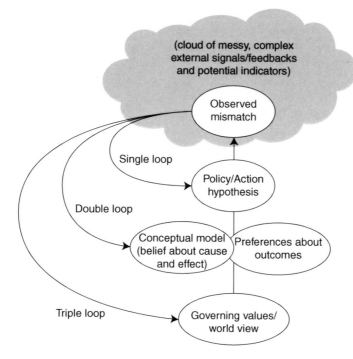

FIGURE 3.1   Single, double and triple loop learning in the context of sense-making and development of ecological and natural resource management knowledge. Modified from Argyris and Schon, 1978

shift in values or goals. Decision making becomes more transparent, and even where processes fail and/or political jockeying and special interests prevail, unjust or unfair decisions are more likely to be challenged. This redesign of natural resource management institutions will shift emphasis from reductionist to holistic and integrative forms of scientific inquiry, essential to ensuring that natural resource systems are viewed not just as a means of commodity production, but also as the natural capital of the economy and a provider of environmental services, biodiversity, cultural and spiritual values.

Embracing adaptive management

Finally, resource management must embrace monitoring, assessment and anticipatory adaptive management. Adaptive management and collaboration are both methodological approaches and processes that have emerged independently as potential solutions to the failures and conflicts generated by traditional natural resource management decision making. Adaptive management emphasizes the conceptual blind spots of expert-driven management – the failure to recognize complexity and dynamism in ecosystems. Adaptive management is predicated on the notion that understanding of ecosystems is always provisional; our knowledge of how ecosystems work is imperfect and uncertainty dominates our interactions within them. The central tenet of adaptive management is that current policy and knowledge is inadequate, and must remain open to learning and revision.

Tools such as adaptive management, collaboration, scenario development and adaptive environmental assessment and management have been developed to facilitate learning-by-doing (Blumenthal and Jannink, 2000; Borrini-Feyerabend *et al.*, 2008; Light, 2001) and to create more anticipatory and flexible solutions to complex problems that do not foreclose future options. Adaptive natural resource management requires clearly defined goals and conceptual management models, selection of performance measures and indicators and monitoring and evaluation methods that are modified as learning proceeds (see Holling, 1978). Building social capacity for adaptive management in Western societies is essential to achieving greater understanding of the value of biodiversity to society, and for developing and diffusing ideas and innovations capable of meeting the challenge of sustainability.

Conclusion

Part of making the quantum shift towards sustainability is recognizing and challenging conventional ways of thinking and acting. It takes time to shed old paradigms, institutions and habits. The new philosophies and frameworks for ecosystem management and biodiversity protection have yet to be fully developed or implemented. Citizen sciences, participatory and collaborative learning methodologies, ecosystem and adaptive management, sustainability assessment and other theories and frameworks are all part of converging and emerging approaches that require iterative adaptation. No single framework is appropriate for every geographic, ecological or cultural setting. Productive interactions between different ways of knowing and between differently situated and partial knowledge need to be established, not in order to combine or translate knowledge, but to permit mutually beneficial dialogue, synthetic understanding, uncertainty analysis and the emergence of new solutions. The task for ecosystem resource managers is to create the conditions for adaptive social learning and the restoration of ecological resilience.

The ultimate test of humanity's ecosystem knowledge is whether it contributes to our survival and quality of life. Human nature, perception and social organizations have evolved in very different ecological and social environments to those in which we now find ourselves. Will we be able to escape the juggernaut of the institutions, pathologies and paradigms that have led us into our current predicament and thus far prevented attempts at reform? Will the new ecosystem and biodiversity knowledge help resolve our current environmental predicament, given the scale and pace of ecological changes our past actions and institutions have produced? Can our knowledge guide effective action under uncertainty? Can we develop new adaptive ways of thinking and acting in time to avoid large-scale irreversible ecological or social collapse? These questions remain poignantly open. The challenge is ours.

> And so long as you haven't experienced this:
> To die and so to grow
> You are only a troubled guest
> On the dark earth
> > *Johann Wolfgang von Goethe (1814)*

Note

1   Based on the seminal work on the Santiago theory of cognition by Humberto Maturana and Francisco Varela (1992), and on Arthur (1994a, b), Arrow (1994) and Capra (1996).

References

Adams, M. S., J. Carpenter, J. A. Housty, D. Neasloss, P. C. Paquet, C. Service, J. Walkus and C. T. Darimont (2014) 'Toward increased engagement between academic and indigenous community partners in ecological research', *Ecology and Society* 19(3): 5.

Argyris, C. and D. Schon (Eds) (1978) *Organizational Learning*, Addison-Wesley Publishing Company, New York.

Arrow, K. J. (1994) 'Methodological individualism and social knowledge', *American Economic Review* 84(2): 1–9.

Arthur, W. B. (1994a) 'Inductive reasoning and bounded rationality', *American Economic Review* 84(2): 406–11.

Arthur, W. B. (1994b) 'On the evolution of complexity', In Cownan, G., D. Pines and D. Meltzer (Eds) *Complexity: Metaphors, Models and Reality*, Addison-Wesley, Reading, MA.

Baker, L., M. Dove, D. Graef, A. Keleman, D. Kneas, S. Osterhoudt and J. Stoike (2013) 'Whose diversity counts? The politics and paradoxes of modern diversity', *Sustainability* 5(6): 2495–2518.

Berkes, F. and H. Ross (2013) 'Community resilience: Toward an integrated approach' *Society & Natural Resources* 26(1): 5–20.

Blumenthal, D. and J. L. Jannink (2000) 'A classification of collaborative management methods', *Conservation Ecology* 4(2): 13.

Borrini-Feyerabend, G., M. P. Pimbert, M. T. Farvar, A. Kothari and Y. Renard (2008) *Sharing Power. A Global Guide to Collaborative Management of Natural Resources*, Routledge, London.

Botkin, D. B. (1990) *Discordant Harmonies: A New Ecology for the Twenty-first Century*, Oxford University Press, New York.

Capra, F. (1996) *The Web of Life*, Bantam Doubleday, New York.

Chesapeake Bay Program Office (1999) *Snapshot of Chesapeake Bay: How's it doing?* Chesapeake Bay Program Office, Annapolis.

Clausewitz, K. von (1976) *On War*, translated by Michael Howard and Peter Paret, Princeton.

Collier, M. and R. Webb (2002) *Floods, Droughts and Climate Change*, The University of Arizona Press, Tucson.

Cooke, B., S. West and W. J. Boonstra (2016) 'Dwelling in the biosphere: Exploring an embodied human-environment connection in resilience thinking', *Sustainability Science* 11(3): 1–13.

Costanza, R. and J. Geer (1995) 'The Chesapeake Bay and its watershed: A model for sustainable ecosystem management?', In Gunderson, L., C. S. Holling and S. S. Light (Eds) *Barriers and Bridges to Renewal of Ecosystems and Institutions*, Columbia University Press, New York.

Cummins, K. W. and C. N. Dahm (1995) 'Introduction: Restoring the Kissimmee', *Restoration Ecology* 3(3): 147–8.

Daly, H. E. (1991) 'From empty world economics to full world economics: Recognizing an historic turning point in economic development', In Goodland, R., H. Daly and S. El Serafy (Eds) *Environmentally Sustainable Development: Building on Brundtland*, The World Bank, Washington, DC.

Dietz, T. and P. Stern (1998) 'Science, values, and biodiversity', *Bioscience* 48(6): 441–5.

Eakin, H., S. Eriksen, P. Eikeland and C. Øyen (2011) 'Public sector reform and governance for adaptation: Implications of new public management for adaptive capacity in Mexico and Norway', *Environmental Management* 47(3): 338–51.

Fahrenthold, D. A. (2008) 'Oyster-saving efforts a wash in Chesapeake: Fewer bivalves in the bay after $58 million campaign', *Washington Post*, 2 June, Washington, DC.

Fischer, J., T. A. Gardner, E. M. Bennett, P. Balvanera, R. Biggs, S. Carpenter, T. Daw, C. Folke, R. Hill, T. P. Hughes, T. Luthe, M. Maasse, M. Meacham, A. V. Norstrom, G. Peterson, C. Quieroz, R. Seppelt, M. Spierenberg and J. Tenhunen (2015) 'Advancing sustainability through mainstreaming a social-ecological systems perspective', *Current Opinion in Environmental Sustainability* 14: 144–49.

Folke, C., Å. Jansson, J. Rockström, P. Olsson, S. Carpenter, F. Chapin, A. S. Crépin, G. Daily, K. Danell, J. Ebbesson, T. Elmqvist, V. Galaz, F. Moberg, M. Nilsson, H. Österblom, E. Ostrom, Å. Persson, G. Peterson, S. Polasky, W. Steffen, B. Walker and F. Westley (2011) 'Reconnecting to the biosphere', *Ambio* 40(7): 719–38.

Funtowicz, S. O. and J. Ravetz (1993) 'Science for the post-normal age', *Futures* 25(7): 740.

Gasparatos, A., El-Haram, M. and M. Horner (2009) 'The argument against a reductionist approach for measuring sustainable development performance and the need for methodological pluralism', *Accounting Forum* 33(3): 245–56.

Goethe, J. W. von (1814) 'The Holy longing', In Bly, R. (ed.) (1980) *News of the Universe: Poems of Twofold Consciousness*, translated by Robert Bly, Sierra Club Books, San Francisco.

Gómez-Baggethun, E. (2013) 'Scale misfit in ecosystem service governance as a source of environmental conflict', *Society & Natural Resources* 26(10): 1202–16.

Gómez-Baggethun, E. (2015) 'In search of lost time: The rise and fall of limits to growth in international sustainability policy', *Sustainability Science* 10(3): 385–95.

Gómez-Baggethun, E., V. Reyes-García, P. Olsson and C. Montes (2012) 'Traditional ecological knowledge and community resilience to environmental extremes: A case study in Doñana, SW Spain', *Global Environmental Change* 22(3): 640–50.

Gunderson, L. H., C. S. Holling and S. S. Light (Eds) (1995) *Barriers and Bridges to the Renewal of Ecosystems and Institutions*, Columbia University Press, New York.

Hilborn, R. and S. C. Stearns (1982) 'On inference in ecology and evolutionary biology: The problem of multiple causes', *Acta Biotheoretica* 31(3): 145–64.

Holland, J. (1998) *Emergence: From Chaos to Order*, Addison Wesley, Reading MA.

Holling, C. S. (1978) *Adaptive Environmental Assessment and Management*, John Wiley and Sons, New York.

Holling, C. S. (1986) 'The resilience of terrestrial ecosystems: Local surprise and global change', In Clark, W. C. and R. E. Munn (Eds) *Sustainable Development of the Biosphere*, Cambridge University Press, Cambridge.

Holling, C. S. (1995) 'What barriers? What bridges?', In Gunderson, L. H, C. S. Holling and S. S. Light (Eds) *Barriers and Bridges to the Renewal of Ecosystems and Institutions*, Columbia University Press, New York.

Horgan, J. (2015) *The End of Science: Facing the Limits of Knowledge in the Twilight of the Scientific Age*, Basic Books, New York.

Hulme, M. (2010) 'Problems with making and governing global kinds of knowledge', *Global Environmental Change* 20(4): 558–64.

Huslin, A. (2003) 'Reports cloud picture of bay's health', November 12, 2003, *The Washington Post*, Washington, DC.

Irwin, A. (1995) *Citizen Science: A Study of People, Expertise, and Sustainable Development*, Routledge, London.

Kahn, J. R. and R. V. O'Neill (1999) 'Ecological interaction as a source of economic irreversibility', *Southern Economic Journal* 66(2): 391–402.

Kauffman, S. A. (2000) *Investigations*, University of Oxford Press, Oxford.

Kay, J. J. and E. Schneider (1994) 'Embracing complexity: The challenge of the ecosystem approach', *Alternatives* 20(3): 3–9.

Klein, N. (2015) *This Changes Everything: Capitalism vs. the Climate*, Simon and Schuster, New York.

Kloppenburg, J. (1991) 'Social theory and the de/reconstruction of agricultural science: Local knowledge for an alternative agriculture', *Rural Sociology* 56(4): 519–48.

Lee, K. N. (1993) *Compass and Gyroscope: Integrating science and politics for the environment*, Island Press, Washington, DC.

Lawhon, M. and J. T. Murphy (2012) 'Socio-technical regimes and sustainability transitions insights from political ecology', *Progress in Human Geography* 36(3): 354–78.

Levins, R. (1998) 'Dialectics and systems theory', *Science & Society* 62(3): 375–99.

Levins, R. and R. Lewontin (1985) *The Dialectical Biologist*, Harvard University Press, Cambridge.

Light, S. (2001) 'Adaptive ecosystem management and assessment: The path of last resort?', In Jensen, M. E and P. S. Bourgeron (Eds) *A Guidebook for Integrated Ecological Assessments*, Springer Verlag, New York.

Light, S. S. (ed.) (2004) 'The role of biodiversity conservation in the transition to rural sustainability', *NATO Science Series: Science and Technology Policy* Vol. 41, The Netherlands.

Loftin, M. K., L. A. Toth and J. T. B. Obeysekera (1990) *Kissimmee River Restoration: Alternative Plan Evaluation and Preliminary Design Report*, South Florida Water Management District, West Palm Beach.

Ludwig, D., R. Hilborn and C. Walters (1993) 'Uncertainty and resource exploitation: Lessons from history', *Science* 260(5104): 17–36.

McCann, K. S. (2000) 'The diversity-stability debate', *Nature* 405(6783): 228–33.

Maturana, H. R. and F. J. Varela (1992) *The Tree of Knowledge: The biological roots of human understanding*, Shambala Publications, Boston.

Meine, C. (1988) *Aldo Leopold: His life and work*, The University of Wisconsin Press, Madison.

Munn, R. E. (1987) 'Environmental prospects for the next century: Implications for long-term policy and research strategies', Working paper RR-87-15, International Institute for Applied Systems Analysis, Laxenburg, Austria.

NAS (2003) *Non-Native Oysters in Chesapeake Bay*, National Academies Press, Washington, DC.

National Research Council (1996) *Upstream: Salmon and Society in the Pacific Northwest*, National Academy Press, Washington, DC.

National Research Council (1999) *Our Common Journey: A transition toward sustainability*, National Academy Press, Washington, DC.

Newell, R. (1988) 'Ecological changes in Chesapeake Bay: Are they the result of overharvesting the American oyster (*Crassostrea virginica*)?', In *Understanding the Estuary – Advances in Chesapeake Bay Research*, Proceedings of a conference, 29–31 March 1988, Chesapeake Bay Program, Publication 129, Chesapeake Research Consortium, Baltimore.

Ostrom, E., T. Dietz, N. Dolsak, P. C. Stern, S. Stonich and E. U. Weber (Eds) (2002) *The Drama of the Commons*, National Academy Press, Washington, DC.

Pastor, J., S. Light and L. Sovell (1998) 'Sustainability and resilience in boreal regions: Sources and consequences of variability', *Conservation Ecology* 2(2): 16.

Phillips, K. (2002) *Wealth and Democracy*, Broadway Books, New York.

Raskin, P., T. Banuri, G. Gallopin, P. Gutman, A. Hammond, R. Kates and R. Swart (2002) *Great Transition: The Promise and Lure of the Times Ahead*, Global Scenarios Group, Stockholm.

Rees, M. (2003) *The Final Hour*, Basic Books, New York.

Regier, H. A. and G. L. Baskerville (1986) 'Sustainable redevelopment of regional ecosystems degraded by exploitative development', In Clark, W. C. and R. E. Munn (Eds) *Sustainable Development of the Biosphere*, Cambridge University Press, Cambridge.

Risvoll, C., G. E. Fedreheim and D. Galafassi (2016) 'Trade-offs in pastoral governance in Norway: Challenges for biodiversity and adaptation', *Pastoralism* 6(1): 1–15.

Robbins, P. (2011) *Critical Introductions to Geography: Political Ecology*, second edition, Wiley-Blackwell, Oxford.

Röling, N. and J. Jiggins (1998) 'The ecological knowledge system', In Röling, N. G. and M. A. E Wagemakers (Eds) *Facilitating Sustainable Agriculture: Participatory Learning and Adaptive Management in Times of Environmental Uncertainty*, Cambridge University Press, Cambridge.

Röling, N. and J. Jiggins (2001) 'Agents in adaptive collaborative management: The logic of collective cognition', In Buck, L. E., Geisler, C. C., Schelhas, J. and E. Wollenberg (Eds) *Biological Diversity: Balancing Interests through Adaptive Collaborative Management*, CRC Press, Boca Raton.

Röling, N. (2002) 'Beyond the aggregation of individual preferences: Moving from multiple to distributed cognition in resource dilemmas', In Leeuwis, C. and R. Pyburn (Eds) *Wheelbarrows Full of Frogs: Social Learning in Rural Resource Management*, Koninklijke Van Gorcum, Assen, The Netherlands.

Ruiz-Mallén, I. and E. Corbera (2013) 'Community-based conservation and traditional ecological knowledge: implications for social-ecological resilience', *Ecology and Society* 18(4): 12.

Sagoff, M. (2000) 'Concepts of ecosystem design in historical and philosophical context', In Pimentel, D., D. Westra and R. Noss (Eds), *Ecological Integrity in Environmental, Agricultural, and Health Systems*, Island Press, Washington, DC.

Sawyer, R. D. (transl) (1994) *Sun Tzu, the Art of War*, Barnes and Nobles, New York.

Scheffer, M. and S. R. Carpenter (2003) 'Catastrophic regime shifts in ecosystems: Linking theory to observation', *Trends in Ecology and Evolution* 18(12): 648–56.

Scheffer, M., S. Carpenter, J. A. Foley, C. Folke and B. Walker (2001) 'Catastrophic shifts in ecosystems', *Nature* 413(6856): 591–96.

Scoones, I. and J. Thompson (1994) *Beyond Farmer First: Rural People's Knowledge, Agricultural Research, and Extension Practice*, Intermediate Technology Publications, London.

Stakhiv, E., R. Cole, P. Scodari and L. Martin (2003) *Improving Environmental Benefits Analysis in Ecosystem Restoration Planning*, Institute for Water Resources, Policy Study Program, Alexandria, VA.

Stern, M. J. and K. J. Coleman (2015) 'The multidimensionality of trust: Applications in collaborative natural resource management', *Society & Natural Resources*, 28(2): 117–32.

Stiglitz, J. (2002) *Globalization and Its Discontents*, WW Norton & Co., New York.

Sullivan, S. (2013) 'Banking Nature? The Spectacular Financialisation of Environmental Conservation' *Antipode* 45(1): 198–217.

TEEB (2010) *The Economics of Ecosystems and Biodiversity: Ecological and Economic Foundations*, Earthscan, London.

Toomey, A. (2016) 'What happens at the gap between knowledge and practice? Spaces of encounter and misencounter between environmental scientists and local people', *Ecology and Society* 21(2): 28.

Toth, L. A., D. Albrey Arrington and G. Begue (1997) 'Headwater restoration and reestablishment of natural flow regimes: Kissimmee River of Florida', In Williams, J. E, Wood, C. A. and M. P. Dombeck (Eds) *Watershed Restoration: Principles and Practices*, American Fisheries Society, Baltimore.

Turnbull, D. (1997) 'Reframing science and other local knowledge traditions', *Futures* 29(6): 551–62.

Walker, B. and D. Salt (2012) *Resilience Thinking: Sustaining Ecosystems and People in a Changing World*, Island Press, Washington, DC.

Walker, B. and N. Abel (2002) 'Resilient rangelands: Adaptation in complex systems', In Gunderson, L. and C. S. Holling (Eds) *Panarchy*, Island Press, Washington, DC.

Wesselink, A., K. S. Buchanan, Y. Georgiadou and E. Turnhout (2013) 'Technical knowledge, discursive spaces and politics at the science-policy interface', *Environmental Science & Policy* 30(1): 1–9.

Wilson, J. (2002) 'Scientific uncertainty, complex systems, and the design of common-pool institutions', In Ostrom, E., Dietz, T., Dolsak, N., Stern, P. C., Stonich, S. and E. U. Weber (Eds) *The Drama of the Commons*, National Academy Press, Washington, DC.

Worster, D. (1977) *Nature's Economy: A History of Ecological Ideas*, Cambridge University Press, Cambridge.

Wu, J. (2013) 'Landscape sustainability science: Ecosystem services and human well-being in changing landscapes', *Landscape Ecology* 28(6): 999–1023.

Zimmerer, K. S. (2011) 'The landscape technology of spate irrigation amid development changes: Assembling the links to resources, livelihoods, and agrobiodiversity-food in the Bolivian Andes' *Global Environmental Change* 21(3): 917–934.

4

ON NON-EQUILIBRIUM AND NOMADISM

Knowledge, diversity and global modernity in drylands

Sian Sullivan and Katherine Homewood

Introduction: conceptualizing drylands and nomads[1]

> [I]mages of poverty and ... pastoralism have in recent years become inextricably bound up together in apocalyptic scenes of drought, famine and warfare. Media representations of swollen-bellied children, skeletal figures in drought-stricken landscapes and pitiful refugee camps are so powerful that, rather than stimulating critical examination of the complex causes of the crisis, they have circumvented it and urged upon planners the simplest of diagnoses and cures ... There is the profoundest possible opposition between the diagnoses and perceptions of the planners and the perceptions of the pastoralists themselves. While planners see the reduction of livestock and moves towards sedentarization and cultivation as the ways to prosperity, pastoralists tend to see these as the very definition of poverty itself.
>
> *(Broch-Due and Anderson, 1999)*

The quote above describes widespread views of pastoral nomads and drylands. In popular perceptions, these are localities and peoples that have been distinguished by their poverty, their environmental fragility, the scourge of degradation and 'encroaching deserts', the eruption of disorder, conflict and banditry and the apparent need for a civilizing intervention that favours settlement, land privatisation and planning (e.g. Hardin, 1968; Lamprey, 1983; Sinclair and Fryxell, 1985; Timberlake, 1988; Grainger, 1992).

In the 1980s and 1990s an alternative discourse emerged which situated the construction of these negative views in historical context, considered the power relationships and often marginalizing policies they support and challenged the evidence and assumptions on which they are based (for the African context see Anderson, 1984; Homewood and Rodgers, 1987; Brockington and Homewood,

1996; Sullivan, 1996c, 2000a; Mortimore, 1998; Oba *et al.*, 2000; Brockington, 2002; Sullivan and Rohde, 2002). For example, numerous studies now question assumptions of agro/pastoralist-induced degradation, whether of rangeland habitat (Sullivan, 1999b; Homewood *et al.*, 2001), soil fertility (Mortimore, 1998; Ramisch, 1999; Hilhorst and Muchena, 2000; Osbahr, 2001), soil erosion/redeposition (e.g. Abel, 1993; Homewood, 1994), deforestation (Leach and Fairhead, 2000) or biodiversity (Homewood and Brockington, 1999; Maddox, 2002; Western and Gichohi, 1993; Homewood *et al.*, 2001). And increasingly, debate and dialogue among range ecologists, development workers, policy makers and practitioners emphasize the ecological and economic rationales behind mobile livestock production systems in drylands (Sandford, 1983; Behnke *et al.*, 1993; Young and Solbrig, 1993; Niamir-Fuller, 1999a; Niamir-Fuller, 1999b).

Beinart (2000) describes these analyses as attempts to build a 'corrective and anti-colonial' discourse that might say as much about the paradigmatic post-colonial framework these scholars have been working within as about the empirical legitimacy of their views. Elsewhere, critique of 'new' thinking and a defence of 'conventional' natural science analyses regarding dryland dynamics indicate the contentious nature of views involved in these debates (e.g. Illius and O'Connor, 1999, 2000; Attwell and Cotterill, 2000; Cowling, 2000). Recently published natural science analyses of long-term climate patterns add further complexity (e.g. Rohde, 1997a, 1997b; Nicholson *et al.*, 1998; Parmesan and Yohe, 2003). For example, contrary to popular assumptions about contemporary Sahelian desertification (Nachtergaele, 2002), and to models linking agro/pastoral land use with rising albedo (increasing land-surface light reflectivity) and falling rainfall (Charney *et al.*, 1975), Nicholson *et al.* (1998) indicate that over the last few decades there has been no progressive change in the boundary of the Sahara, in the vegetation cover of the Sahel, nor in productivity (as defined by water-use efficiency of vegetation cover).[2]

In this chapter we aim to extend discussion not by asking who is 'right' or 'wrong' and why in these debates, but instead by interrogating *why* views regarding the dynamics of drylands, and the knowledge and practices of pastoral nomads, are so contested and seemingly irreconcilable. We focus on existing and emerging debates signified by the two key terms of our title, namely 'non-equilibrium' and 'nomadism'. In a sense, these terms represent all that is and has been problematic for scholars and policy makers regarding both drylands and the mobility and diverse livelihood practices of the variously nomadic peoples who live there. As such, we explore ways in which differences in values and assumptions regarding environmental phenomena in drylands affect the ways in which 'the environment' is used, managed and perceived by people in these contexts.

A significant dimension of these interrelationships relates to how particular environmental discourses can become reified as 'truth', and thereby inform modern policy and planning in ways that may disenfranchise those with different – but perhaps no less 'true' – perceptions regarding the same phenomena. This is not only an outcome of a Foucauldian power/knowledge nexus (Foucault, 1981) – it is also related to ways in which *ignorance*, conscious or otherwise, sustains exclusionary discourses, policies and practices (Gordon, 1998; Sullivan, 2000b). Thus a reining in of the imagination, and an everyday unwillingness to engage with the complex, constructed and contingent nature of ways of knowing (e.g. Belenky *et al.*, 1986) translates into the occlusion of alternative knowledges along axes of difference supported by current power structures (e.g. Richards, 1985; Nader, 1996a; Leach and Fairhead, 2000).

We complement this analysis by drawing on the explanatory power of theories of conceptual and ritual purification, associated with anthropologist Mary Douglas (Douglas, 1996); of the empowered Panopticon[3] society, with its requirements for diffuse and minutely controlled surveillance and regulation (Foucault, 1981); and of the ideological differences between state and nomad science as considered by philosophers Gilles Deleuze and Félix Guattari (1980).

For example, we consider the classic anthropological concept of purity and danger (Douglas, 1966) to be central to understanding both the current situation and long-term trajectories regarding dominant policy trends in drylands. Mary Douglas argued that rituals of purity and impurity enacted as components of cultural and religious praxis are central for the maintenance of unifying categories used to classify, conceptualize and construct 'reality' (Douglas, 1996). Importantly, these categories require the avoidance and purification (or eradication) of phenomena representing danger and disorder to their internal order. The elimination of sources of disorder thus becomes 'a positive effort to organize the environment', carried out by 'separating, purifying, demarcating and punishing transgressions' through acts made possible 'by exaggerating the difference between within and without' so as 'to impose system on an inherently untidy experience' (Douglas, 1966). We extend these ideas in considering the ways in which socio-political processes of purification – of knowledges, peoples, spaces and practices – have structured encounters with modernity for drylands and their inhabitants. We argue that this has manifested as the exclusion of phenomena that run counter to the normative frame of reference of a powerful, colonizing and globalizing culture of modernity.

We consider here that three interrelated dimensions in particular constrain understanding of contexts positioned as peripheral to this culture. First, the reduction of complex and diverse phenomena to bounded and reified categories that act as homogenizing reference points, transferable across time and space (Latour, 1987; Smith, 2001). Second, the construction of a rationalist and positivist procedure for knowledge acquisition which, through separating, partitioning and abstracting phenomena from their social and moral contexts, makes possible their use in technological, industrial and militaristic arenas with significant social and ecological consequences (cf. Nader, 1996a). And third, the particular and constraining gender constructs embodied by modern patriarchy (Belenky *et al.*, 1986; Hodgson, 1999; Hodgson, 2000). Organically and pragmatically, these structuring norms have underscored a number of familiar and globalizing phenomena, including:

- centralized state-planning and the ordering of spatial contexts (e.g. Corbin, 1986; Smith, 2001), building on the codification, via surveying and mapping, of territories and peoples (cf. Peluso, 1995; Scott, 1998; Hodgson and Schroeder, 1999; Hughes, 1999; Abramson, 2000). To use Foucault's words, space is thus managed and controlled by becoming 'segmented, immobile and frozen', making possible the 'constantly centralized' surveillance, registration and regulation of the dangerous, contaminating 'other' – the 'pathological' (Foucault, 1977);

- the instrumentalization, commodification and militarization of a reified western 'technoscience' (Nader, 1996a) with the ability to 'act at a distance' from the locales of its formalization (Latour, 1987, 1993; Murdoch and Clark, 1994);
- the standardizing and commercializing of production practices, coupled with prescriptive regulation of both production and reproduction (e.g. Greer, 1984);
- and an inflexible gendering of public and private domains coupled with the 'othering' of woman by a normative frame that takes man as the human generic (e.g. Irigaray, 1996).

Thus the *power* of state science, planning and regulation is maintained precisely by the delegitimizing and dehumanizing of concepts, practices and peoples that, in Mary Douglas' terms, pose *danger* to hegemonic structures and categories (Douglas, 1966). As Foucault writes of the extensive disciplinary power over all individual bodies desired by 'the utopia of the perfectly governed city', state-making requires 'a whole set of techniques and institutions for measuring, supervising and correcting the abnormal,' including 'people who appear and disappear,' i.e. 'nomads' (Foucault, 1977). It is this empowered conceptual process, together with the technological phenomena utilized in its support, that in part enables the acts of assimilation and colonization – not to mention the purification and eradication – of 'the other' that we know only too well from history.

Clearly, the 'edge' (Jacobs, 1996) of the meeting between 'the modern colonial imperative and the colonized periphery' has manifested differently in the geographically distant drylands that form the focus of this chapter. We distinguish here between 'Old World' drylands of the Middle East, Africa, Asia and Europe, and a pastoral 'New World' of the Americas, Australia and Southern Africa (Behnke, 1983). In the former, pastoralism has existed for millennia and, in relative terms, modern (European) colonialism was based on resource extraction and labour administration as opposed to large-scale European settlement. In the pastoral 'New World' of the Americas, Australia and Southern Africa, European settlers unrolled a cattle ranching system and a cowboy culture harking back to medieval Spain during the eleventh and twelfth centuries, when the Christian Reconquista frontier forced back the Moors (Behnke, 1983). European colonists displaced earlier inhabitants across the vast part of the 'New World' drylands through genocidal dispossession at the colonizing frontier (seventeenth century in South America, nineteenth century in North America, Australia and Southern Africa), and by eventual incorporation of indigenes as landless stockmen/herders, labourers and servants. In these drylands a European settler imperative focusing on commercial livestock production based on introduced species generated the continual requirement for new land, becoming associated with the extreme violence and 'genocidal moment' of the frontier (Dirk Moses, 2000).[4]

Although these dryland contexts represent major differences in the specificities of how the modern colonial encounter played out, in simple terms we maintain that the rationality underscoring these processes has been the same, contributing to broadly similar outcomes in terms of the management and administration of both environment and people. Thus landscapes have been carved into fenced holdings with defined livestock carrying capacities, while people have been encouraged and coerced to settle, often in bounded reservations and following ethnicide (e.g. Trafzer, 2000), or as an underclass and labour pool (Behnke, 1983; Holmes and Mott, 1993; Gordon and Sholto Douglas, 2000). 'Wild lands' have been purified of undesirable beasts – from wild dog to tsetse fly – only to later become the desired and imagined spaces of 'untouched Edenic Nature', or the locales of various 'community-based conservation' schemes designed to emphasize wildlife and wild landscapes over, or as well as, other livelihood practices (e.g. Alexander and MacGregor, 2000; Duffy, 2000; Brockington, 2002). Women have been excluded from decision-making processes (e.g. Sullivan, 2000a) and undermined by the commercializing and formalizing of production practices (Hodgson, 1999). As Smith (2001) argues in commenting on the (anti-)social space of modernity, '[t]his repetitively patterned space consumes and regulates the differences between places and people: it encapsulates a normalizing morality that seeks to reduce all differences to an economic order of the Same'.

In extending discussion we also draw into debate insights from the brilliant (if sometimes frustratingly obscuring) work of philosophers Gilles Deleuze and Félix Guattari (1980), particularly their conceptions of the differences and relationships between 'state' and 'nomad' science. We argue that these correspond well to expositions and arguments regarding equilibrium and non-equilibrium ecological dynamics respectively (discussed further below), as well as to the similar distinctions drawn between modern, commercial, privatized and settled systems of production on the one hand, and traditional (customary) subsistence, communal (common property) and mobile production practices on the other. Work by Deleuze and Guattari among others can extend our frame of reference and analysis beyond the somewhat crude and even environmentally deterministic equilibrium/non-equilibrium divide that has dogged recent debate regarding drylands and pastoral production practices. In particular, this work can contribute to understanding that non-equilibrium thinking and nomadic practices are problematic *precisely because they are qualitatively and conceptually different* to the cross-cutting phenomena of formal science, the categorizing rationality of modernity and centralized state planning and governance.

From this perspective, non-equilibrium thinking and nomadic practices can be seen as comprising conceptual and pragmatic challenges to the norms delineated and required by the logic of state-centrism and rationality. In resisting what amounts to a paradigmatic contestation of a colonizing, hegemonic and state-centric modernity, supported by a mechanistic, linear and equilibrium-oriented technoscience, these categories thus have been physically and/or conceptually suppressed (purified), incorporated and transformed (colonized), or peripheralized (marginalized). In other words, the problems of legitimacy faced by drylands, pastoral nomads and perhaps even by scientists adhering to a non-equilibrial conception of dynamics (see below), attain sharper relief when set within a broader socio-political and historical context: namely, a context associated with an emerging and global hegemony of a particular and constructed humanity – from which difference is erased, whether by persuasion, suppression, coercion or violence.[5]

We are coming to a view that these differing perspectives may never be reconciled. This is because in many ways the binary oppositions on which they are built – equilibrium/non-equilibrium thinking, state/nomad science, settled/mobile practices, modernity/postmodernity – are ideological in nature, extending from fundamentally different ways of imagining, evaluating and being in the world, as well as from different ways of realizing power. Indeed, it is salutary to remember that even for the so-called objective or 'hard' sciences, 'a scientific question cannot be completely separated from the question of *values*' (Saner, 1999, emphasis added), such that natural science data themselves can be construed as inference-laden signifiers that represent choice, perception, interpretation and scientific *habitus* (Bourdieu, 1980) in the building of an empirically verifiable and variously technologically useful world 'out there' (Orestes *et al.*, 1994, in Baumann, 2000).

If it is indeed the case that an adherence to equilibrium/linear or non-equilibrium/non-linear thinking speaks more of ideology than 'reality', then how might we be able to take debate forward? In particular, how do we find ways of engendering a conversation across this ideological divide?

One possible path might be to move towards an explicit view that these categories – i.e. equilibrium and non-equilibrium thinking, state-centric (settled) and nomadic practice – do not exist in isolation from each other, but *in relationship with* each other (cf. Nader, 1996b). To paraphrase Deleuze and Guattari (1980), they function as pairs, each containing the seed or definition of the other (Kumar, 2002), in the same way that *yang* contains *yin* and vice versa in the familiar and powerful symbol of Chinese Taoist thought (e.g. Hooker, 1996). Conceptually this approach represents a movement away from entrenched and static either/or dichotomies and binary oppositions, towards an understanding that accepts the empirical reality of dualities, but sees dualistic categories as relational, dynamic and essential for each other's existence. It is an approach that opens up greater possibilities for thinking beyond linear framings of the equilibrium/non-equilibrium dynamical relationship as being located on a continuum from one extreme to the other (e.g. Wiens, 1984; Illius and O'Connor, 1999; Sullivan and Rohde, 2002). For example, a more explicitly relational view might be better able to embrace the cross-cutting interrelationships of temporal and spatial scale with those biological and abiotic dynamics that have become known as equilibrium and non-equilibrium dynamics respectively (also see Briske *et al.*, 2003; Oba *et al.*, 2003). In this sense, the empirical variability and complexity of dryland environments and pastoral practices – associated with temporal and spatial scale, varied species and suites of species and diverse socio-cultural practices – might be more critically and effectively conceptualized and analysed in ways that move beyond and reflect back on simple defences of entrenched positions.

In terms of the empirical issues and knowledge debates that form the focus of this chapter, the question therefore is not whether equilibrium or non-equilibrium thinking, settlement and nomadism, are 'right' or 'wrong', true or false. Instead, both intellectually and pragmatically the relevant questions relate more to distinguishing in what contexts, at what scales, and under what conditions, might these different dynamics and practices arise. Further, what might be learned or elucidated by their relationships to each other in these contexts? Hegemony of one component of an oppositional pair, and purification or marginalization of the other, for example, would denote a relationship that is out of balance. With exceptions, we maintain that this is precisely what can be observed with regard to the linear, equilibrium thinking that has underscored the conceptualization and management of drylands and pastoral peoples under modernity. If this is indeed the case, then engagement with non-linear and non-equilibrium concepts may be critical if we are to understand the processes that generate the misunderstandings and detrimental outcomes described below and summarized in the quotation with which we began this chapter.

The chapter proceeds in three sections. In the next two sections we review debates regarding knowledge production and policy intervention in dryland environments and in relation to pastoralist/nomadic peoples. For ease of organization and readership we focus first on ecological debates, and second on socio-cultural aspects with the important provision that these domains are overlapping and cross-cutting in all areas of discussion and 'reality'. We then move on to focus on policy and intervention, emphasising ways in which these have been shaped by the conceptual frameworks we discuss in the previous sections. We refer to case material throughout the chapter, with a particular emphasis on African contexts, which is where we both have primary fieldwork experience.

Non-equilibrium and drylands

The term 'non-equilibrium' has fast become shorthand for ways of thinking in dryland ecology that emphasize the abiotically driven, variable productivity of arid and semi-arid environments. As such, this 'new rangeland ecology' challenges conceptual 'norms' in ecology, population biology and rangeland science that emphasize the emergence of density-dependent dynamics from the producer-consumer relations that exist between species, and particularly between vegetation and herbivores. What this means in practice is a growing scepticism towards statements of irreversible environmental degradation caused by the herding practices of pastoralists, specifically the impacts of livestock on soils and vegetation (see references below). Non-equilibrium views affirm instead that tight links between variable rainfall and primary productivity, particularly in more arid environments, may mitigate degradation processes (caused by the impacts of livestock on vegetation) by weakening and/or disrupting the relationship between herbivores and forage.

In order to avoid duplication, we do not intend to provide here either an extensive review of the ecological specificities of equilibrium and non-equilibrium conceptions of dynamics and ecological functioning in drylands, or a detailed critique of the arguments for and against each aspect of these conceptions. Table 4.1, however, presents a brief typology of both, with the *proviso* (as above) that the descriptions noted in each column are each defined by, and exist in dynamic relationship with, the other. Readers who wish to access in more detail the specific components of recent debate are advised to turn to the excellent overview by Oba *et al.* (2000), as well as to recent in-depth reviews in Illius and O'Connor (1999), Scoones (1999) and Sullivan and Rohde (2002). The latter paper is a detailed response to Illius and O'Connor (1999) and as such, these two papers go some way to presenting equilibrial (Illius and O'Connor) and non-equilibrial (Sullivan and Rohde) views respectively. A particular technical focus of debate has been on whether or not sporadic and weak density dependent effects (Scoones, 1993), or density dependent effects operating in restricted but key resource parts of ecosystems (Homewood, 1994) are (1) of such significance that 'a system' can effectively be better understood as an equilibrium system; or (2) allow survival/maintenance

TABLE 4.1 An overview of ecological dynamics associated with equilibrium and non-equilibrium conceptions of rangelands, together with related policy and economics outcomes

Equilibrium	Non-equilibrium
Ecology:	
Climate stability	Unpredictable climatic variability
Stable interannual primary productivity	Unpredictably variable primary productivity (tightly linked to rainfall)
Livestock population strongly coupled with vegetation (density-dependent)	Livestock population density-independent
Change in stocking density creates predictable changes in plant assemblages	Livestock track unpredictably varying forage production
Policy and economics:	**Policy and economics:**
Potential carrying capacity can be predicted	Calculations of carrying capacity not useful
Stocking density can be regulated according to carrying capacity	Opportunistic grazing practices employing mobility are more appropriate
Land and resources managed under private/freehold tenure	Land and resources held and managed as common property, and/or under communal tenure regimes in southern African reservations
Goals:	**Goals:**
Strongly commercial/financial; benefits/ profit vested in cash and capital	Subsistence; reproduction of herd; profit vested in social relationships (reinforced by prohibitions on indigenous participation in emerging capitalist economies)

Source: Drawing on but modifying Oba *et al.* (2000, p37).

of enough grazers to *exceed* the 'carrying capacity' of surrounding wet season dispersal areas, thereby causing degradation (as argued in Illius and O'Connor, 1999, 2000). Contained within the synthesis papers above, as well as elsewhere in this chapter, are references to specific aspects of the debate and to detailed and location-specific case-studies.

Instead, our aims in this section are threefold. First, we consider some ways in which non-equilibrium approaches have been, and are being, discredited by proponents of what we might frame as 'mainstream' or 'orthodox' approaches in ecology, population biology and rangeland science. Second, we describe some ways in which key general assumptions in equilibrium and non-equilibrium conceptions of ecology dynamics differ. And finally, we indicate some reasons why non-equilibrial approaches are actively discredited, suggesting (as above) that this to some extent is a logical outcome of the cultural, ideological and institutional contexts within which each position is located, as well as of the policies and power relationships they legitimate.

Having to some extent ridden the crest of a wave of paradigmatic change in the 1980s and 1990s, non-equilibrium ideas in dryland ecology have been undergoing intense scrutiny by ecologists, particularly in Southern Africa (e.g. Illius and O'Connor, 1999, 2000; Campbell *et al.*, 2000). Many of the tenets and precepts of non-equilibrium ideas have been asserted as both 'challenged' and 'falsified' (Cowling, 2000), and many of these challenges have themselves been disputed (Sullivan and Rohde, 2002). What is of interest to us here is that beyond the playing out of this academic debate through discussion and critique of theory, empirical analyses and interpretation, there has been a noticeable attempt to discredit non-equilibrium concepts and analyses on the basis that these are somehow focused in publications that do not comprise 'rigorous', 'real' or 'primary' science. Take, for example, a review by South African ecologist Cowling (2000), in which he asserts his scepticism for all things non-equilibrial by stating that 'very little of this "new" science has appeared in the primary literature' or 'been subjected to rigorous peer-review'. This seems to us to be a misrepresentation of the situation.

To illustrate our point, let's take a closer look at these statements in relation to a selection of contributions to a non-equilibrial framing of dynamics in ecology, and particularly African dryland ecology, over the last three decades (Table 4.2). Among these references are a score of articles drawing on non-equilibrial dynamical concepts that have appeared in major peer-reviewed journals, including the three highest impact general science periodicals (*Nature, Science* and *Proceedings of the National Academy of Sciences* (of the USA) as well as specialist publications. It seems unreasonable to dismiss these articles, as well as contributions made by their authors and others as peer-reviewed chapters in edited volumes brought out by academic presses, as academically and scientifically irrelevant.

It seems to us pertinent to consider the knowledge politics signalled by Cowling's (inaccurate) dismissal of non-equilibrium perspectives as not having received the credentials conferred by appearance in the primary, and rigorously peer-reviewed, literature. The word 'primary' here says a lot about the assumed and imputed relationship between non-equilibrium and equilibrium concepts in dryland ecology

TABLE 4.2 Publications regarding rangeland dynamics that draw on non-equilibrium ideas in ecology

Journal	Author/date
Annual Review of Ecology and Systematics	Holling, 1973; Noy-Meir, 1973
Journal of Ecology	Noy-Meir, 1975
Science	Coughenour *et al.*, 1985
The Journal of Applied Ecology	Belsky *et al.*, 1989, 1993a
The Journal of Arid Environments	Belsky, 1989; Coughenour *et al.*, 1990; Scoones, 1995; Ward *et al.*, 1998; Turner, 1999
The Journal of Range Management	Ellis and Swift, 1988
The Journal of Animal Ecology	Dublin *et al.*, 1990
Forest and Conservation History	Dublin, 1991
Nature	Mace, 1991
Land Degradation and Rehabilitation	Scoones, 1992
Agroforestry Systems	Belsky *et al.*, 1993b
Ecology	Belsky, 1994
BioScience	Belsky and Canman, 1994; Oba *et al.*, 2000
The Journal of Biogeography	Sullivan, 1996c; Turner, 1998a, 1998b; Sullivan and Rohde, 2002
The Geographical Journal	Scoones, 1997
Conservation Ecology	Holling, 1998
Global Ecology and Biogeography	Sullivan, 1999b
Proceedings of the National Academy of Sciences	Homewood *et al.*, 2001
Global Environmental Change	Lambin *et al.*, 2001
	For chapters in edited volumes published by academic presses see Wiens, 1984; Caughley *et al.*, 1987; Homewood and Rodgers, 1987; Belsky, 1995; Scoones, 1993; Behnke *et al.*, 1993; Ellis *et al.*, 1993; Ellis, 1994; Sullivan, 2000a/b; Turner, 1999.

Listed by journal of publication. For full details of the citations see the list of references.

(and beyond): namely, that non-equilibrial analyses and approaches are somehow *secondary*; that they exist in a peripheral relationship to a *hardcore* of conventional (equilibrial) rangeland science; that this core retains its functions as providing the key conceptual reference points against which all else is measured and revealed; and that these relationships make work drawing on 'eccentric' (Deleuze and Guattari, 1980) non-equilibrium ideas somehow less rigorous, less accurate, less *hard*. If this is the case, then what are the underlying differences between this core and the periphery in ecology, and what is gained by the entrenching of these positions, and by defending ecological orthodoxy?

Scientific ecology emerged in a particular historical, cultural and environmental context. In the simplest of terms, this was fuelled by the imperatives and assumptions of European, and particularly British, capitalism and imperialism (Scott, 1998;

Anker, 2002). Scientific ecology in this reading thus acted to entrench on a global scale the situations of inequality arising from a particular mode of economic behaviour: the latter associated with the mechanization and homogenization of production; the grabbing and mining of previously uncommodified resources; the desirability of continual economic growth; the monopolization of profit; the capture of peoples' effort as abstracted labour; and lucrative collaboration between government and business. In both general and systematic terms, a further empire-building necessity was of the possibility and desirability of the distribution of sameness – namely, of what were considered to be superior Christian-European, patriarchal, modern and scientific values and practices (Nader, 1996b). These desires were pursued variously through the assimilation, colonization, and extermination of local peoples, assisted by the surveying and mapping, and hence control, of geographical spaces (Scott, 1998), combined with the rationalization and measurement of time-keeping (Corbett, 2003), and the measurement and purification of smell (Corbin, 1982). Anker (2002) has argued cogently that ecology, a new science, expanded rapidly in this context to provide expertise in establishing tools for the extraction and management of natural resources, and for informing the planning and management of human settlement and land-use practices.

Without affirming a simplistic environmental determinism, it is significant that these desires and assumptions emerged within, and were/are influenced by, a temperate environmental context. This is relevant because it meant that the academic discipline of ecology, and its practical application in terms of resource management, emerged where abiotic conditions and productivity were relatively constant within the timescales of relevance to economic productivity and decision making (interannual and over several decades). This is not to say that variability in productivity and unpredictable abiotic and biotic events were or are unimportant in these contexts – in Britain the low temperatures of the mini-Ice Age in the nineteenth century, the extreme drought of 1976 and the devastations of foot and mouth disease in 2001 (the latter exacerbated by administrative and political reactions over and above biophysical processes) demonstrate that this is not the case. What it does suggest, however, is that this relatively predictable and even stable environmental context supported a particular *modus operandi* in the natural sciences that operated from core values and assumptions about the nature of nature, and was fuelled by the successful delineation of laws and models to describe the dynamics of physical systems at particular scales of observation and in experimental contexts abstracted from the 'real world' (as discussed in detail for scientific forestry as a state-building activity in Chapter 1 of Scott, 1998). This further supported a particular and instrumental relationship with nature (Merchant, 1980), and was entangled with the structuring and maintaining of power relationships and processes of territorial expansion through the rationalization of landscape administration (Mukerji, 1997; Scott, 1998). As asserted by Deleuze and Guattari (1980), through such techniques 'nomad science' is variously submitted 'to civil and metric rules that strictly limit, control, localize'; under what Paul Virilio refers to as the 'geometrical imperialism of the West' (1975, in Deleuze and Guattari, 1980).

We outline below some key assumptions underlying the core principle of equilibrium, and indicate ways in which non-equilibrium ecology departs from these assumptions. Table 4.3 summarizes these differences and indicates correspondences with the concepts of equilibrium dynamics and state science on the one hand, and non-equilibrium dynamics and nomad science on the other.

Equilibrium community state

In modernity, a key assumption is that all 'systems', whether ecological, social or economic, have a natural and fundamental state or stable 'equilibrium'. In classical ecology, this is the original, primary or climax community (*synusiae*), i.e. the stable community that exists in the context of its abiotic environment (comprised of edaphic or soil, climatic, topographic and fire factors). For analytical purposes, these are treated as stable, and as exogenous to biotic community factors. Thus community equilibrium exists when all else is equal, with each species – or member of the community – functioning as part of the whole to maintain this equilibrium state. In anthropology and sociology, an analogous framing of human communities and societies is that of structural functionalism, whereby all socio-cultural phenomena are interpreted as performing a function in maintaining the stable structure of a society (as noted by Richards, 1996, and discussed in Fairhead, 2000). In ecology, the equilibrium community state frequently has been delineated with respect to vegetation parameters (Richards, 1996): a tendency thus has been to analyse animals and people in terms of their *impacts on* this primary community, rather than their *contributions to* the emergence of observed and desired communities.

From a non-equilibrial perspective, key understandings are that biotic and abiotic phenomena are integrated in their dynamical behaviour (i.e. making the delineation of endogenous and exogenous variables problematic if not impossible). As framed in ancient times by the Greek philosopher Heraclitus (Stott, 1998) – as well as throughout Oriental/eastern philosophy – flows and flux are considered 'reality itself' (Deleuze and Guattari, 1980), such that change is the only consistency and it is impossible for all else ever to be equal. Thus, 'the system itself is a moving target' (Holling, 1998), with surprise, uncertainty and unpredictability emerging from both biotic and abiotic sources and with effects that differ according to scale of observation. Pattern and order emerge, whilst phenomena are never absolutely repeatable – exactly the same – through time and space. Patterns and persistence are better imagined as qualitative system trajectories in n-dimensional phase space, drawn to basins of attraction but sometimes shifting from these, and always pursuing pathways that differ quantitatively to varying degrees through time and space. Analytically and dynamically a state of equilibrium in a living system (or complex) can signify only one thing, namely, death (Jantsch, 1980; Waldrop, 1993; Cilliers, 1998). Thus, Waldrop (1993, following computer scientist John Holland, e.g. 1992, 1998, 2000), states that, 'it's essentially meaningless to talk about a complex adaptive system being in equilibrium: the system can never get there. It is always unfolding, always in transition. In fact, if the system ever does reach equilibrium, it isn't just

TABLE 4.3 Correspondences for the intertwined notions of equilibrium dynamics and state science, and non-equilibrium dynamics and nomad science

State science	Nomad science	Key references
Science practice	*Science practice*	*Science practice*
Analyst	Synthesist	Baumann, 2000, p3
Atomism/reductionism – a science of parts	Holism – a science of wholes	Rosenberg, 1995
Mechanistic	Living	Deleuze and Guattari, 1988, p372
Rationalist/materialist	Spiritualist/existential	Deleuze and Guattari, 1988, p361
Equilibrium	Non-equilibrium	
Quantitative	Qualitative	
Extraction of constants/laws/absolutes – universalist and globalising	Engagement with continuous variation of variables	
Focus on solid forms and linear analytics	Focus on flows, vortices and spirals and nonlinear analytics	
Manifestations	*Manifestations*	*Manifestations*
Technoscience (associated with instrumental outcomes; commodification and militarization of knowledge) a science of ends/goals and of experts	Ethnoscience	Levi-Strauss, 1966, in Nader, 1996, p6
	a science of means/processes and of folk/citizens	Deleuze and Guattari, 1988, p372
City and Polis (government)	Outskirts/country and Nomos (governance)	Deleuze and Guattari, 1988, p80
Planning from the centre	Knowledge distributed through networks	
Managerial/state centric	Devolved/distributed decision making	
	Bricolage	
Engineering Reproduction of sameness	Following/tracking of variability and change	
Static/settled	Moving/mobile/mobilizing	
Knowledge	*Knowledge*	*Knowledge*
Information-based	Practice-based, *habitus*	Bourdieu, 1990 (1980)
Doctrinal	Gnostic (self-knowledge; intuitive wisdom)	Pagels, 1979
Symbolically conservative/impoverished	Symbolically imaginative/rich	

continued

TABLE 4.3 Continued

State science	Nomad science	Key references
Models and metaphors of organization	*Models and metaphors of organization*	*Models and metaphors of organization*
Top-down, strong hierarchies	Bottom-up, agent-based, loose/temporary hierarchies or nodes	Baumann, 2000, p3
Tree	Rhizome	Deleuze and Guattari, 1988
Formal	Informal/'underground'/dissident/'illegitimate'	De Certeau, 1984
Geographies	*Geographies*	*Geographies*
Space (abstract and homogenous)	Place (differentiated meaning, heterogeneity, diversity)	Tilley, 1994
Land = parcelled/enclosed/delimited/privatised/allocated/striated	Land = open/unenclosed/managed in common/distributed/smooth	Deleuze and Guattari, 1988, p380, 557
Power	*Power*	*Power*
Power over	Power to	Foucault (1998 (1976) after Nietzsche)
Orthodox (apostolic)	Heretic (gnostic) = persecuted and purified	Pagels, 1979, 382
Centred	Acentred	
Associated gender	*Associated gender*	
Male	Female	

stable. It's dead'. Similarly, Cilliers (1998) asserts that 'to yearn for a state of complete equilibrium is to yearn for a sarcophagus'.

Disturbance (from equilibrium)

In equilibrium thinking, movement or perturbation away from the predetermined and functional equilibrium indicates disturbance. It is generally framed as negative, i.e. as degradation. Disturbance in ecology might be seen as something akin to falling from grace as framed in the Christian apostolic tradition. The conceptual acceptance of a baseline or original condition tends to frame analyses of species assemblages in terms of what they may have been in the past, with present and future circumstances seen as deviations from this. In classical ecology this has manifested in some key organizing ideas. The concept of ecological succession (primarily associated with Clements, 1916), for example, analyzes changes in assemblages occurring due to disturbance in terms of their repeatable (and predictable) recovery or return to the baseline or 'climax' assemblage via a number of stages, which may themselves attain some temporal and/or spatial stability before succeeding to the next stage (as predicted by state and transition, and multiple equilibria models, May 1977).

This has since been challenged and reformulated by concepts that affirm the possibilities for contingency, indeterminacy and irreversibility. Such dynamical behaviour is introduced, for example, by:

- path dependency (i.e. history; de Rosnay, 1979, in Saner, 1999) and the possibility of there being a multiplicity of possible paths (e.g. Turner, 1998a, 1998b, 1999);
- patch dynamics conferred by location-specific events and interactions (e.g. Belsky *et al.*, 1993a);
- the impacts of biotic ecosystem components on abiotic factors, as, for example, with the influence of tree canopies on physical and chemical soil properties (Belsky *et al.*, 1989; Belsky *et al.*, 1993a, 1993b) and the long-term effects of animals on substrate factors (e.g. Turner 1998a, 1998b, 1999);
- and the possibility for positive (non-linear) feedback relationships between species (i.e. biotic-biotic relationships), as observed by Belsky *et al.* (1989), who found changes in the nutrient content of understorey grasses occurring under the tree canopies of selected species.

All of these types of interrelationships contribute to the dynamic mosaics of species observed empirically (Aubréville, 1938). Ingersell notes, therefore, that ecology in the latter part of the twentieth century has shifted 'from seeing nature as composed of stable, self-perpetuating and self-balancing ('equilibrium') natural communities or systems, to seeing nature as always in flux, and studying natural systems and landscapes as the products of unique events and histories' (Ingersell, n.d.).

Successional dynamics nevertheless remain an important conceptual organizing principle in the design and interpretations of ecological field studies, and in driving conservation goals and policies. For example, a frequent feature of ecology case studies in drylands is to interpret species assemblages, and the presence or absence of particular 'indicator species', as evidence for degradation from, or closeness to, a desired ecosystem state, i.e. one that is conceived as relatively undisturbed and therefore undegraded (for a range of references in relation to the Southern African context, see Sullivan and Rohde, 2002).

Underlying the somewhat Edenic notion of an equilibrial baseline community or ecological 'deep structure', is a dominant organizing and philosophical metaphor in western thought, namely that of the tree (Deleuze and Guattari, 1980). This refers to a tendency to think in terms of the primary legitimacy of a root or foundation of things from which all else is distinguished or separated following the logic of dichotomies and binary splitting (one to two to four, etc.). With regard to knowledge production across disciplines this supports a view that 'truth/reality' – first principles – can be uncovered and revealed through reductionist analytics (Holling, 1998) combined with processes of excavation and experimentation involving the tracing back of genealogies and lineages. In terms of organization, the metaphor of the tree is well-known to us in the establishment of hierarchical (or 'arborescent') structures in which authority is invested. Arguably, the assumption that there is always a 'deep' structure, with a 'right', 'true' or 'primary' baseline that can be traced given the appropriate tools and conceptual framework, is what legitimates both the assumption of 'expert' knowledge on the part of ecologists, planners, policy makers and other professionals, as well as the hierarchical organizational structures from which they are able to divulge their expertise. In other words, it legitimates the hegemonic relationships at the receiving end of which pastoralists frequently find themselves.

Equilibrium and economics

The acceptance that system behaviour is underlain by a condition of equilibrium makes tractable the building of economic predictions and models in relation to resource and environmental productivity. For example, the maximum sustainable yield of a product can be defined, thereby theoretically marrying the desires for maximum income on the part of a harvester or farmer with the need to maintain environmental integrity so as to sustain further harvests (cf. Scott, 1998). Similarly, the carrying capacity – i.e. the number of animals that can be sustained through time by a particular area of land – can be calculated, and used as a benchmark from which to regulate and enforce production and land-use practices, and to decide who might be free-riding in relation to these calculations.

In other words, there are instrumental reasons for assuming equilibrium dynamics. They enable harvesting rates to be set and profits to be predicted; they justify policy, planning and intervention from the centre; they empower the expert by generating an impression of being able to provide solutions to pressing issues; and they make elegant analyses possible. As Levins and Lewontin assert, however, a tautologous situation can arise such that analyses are constrained to the problems and methods that are amenable to analysis (Levins and Lewontin, 1985, in Baumann, 2000). The use of equilibrium acts to 'narrow uncertainty' in both conceptual and applied domains (Holling, 1998), contributing a 'normal science' framework (cf. Kuhn, 1962) that dictates possibilities for the types of questions asked and the analytical methods applied, whilst perhaps enabling scientists to maintain an aura of certainty and expertise in 'today's institutional[ized] science regime' (Baumann, 2000). Further, the naturalizing of a dominant normal science and the perspectives arising from it, makes possible the maintenance of expert opinions *in the absence of natural science data*, as frequently has been the case where pastoralists have been accused of degrading pastures (discussed in Brockington and Homewood, 1996; Sullivan, 2000b; Brockington and Homewood, 2001; Homewood *et al.,* 2001; Brockington, 2002).

As hypothetically posed by several authors (e.g. Seddon, 1997; Stott, 1997, 1998), if the science of ecology had emerged in a different environmental context – the more explicitly variable environments of drylands, for example – its key norms and signifiers might have been very different. The debates outlined above, then, are significant because they carry with them political currency shaped by specific contexts, and as such translate into impacts on the lives of people who do not share their foundational assumptions. As we have indicated, non-equilibrium ideas are resisted in some scientific quarters, and also pose challenges and problems for developing and implementing appropriate policy (discussed further below). Non-equilibrium thinking perhaps generates resistance – not so much because of inaccuracies, but for the following reasons: it demotes the superior positioning of 'experts' by emphasizing 'unknowability' in terms of predicting the behaviour of complex systems; it creates problems for conservationists wishing to clear (purify) landscapes of people and livestock in order to return these spaces to a desired, imagined original undisturbed state of nature; and it emphasizes the significance of local and historical specificities in creating currently valued landscapes.

Nomadism: 'Not all those who wander are lost'[6]

This brings us to the second key term of our title – 'nomadism' – and to the ways in which mobile lifestyles and livelihood practices have been denigrated and displaced by modernity. In this nexus of interrelations – between peoples, cultures, ideas and practices – pastoralists are misunderstood and marginalized *because* of the different practices and freedoms they represent as mobile peoples in contrast to the settled and more easily administered (and controlled) peoples of the city and of settled agriculture. Such circumstances are heightened when mobile pastoralists require access to land areas that also support natural resources critical to colonial and current empire-building, capital accumulation and profit in recent times – as has been the case for Bedouin pastoralists throughout the oil-rich drylands of the Middle East (e.g. Rae, 1999; Chatty, 2003). As Deleuze and Guattari describe in their juxtaposing of nomad science with state or royal science, '[a]ll of this movement is what royal science is striving to limit. ... nomad science is continually "barred", inhibited, or banned by the demands and conditions of State science' (Deleuze and Guattari, 1980).

The corresponding suppression of pastoralist knowledges that has occurred with the imposition of state-centric and/or modern administrative and production practices thus is understandable as part of a broader hegemonic process of social and spatial rationalization. Given the legislative and assumed primacy or interiority (i.e. 'habit') of the state (Deleuze and Guattar, 1980), those on the margins are either gradually or forcibly brought into its fold, or pushed more and more into the frontier and into the lifestyle of the outlaw – literally of someone outside the rule of law. In combination with the constricting and fragmenting effects of imposed nation state frontiers (e.g. Galaty and Bonte, 1992; Oba, 2000), nomadic peoples have been both marginalized and placed at the frontlines of international conflicts between neighbouring states and in relation to more global geopolitical tensions. Combined with customary expressions of conflict and power within and between pastoralist peoples (Kurimoto and Simonse, 1998), and with the exponential spread of firearms and automatic weapons (Hogg, 1997), this has acted to make whole regions vulnerable to escalating banditry and warlord rule (see, for example, Markakis, 1966, 1993; Lewis, 2001).

In this section we attempt to distinguish some key elements constituting the sciences and knowledges of nomadism that inform pastoralist practice in drylands, and to clarify why these pose a challenge to the rationality of 'state science', making them subject to modification, constraint and processes of purification. Again, refer to Table 4.3 for an overview of relevant components and correspondences of both 'state' and 'nomad' science. Here we focus on three overlapping domains of practices and the knowledges by which variously nomadic pastoralism is informed: first, the material realities of herd and livelihood management strategies, incorporating geographical mobility and the maintenance of diversity in both knowledge and practice; second, the significance of socio-cultural networks in contributing to the maintenance of both physical and social well-being; and third, an overview of customary arrangements in facilitating access to, and management of, land and other resources.

Making a living and nomad knowledges

In perhaps idealized terms, pastoral/nomadic living affirms, manages and responds to the variable productivity of drylands through maintaining heterogeneity and diversity in socio-economic practices. Herds are managed for species, breed and product diversity rather than (only) for single products with value on commercial commodity markets (e.g. Evans-Pritchard, 1940; Sandford, 1983; Coughenour *et al.*, 1985). Members of livestock-keeping 'households' distinguish multiple and different rights to animal products, with individuals, households and families deployed in varying productive capacities across social groups through time and animals distributed and dispersed throughout herding kinship networks (e.g. Talle, 1987, 1988, 1990). Depending on opportunities and constraints, individuals and families may move between different livelihood practices and knowledges, complementing livestock-herding with various combinations of 'wild' product gathering and hunting (e.g. Sullivan, 1999a, 2000b, 2005 and references therein; Sullivan and Homewood, 2004), cultivation (e.g. Thompson and Homewood, 2002), trade (e.g. Zaal and Dietz, 1999), and remittances from wage labour (e.g. Pantuliano, 2002). And women, contrary to assumptions of the 'patriarchal pastoralist' (as critiqued in Hodgson, 2000), frequently hold positions of authority and responsibility as managers and decision makers. This is in relation to the milking of animals and the distribution of this primary subsistence item; the means by which women have ownership over animals; and their authority, as 'heads of houses', over consumption, production and social and biological reproduction (e.g. Broch-Due and Anderson, 1999; Grandin, 1988; Dahl, 1987; Talle, 1987, 1990; Joekes and Pointing, 1991; Jowkar *et al.*, 1991; and chapters in the volume edited by Hodgson, 2000).

Underpinning this dynamism and flexibility in livelihood practices are both a conceptual acceptance (and practice) of the validity and necessity of physical movement through time and space, and the maintenance of a diversity of relevant knowledges to support and make possible such practices. Numerous studies document the mobility practices of pastoralist societies – these will not be described in depth here (e.g. Niamir-Fuller, 1999a, 1999b; Hampshire and Randall, 1999, 2005). What arises from these studies is an appreciation of the ways in which the physical mobility of herds through time and space is essential to enable livestock to access forage resources whose availability varies according to abiotic conditions. It is through these practices that herders access the full repertoire of available herding opportunities (from wet season grasslands, to browse and leguminous pods, as well as swamps or vleis/dambos in dry seasons; e.g. Scoones, 1991).

This material necessity of mobility practices means that in many circumstances nomadism is maintained through disobedience of state rules and across landscapes that now are demarcated into fenced holdings under various forms of individual or private tenure (see below). In the former 'homeland' of Damaraland, northwest Namibia, for example, and despite a rather static geography of delineated and fenced farms plus an administrative and apartheid context that was not amenable to movement by local people, migration histories for indigenous herders indicate that complex movements of people, livestock and other traded commodities across farm boundaries have characterized the area since its demarcated farms were redistributed to indigenous herders in the 1970s (Sullivan, 1996a). In fact, even in contexts where European settler livestock farmers have exclusive use of huge ranches under freehold tenure (such as in this area prior to the 1970s), it is apparent that herders need to move livestock across ranch boundaries, and sometimes over large distances, in order to maintain herd numbers in the face of variable forage productivity (Sullivan, 1996a; Beinart, 2003). Similarly, several case studies report that herd mobility remains essential where pastoralists have been settled on delineated group ranches, as is the case for group ranches in Kenya (Grandin and Lembuya, 1987). These studies suggest that where access to extensively distributed resources is important, as is the case for dryland environments, it might be inappropriate to assume that individualized land tenure holdings are essential for economic productivity and welfare.

But as well as this, and as framed by authors as varied as Bruce Chatwin (in his 1987 bestseller *The Songlines*) and Deleuze and Guattari (1980), abiding in a *habitus* of nomadism carries with it a rationality or 'pool' of collective subjectivities that positions mobile pastoralists – those accessing and using the dispersed resources of drylands – as counter or peripheral to the centre-oriented interiority of the settled state. As Chatwin (1987) describes:

> To survive at all, the desert dweller – Tuareg or Aboriginal – must develop a prodigious sense of orientation. He [or she] must forever be naming, sifting, comparing a thousand different 'signs' – the tracks of a dung beetle or the ripple of a dune – to tell him where he is; where the others are; where the rain has fallen; where the next meal is coming from; whether if plant X is in flower, plant Y will be in berry, and so forth.
>
> *(Chatwin, 1987, pp222–23)*

Chatwin's 'desert dwellers' in the above quote again are somewhat idealized. Depending on wealth and other opportunities (and constraints), today's pastoralists are as likely to make livelihood decisions via their mobile phones, or to have been drawn into 'food for work' programmes established for those dropping out of the system due to varying combinations of drought, land appropriation and warfare. But what Chatwin does convey is a sense of the importance of retaining openness in the *process* of enacting knowledge. Knowledge thus is called upon as and when necessary – in relation to the flow of changes in circumstances that occur through time – such that we might think of nomad knowledge, or of 'citizen science' or ethnoscience more generally, as integrative through its practice of collating and using multiple sources of knowledge, evidence or information (Holling, 1998). The phenomenologist Edmund Husserl describes this as a 'vagabond nomadism', for which knowledge is 'essentially and not accidentally inexact' (cited in Deleuze and Guattari, 1980). Thus, classificatory categories have loose boundaries, names (e.g. for species) vary through time and space and according to the lineage and history of the person doing the naming (Sullivan, 1999a), and knowledge expertise and specialization, in relative terms, are distributed throughout collectives of people. This way of knowing is flexible and open, and is inseparable from heterogeneity and inexactness because 'it' also is inseparable from the unique experience, ideology and power of the knowledge-holder/producer (cf. Negri, 2002).

Networking

Pastoralist welfare is bound intimately with concepts and practices of exchange and reciprocity between and within 'groups', which thereby facilitate broader social networks that are activated and maintained by these practices. In East Africa, for example, pastoralists engage in complex 'cross-sectional and cross-ethnic bond-friendships' (Lind, 2003, following Sobania, 1991) which act to 'sort out' the particular attributes and niches of different 'groups' to help minimize conflict, and to act as the 'glue' that binds groups into broader regional and societal networks. Malleable and ambiguous ethnic identities also have enabled people to move in and out of 'groups' and to accommodate others when appropriate (e.g. Waller, 1985).

Negotiation, between groups and individuals, is critical in enabling exchange and reciprocity, as is an ability to recognize potential alliances through the process of reckoning relationships. The key to negotiation is kinship; in particular, a conception of kin relationships as reciprocal networks that can be continually modified or reorganized on the strength of new interactions between individuals (e.g. Lancaster and Lancaster, 1986). To take a regional example, kinship among KhoeSān peoples inhabiting Southern African drylands provides what Fuller describes as a superbly enabling framework 'for the expansion and contraction of the network of relatives with whom one maintains reciprocal obligations' (Fuller, 1993). This occurs primarily through parallelism in parents and same-sex cousins, a high incidence of fostering and adoption and flexible definitions of those constituting family. Of particular significance is the potential for network expansion, embodied by a kinship frame that is 'constituted by relations of incorporation rather than exclusion, by virtue of which others are "drawn in" and not "parcelled out"' (Ingold, 1992). Fuller further maintains that this is linked with the exigencies of an uncertain environment: thus, '[t]he intimate connection between kin and the social imperatives of economic survival leads to an imprecision in the definitions of who and who is not kin because the imperatives of economic survival are themselves constantly changing. ... A wide net of kin increases the area over which one could utilize resources thus counterbalancing the periodic localized droughts that occur' (Fuller, 1993; cf. Gordon, 1972). It is this in-built flexibility that confers buoyancy to any network. In this instance it means that the potential inherent within the social network for future linkages and reciprocity is not limited to the connections between individuals (and/or groups) that are activated at any one time. Viewed in this way, it is easy to conceptualize the multi-layering of social and kin networks, and the 'contractual alliances' on which they are based (Knight, 1991), as literally providing a 'safety net' for the individuals and families constituting its 'members' (although by the same token, 'extended family relationships' also may be 'fraught with conflicting demands and opportunities' (Rohde, 1997a; cf. Fuller, 1993).

The colonial administrative imperative ushered in an era that fetishized the ordering of land allocation and the registration of individuals within localities for administrative purposes. By fragmenting both land and social groupings and extending the arm of the state over both, this arguably has undermined local and autonomous welfare and livelihood practices. Nevertheless, kin relationships and the dynamic and fuzzy logic of kin and social networks retain significance in guiding the negotiations that make herd mobility and other welfare decisions possible, again frequently in contexts where such mobilizations occur through disobedience against imposed administrative constraints. A problematic ramification, however, has been a tendency for wealthier individuals and families to draw both on their position within local kin and social networks, and their access to and influence over formal processes of land registration, to consolidate ownership of land and resources while poorer land-users are excluded (e.g. Thompson and Homewood, 2002, discussed further below).

Customary tenure arrangements[7]

As a general rule, and especially pre-colonialism, the more arid and infertile the land, and the more seasonally and annually variable its productivity and ensuing use, the more likely it is that the area and its resources will be under communal control rather than individualized tenure. This makes common property regimes typical of pre-colonial drylands where movement is essential in order to access forage and other resources.

Box 4.1 provides a detailed case example of the workings of the overlapping forms of tenure that may comprise common property regimes in dryland environments. Common components include:

- management of a dry season grazing area, often with a committee of elders who decide when and where to reserve, or allow access, to dry season grazing (for a detailed case example regarding Tanzanian Maasai, see Potkanski, 1994; Brockington and Homewood, 1998; and Brockington, 2002);
- sophisticated collaborative management, of both the timing of herd access and the co-ordination of labour, to enable group access to shared water sources (as among Borana pastoralists of southern Ethiopia; Cossins and Upton, 1987);
- negotiation of group access to other 'key resources' such as local 'hotspots' of productive potential (for example, access to, and inheritance of, riverine tree resources for dry season forage managed by Turkana pastoralists in north Kenya; Barrow, 1988, 1990);
- cultivated fields allocated as a common property resource such that plots are designated to be worked by particular individuals or households for one or more farming seasons, or until the household head has died, after which it reverts to the pool of common land for reallocation (Birley, 1982).

BOX 4.1 OVERLAPPING FORMS OF RESOURCE TENURE AND TENURE CHANGE UNDER AGROPASTORALISM IN SEMI-ARID NORTH-CENTRAL NAMIBIA

Land tenure: Settled and private, unsettled and communal

For Oshiwambo-speaking peoples of north-central Namibia, land can be divided between a wetter central floodplain area, which is permanently settled and allocated under relatively secure tenure, and a peripheral unsettled area which is used and managed communally as wet season pasture for livestock.

In the *wetter, permanently settled central area*, land has been cleared for fields and kraals, and is divided into plots with recognized boundaries. These traditionally are allocated on a lifetime tenure basis to a household head (usually male) following payment of a fee to the chief/headman. The boundaries of these plots remained fixed so that, should a farmer wish to augment the size of his or her holdings, they would either be allocated a second plot in addition to that already inhabited, or would move to a completely different but larger plot. While the 'tenant/owner' did not have the right to alienate his (or occasionally her) allocation of land in the inhabited area, they could consider it as essentially theirs for the duration of their life, as long as it remained suitably productive and was improved during their 'tenancy'. Following their death, or the termination of tenure for any other reason, the farmland would return to the traditional land allocator, i.e. the King, chief or headman. Women did not normally 'own' land but had greater rights to the fields allocated to her by her husband and to the produce from these fields. Since independence in 1990 the Namibian government has recognised this as discriminatory against women and formal policy now makes provision for the ownership of land by women and the inheritance of land by widows.

The *drier unsettled peripheral* areas are used primarily as wet season pastures allowing a pattern of transhumance, i.e. annual livestock movement, between the two categories of land. In the past, fees were not required from users of the uninhabited area. Here, established boundaries for plots did not exist and the only constraints to expansion were labour (for herding) and water availability. The land and its resources were loosely divided between the different Owambo-speaking communities. They were managed by the local 'community' with rights to a particular area, but flexibility in tenurial rights allowed the opportunistic and reciprocal use of pastures by different communities in response to rainfall-driven variability in pasture availability. In periods of severe drought, herds were driven to the sparsely populated pastures of eastern Kaokoland and southern Angola. Since the mid-twentieth century, increasing control by local headmen is indicated by records of payments being made for the establishment of kraals in the uninhabited zone, and the declaration and removal of 'illegal' settlement in this zone.

Water tenure

While the unsettled areas were communally managed, access to water occurring in these areas, without which the pastures could not be used, was controlled by those with recognized rights to an area. This could be an official leader, or if a waterpoint was constructed on the initiative of an individual it would be managed by them and inherited by their family as private property. Other farmers who wished to draw from these wells essentially became the clients of the presiding occupant.

Tree tenure and management

The distinction between land allocated to individuals and land open to access and use by others in the community is complicated by common property rules governing the use and protection of key resources occurring in particular areas. Under traditional communal ownership of land, important tree species, especially those providing edible fruits, were protected by making the cutting of trees without the permission of the local King or his councilors a punishable offence. The marula (*Sclerocarya birrea*), important for its nutritious fruits from which a nourishing beer is made, for example, was among the most valued of tree species and individuals of this species were considered the property of the King, regardless of where they were located. For this and other highly regarded species, chiefs had partial first rights to the fruits. Often rules concerning usufructuary (i.e. use) rights to trees were supported by symbolic values attached to different species and different areas of land. For example, at the edge of each Owambo tribal area was located a sacred portion of land from where tree removal was considered to result in various physical afflictions such as blindness or paralysis.

Traditionally, tree tenure and land tenure thus were separate entities, and allocation of farmland did not necessarily confer 'ownership' of the trees on this land. This was particularly true of fruit trees, to which rights may be preserved by the traditional leader even when they occurred on allocated farmland. Rights to a plot of land within the inhabited areas, however, generally confers rights of first access to other resources on the plot to kraalheads and their families, the most important of these being waterholes and trees bearing edible fruits. Further complicating the system of rights accruing to individual trees are instances where several individuals may have access to different products of a single tree. So, while fruits may be harvested by women, with some distilled into saleable liquor for their own profit, other products might be accessible to the whole household. Cutting of the tree for firewood or other wood products generally requires permission from the household head and neighbours may request permission to harvest excess fruits and/or use the branches for livestock forage.

Indigenous cultural practices whereby particular tree species were protected have been complicated by the recommendation for several species occurring in north-central Namibia to be officially protected under colonial forestry legislation. The palm (*Hyphaene petersiana*) and various fig species, for example, were identified in 1927 as requiring protection by the forestry officer at the Union of South Africa Forestry Department, and the 'birdplum' (*Berchemia discolor*) has been protected since 1975. Protected status meant that permits were required before these trees could be felled by local inhabitants. An unfortunate consequence is that these rules have effectively removed responsibility for trees from local farmers and village headmen, eroding incentives among local farmers to manage these resources for use by themselves. The legislated restoration of limited ownership rights and management responsibility for natural resources by local farmers and village leaders currently is viewed as a way of encouraging appropriate resource management in post-independence Namibia and elsewhere.

(Sullivan, 1996b and references therein)

For so-called 'hunter-gatherers', and despite conventional stereotypes of their relentless mobility and their inability to recognize land and natural resources as belonging to any individual or group, a number of anthropological studies indicate complex conceptualizations of land access and tenure rights (Box 4.2). Again, these are mediated via kin relations and rules guiding inheritance.

In other words, tenure and the regulation of access to resources in drylands have tended to be based on the customary bond rather than the legislated pact/contract, i.e. on 'collective mechanisms of inhibition' (Deleuze and Guattari, 1980). As noted above, these are maintained by the diffuse regulatory understandings and practices found in relatively acephalous (or non-state) societies, which often continue to operate despite the imposition of a codified state and administrative apparatus and power (Deleuze and Guattari, 1980). Thus, mechanisms of constraint are embodied in the 'fabric of immanent relations' (Deleuze and Guattari, 1980) characterizing such societies, i.e. in the flexible and rhizomous (horizontally spreading) networks of kin and social solidarity; in genealogies and the processes of classificatory kinship reckoning (Fuller, 1993; Knight, 1991; Sullivan, 2005); in sharing; and in widely observed mechanisms for the diffusion of wealth. Together, these represent 'another kind of justice' (Deleuze and Guattari, 1980): one that is relatively distributed throughout the 'system', rather than meted out from centres of power that are removed from the localities and individuals concerned; and that is relatively *processual* in relying on the prediction and tracking of opportunities and constraints, rather than the rigid codification of rules of access and ownership (Sandford, 1983; Gordon, 1991; Roe *et al.*, 1998).

BOX 4.2 TRADITIONAL CONCEPTS OF LAND OWNERSHIP AMONG JU|'HOANSI 'BUSHMEN'

Although conventionally thought to have little concept of land tenure or resource ownership (an assumption that has undermined formal claims to land throughout Southern and East Africa) 'hunter-gatherer' populations conceptualize land and natural resources in terms of socially defined access rights determined through kin relatedness and inheritance. Here we review categories of land among the Ju|'hoansi, speakers of a central !Kung language who inhabit the Nyae Nyae area of western Botswana and eastern Namibia. The Ju|'hoansi recognize two types of communal land: the broad category of *gxa|kxo* and the named places of *n!oresi*. These are discussed separately.

Gxa|kxo

This term translates literally as 'face of the earth' and refers to all the land and its resources in Nyae Nyae, to which all Ju|'hoansi have use and habitation rights as individual members by descent. Gxa|kxo thus is not the property of any corporate body within the Ju|'hoansi. The rights of individuals within the gxa|kxo include:

- the right to use major plant-food resources such as the tsi or morama bean (Tylosema esculentum) and g|kaa or mangetti nuts (Schinziophyton rautanenii, formerly Ricinodendron rautanenii);
- the right to hunt and track animal wildlife, such that a hunted animal belongs to the hunter who strikes it, and not to the owners of the recognized territory or n!ore (see below) in which it was hit or in which it dies from the effects of arrow poison;
- the freedom to travel;
- and the right to live at a permanent source of water during drought periods.

N!oresi

N!oresi are named territories without fixed boundaries, usually with important focal resources such as permanent or semi-permanent water-holes and concentrations of valued plant-food species. Individual rights to residence within a n!ore, and to use its resources, are inherited directly from both parents and ownership of a n!ore is recognized only if this traceable descent can be demonstrated. As such, 'ownership' of a n!ore is exclusive to a group related through kin alliances who manage its resources communally. 'Ownership' cannot be conferred on outsiders, even though they may reside within a n!ore for a prolonged period of time with permission of its recognized owners. An individual chooses in adulthood which of their parents' n!ore they wish to claim as their own and, through marriage to someone from outside that n!ore, gain rights of access and resource use to a second n!ore. In this sense, kinship networks underpin in a fundamental way an individual's rights to land and resources.

(Ritchie, 1987; Botelle and Rohde, 1995)

Early analyses of land access and management under common property regimes tended to represent these complex understandings of 'right' use, allocation and management as situations of 'open access', i.e. with resources used on an *ad hoc* and 'free-for-all' basis until 'degradation' occurred and people were forced to move or turn to alternative resources. The most famous exposition of this scenario is Hardin's (1968) 'Tragedy of the Commons'. This model alleges that environmental degradation is inevitable since pastoralists 'free ride' by benefiting from the profits of individual herd accumulation while bearing none of the costs of communal range use and possible degradation. Although still often invoked, this analysis is misleading. It discounts the reality of the possibility for *collective* management and restraint, in favour of an emphasis on individual profit maximizing behaviour necessitating freehold title to land. As discussed further below, this discounting has led to some significant socio-economic and environmental impacts.

Arising from the above overview is an appreciation that the flexible mobility and other practices employed by pastoralists may be better equipped to mobilize the opportunities presented by variable environmental productivity than the various livestock development initiatives introduced to stabilize production in settled locations and thereby reduce the perceived poverty and insecurity of pastoralist livelihoods (see below). We might say that the flexibility these practices embody permits an unfolding of lifestyles and livelihood practices that reflects becoming rather than being, flowing rather than stasis and following/tracking rather than stability, settlement or constancy (cf. Sandford, 1983; Deleuze and Guattari, 1980; Rohde, 1994). Modes of organization are characterized more by the horizontally spreading rhizome rather than the rigidly hierarchical tree (Deleuze and Guattari, 1980), while pastoral production practices are infused with multiplicity, diversity and heterogeneity over specialization regarding products and skills.

These points have implications for analytics. For example, and as Waldrop describes:

> there's no point in imagining that the agents in the system can ever 'optimize' their fitness, or their utility, or whatever. The space of possibilities is too vast; they have no practical way of finding the optimum. The most they can ever do is to change and improve themselves relative to what the other agents are doing. In short, complex adaptive systems are characterized by perpetual novelty.
>
> *(Waldrop, 1992, following computer scientist John Holland, e.g. 1992, 1998, 2000)*

Given that '[t]he concern of the state is to conserve' (Deleuze and Guattari, 1980), i.e. to protect its institutions and organs of power, and to conserve desirable environments and lifestyles, it is not surprising that from the standpoint of the state, 'nomads' – mobile peoples – are portrayed in terms that convey 'illegitimacy' *vis à vis* all that the state stands for (Deleuze and Guattari, 1980). In the next section we elaborate this 'standpoint of the state' through outlining trends in policy and intervention and their impacts on dryland dwellers and environments.

Policy and interventions in pastoral drylands: Herding, agriculture and wildlife conservation

The assumptions of a colonizing and globalizing modernity have influenced state policy and development interventions in drylands. Here we focus on how the rationality underscoring indigenous land use practices has been marginalized in the processes of change associated with colonialism and globalization. The conventional wisdom that rangelands are undergoing environmental degradation and desertification due to climate change combined with overgrazing, overstocking and damaging soil management practices (including nutrient mining) is a strong current running through the international development literature (discussed in Homewood and Rodgers, 1987; Homewood, 2008; Sullivan, 1996c, 2000a; Niamir-Fuller, 1999a, 1999b; Platteau, 2000; Nachtergaele, 2002). As a result, techniques associated with state science and the central control of natural resource management frequently have been emphasized at the expense of local practices and social institutions, with 'western' systems of management and production replacing customary institutions of control (e.g. Leach and Mearns, 1996; Mortimore, 1998; Carswell, 2002). Outside expertise consistently has been ranked above indigenous knowledge. Commercialized production, benefiting national élites, has tended to be subsidized and to take priority over local livelihoods that sustain the majority of the population (Klink *et al.*, 1993; Silva and Moreno, 1993). Associated with this has been repressive regulation of natural resource use by indigenous smallholders, tenant farmers and landless peoples, while environmentally problematic large-scale commercial land uses, whether of crops or livestock, have been favoured (e.g. Lane and Pretty, 1990; Young and Solbrig, 1993; Government of Tanzania, 1997). Post-Soviet steppe drylands have been following other 'Old World' common property drylands down the rhetorical pathways of overgrazing and pastoralist-induced degradation into the realities of rapid and inequitable privatization (Debaine and Jaubert, 2002; Arab World Geographer, 2002).

Below, we focus in more detail on the broad trajectory of policies and interventions in drylands livestock, agriculture and wildlife conservation initiatives. We maintain that four interrelated and globalizing contextual trends have guided these interventions:

1. the increasing commercialization, commoditization and monetization of production practices (e.g. Zaal and Dietz, 1999);
2. the rationalization of both people and landscapes for administrative purposes;
3. increasing statism, i.e. the consolidation of the nation state and state-centric systems of government and management (of production and reproduction);
4. and the interaction of political economies with ideologies of ecological 'truth', which tend to assume that degradation follows from pastoral land use, and which emphasize the need for the conservation of landscapes from which people either are removed or constrained in terms of their access to, and use of, such landscapes.

As discussed below, these contextual trends have tended to support particular interventions with a now well-known litany of problematic outcomes.

Development trends in drylands

Commercializing production

Development intervention in drylands has taken as its normative framework a model of livestock production for commercial markets established in 'New World' drylands: i.e. based on extensive and fenced ranches under freehold tenure, with production focused on single marketable products and management drawing on predictive models assuming equilibrium dynamics. Prior to the 1980s, the emphasis of development in the 'Old World' drylands of the Middle East, Africa, Asia and Europe thus was on introducing high-tech, capital-intensive, exotic systems and breeds to revolutionize agricultural and livestock production. The aims were to generate wealth and kick-start health, education and infrastructural improvements, while bringing greater numbers of citizens into the formal monetary economy. A number of comprehensive reviews highlight the failure of these attempts in Africa in terms of wealth generation, livelihood security and environmental impacts (e.g. Horowitz, 1979; Haldermann, 1985; Adams, 1992). In Indian drylands, the nominally state-controlled but *de facto* open access regime allowed an 'iron triangle' of politicians, bureaucrats and commercial entrepreneurs to manipulate subsidies and corner the benefits of development during the same period (Gadgil, 1993). In particular, massive subsidies facilitated the channelling of artificially cheap timber and wood pulp materials to manufacturing industries supplying urban markets in ways that passed the costs of ensuing woodland degradation onto the rural poor. Access by tribal and landless people dependent on the commons for subsistence grazing, fuel, fibres, construction needs and income from the sale of these products has been progressively marginalized. Similarly, in the Brazilian *cerrados,* heavily subsidized inputs have favoured commercial enterprises and mechanized farming by wealthy landowners (Klink *et al.*, 1993), while in the *llanos* of Venezuela's Orinoco Basin environmental pollution has been an outcome of heavy dependence on petro-chemical-based fertilizers and pesticides (Silva and Moreno, 1993).

Rationalizing land tenure

In the pastoral 'New World', European or Euro-American settlers established private ownership of large ranches on land alienated from indigenes to support commercial enterprises characterized by extensive cattle ranching, low stocking rates per unit area of land and the regular harvest of a surplus 'crop' of young cattle for meat. As noted above, this has become the model for rangeland development interventions worldwide, requiring the codification of land tenure to facilitate the rationalization of livestock management (i.e. based on the setting of carrying capacities, the monitoring of veterinary controls and the administration of people). Fencing thus became a key management tool throughout the settler economies of 'New World' drylands.[8]

In 'Old World' drylands, the imposition of private forms of land tenure, usually accompanied by the delineation of land areas using fencing, has since become a norm guiding development interventions (as described in Box 4.3). Further, by assuming that land is not occupied in times when it is not in use, this view has paved the way for land dispossession due to pressures from elsewhere (e.g. Lane and Pretty, 1990; Birch, 1996). This also is occurring through the *de facto* privatization of land through fencing by wealthy and frequently absentee herders, accompanied by *de facto* private control over key or focal resources such as boreholes and other water points, access to which is crucial in enabling use of the wider landscape (e.g. Graham, 1988; Berkes, 1989; Bromley and Cernea, 1989; Prior, 1994). As capitalist relations of production and the demands of a global 'free' market increasingly penetrate African farming sectors, this land privatizing trajectory becomes ever more likely, even in contexts where land redistribution to poorer farmers on communal land is a stated objective (as, for example, in the post-apartheid contexts of Zimbabwe, South Africa and Namibia). A systematic outcome has been the impoverishment of those not able to access and capitalize on these opportunities, e.g. women, poorer individuals/families and sometimes particular ethnic groupings (as documented in Talle, 1988; Galaty, 1999; Igoe and Brockington, 1999).

BOX 4.3 LAND TENURE AND SUBDIVISION ON MAASAI GROUP RANCHES, KENYA

Lemek group ranch near the Maasai Mara in Kenya (745 km²) was established in 1969. The group ranch chairman and land adjudication committee allocated land to educated or influential Maasai in a belt along the western portion of the group ranch boundary bordering the Mara River. These allocations were cemented under private ownership with the issuing of title deeds, the process being facilitated by the local administrative chief and land registry staff. Ostensibly to guard against the continued westward movement of non-Maasai cultivating groups onto Maasai lands, beneficiaries included Maasai administration chiefs, MPs, councilors, county council officials and a police inspector. Ironically, many of these new landowners rapidly sold land on a piecemeal basis to the same in-migrant cultivating groups apparently causing concern to Maasai pastoralists.

On the northern portions of Lemek and since 1984, outside entrepreneurs have been approaching the administration chief and group ranch chairman to cultivate wheat on leases of upwards from 2,000–4,000 acres per contractor. In addition to arranging these leases for their own benefit, the administration chiefs and chairmen have been giving responsibility to other group ranch committee members, councilors and associates to arrange leases with contractors. On sub-divided land on Lemek, each registered member was supposed to be entitled to receive 100 acres of land (in fertile places) or 128 acres on steeply sloping or marshy areas. The process of registering involves all circumcised men deemed to have been resident on the Group Ranch by the land adjudication committee prior to the closing of the register in 1993. According to the Narok County Council there were 1,021 registered members on Lemek. Initial attempts by local élites to allocate larger shares to themselves were thwarted in 1995 when, under the supervision of the District Commissioner, a revised survey was undertaken to ensure plots were of equal size.

Despite this survey, locally influential people (with access to the register and map providing the location of the plots) have still been able to exercise control for personal benefit of the land sub-division process. Examples include:

i) Those previously involved in leasing land for wheat cultivation using the considerable sums generated to buy the permanent/modern houses constructed by contractors. Once owners of the permanent housing, their stake to the land on which the house is located is secure, thus ensuring a position in the lucrative wheat-leasing belt.

ii) Those involved in leasing out the land for wheat farming use the money accrued to buy out poorer neighbours' shares in land. Once agreement has been reached (usually a hand-written confirmation signed or marked with a fingerprint) the position of the selling party's land is changed to ensure it is located on the wheat belt.

iii) Influential people register their younger (uncircumcised) sons and ensure that the shares are located adjacent to each other in the wheat belt. In this way, farms of up to 1,000 acres in extent are established. All of these mechanisms facilitate the further consolidation of land in the hands of the wealthy, while excluding poorer land users whom the subdivision process is ostensibly intended to benefit.

(Thompson and Homewood, 2002)

Formal land tenure reform at the level of national policy has also tended to be based on assumptions guiding farming practices for commercial export markets (e.g. Birley, 1982; Rohde *et al.*, 2001). The assumption here is that inalienable title to land will increase investment in agriculture and thereby increase commercial productivity (although this is not necessarily what does ensue – e.g. see Haugerud, 1989). For example, the Government of Tanzania's Livestock and Agriculture policy specifically stresses that 'shifting agriculture and nomadism will be discouraged'; transhumant movements are to be 'modernized' and regulated; 'pastoralists and agriculturalists … will be educated on good land management'; and free movement of pastoralists with their cattle is to be regulated to limit conflict and degradation (Government of Tanzania, 1997). This then is a clearly stated policy to convert an indigenous livestock production system to western style commercial ranching by means of demarcation of land, fencing, pasture improvement, breed improvement, intensification of fodder production and veterinary inputs (although little of this has been evident in practice). Similar tenets structure the Nigerian agriculture and livestock development policies (Fraser, 2003). Overgrazing, overstocking and environmental degradation myths remain central to these policy documents, as does the persistent assumption that local dryland agricultural practice in Africa is detrimental to soil structure, soil fertility, water relations and productivity generally (critiqued by Mortimore, 1998). The demarcation, subdivision and privatization of

formerly communally held and managed lands is a consistent feature of these policies, as is the pressure to move from more mobile to more settled lifestyles. At the
same time, herders who are unable to qualify for, or otherwise maintain access to,
privatized pastures and the other natural resources occurring on these lands, tend
to experience disproportionately adverse effects due to privatization and the application of monetarist macro-economic policy. This has been noted, for example, in
Venezuela where the outcome of 'land reform' was in fact to concentrate private
land in the hands of the wealthiest owners and further reduce the commons (Silva
and Moreno, 1993; also Galaty, 1999; Toulmin and Quan, 2000; Thompson and
Homewood, 2002; Homewood *et al.*, 2004).

In many 'New World' drylands, land reform also is occurring in response to
the challenge to reinstate land rights to indigenous inhabitants. Again, problems
emerge due to the radically different conceptions of, and relationships with, land
associated with a settler European farming culture and with indigenous peoples.
Broadly speaking, this can be summarized as the differences that emerge respectively between 'people owning the land' and 'the land owning people'; corresponding to the conceptual and experiential differences arising between land as
object to be transformed into profit, and as partner in affective relationships with
cultural practices (for elaboration see Bender, 1993; Tilley, 1994; Abramson and
Theodossopoulos, 2000; Ingold, 2000). For example, Australia's major programme
of restoring Aboriginal land rights has come under criticism as a result of the seem-
ing impossibility of genuinely accommodating ideational, affective and dynamic
relationships with landscape within the cadastral logic of a legal system based on the
necessity of formalizing, structuring and attaching these relationships to a separately
surveyed and mapped landscape:

> [A]ward of lands is constrained by the historical accident of land availability,
> either as Aboriginal reserve or as vacant public land or national park, rather
> than through any informed appraisal of the balance between Aboriginal and
> other interests...the outcome is often very inequitable. ... the title being
> issued is simultaneously more powerful and more restrictive than those avail-
> able to non-Aboriginal people ... [reinforcing] ... the dualism between
> Aboriginal and non-Aboriginal lands and the associated social divisions. ...
> [and strengthening] ... resistance towards recognition of further Aboriginal
> land claims....Accordingly it may reinforce and perpetuate inequitable out-
> comes for Aboriginals, and preclude multiple or joint land use options that
> require shared decision-making between Aboriginal and non-Aboriginals
> representatives.
>
> *(Holmes and Mott, 1993; see also Morphy, 1993; Jacobs, 1996)*

The root of the difficulty lies in the different cultural significance with which
land and landscape are imbued: thus, 'to Aborigines, land is not merely a "factor
of production ... (but) ... a factor of existence ... (providing) ... religious significance, cultural integrity and social identification" as well as a resource base for

traditional activities' (Coombes *et al.*, 1990, cited in Holmes and Mott, 1993). Non-transferable, freehold, communal land titles which attempt to deal with these differing conceptions of land have to be radically different to forms of land title which treat land as a commodity and as transferable property: shifting between these two forms is unlikely to be seamless.

Trends in wildlife conservation

In 'Old World' drylands the drive to substitute commercial ranching on private land for indigenous livestock production on communal rangelands has been mirrored by a drive to substitute wildlife-based systems for agropastoralism through establishing protected areas on otherwise agropastoral land (Simpson and Evangelou, 1984; Alexander and MacGregor, 2000; Kristjansen *et al.*, 2002). This perhaps has been most marked in African drylands, which retain a spectacular large mammal wildlife, but is also clear in the Middle East (Debaine and Jaubert, 2002; Chatty, 2003), in Mongolia, and in India (the latter dominated by forest 'conservation' which is in practice tied to commercial exploitation, e.g. Gadgil, 1993; Rangan, 1996).

Biodiversity is perceived widely as declining in drylands (e.g. Grainger, 1999), although this perception is not always well supported (Shackleton, 2000; Homewood and Brockington, 1999; Maddox, 2002). The dominant explanatory model underlying biodiversity conservation policies has been that local land-use practices are detrimental to soil, water, vegetation and habitat in general (Grainger, 1999; Hartmann, 2002). This is seen as an accelerating threat due to a growing human population and, particularly in sub-Saharan rangelands, in relation to expanding agropastoral land use leading to habitat conversion (Grainger, 1999). Mammal species survival is viewed as threatened by increases in local hunting, especially where urban demand gives rise to trade in valued species (Campbell and Borner, 1995; Campbell and Hofer 1995; Caro, 1999a, 1999b). Ironically, the erosion of land management practices bound with culturally informed and praxis-oriented knowledges of the landscape also has been noted to have had undesirable ecological effects. This is the case, for example, with the restrictions placed on 'traditional' early dry season fire management practices in Australia, which has increased the incidence of late, frequently destructive and uncontrollable burns (e.g. as noted by CSIRO researcher Cheney, cited in Pockley, 2002; Dennis, 2003).

However, narratives of biodiversity (and particularly large mammal) decline arose in a context where a colonial European culture, identifying hunting using firearms with the leisure pursuits of the aristocracy, had enormous impacts on animal wildlife, while criminalizing local hunting for subsistence as poaching (MacKenzie, 1987; Escobar, 1996a; Neumann, 1996). The associated narrative has been so strong that in some cases it has distorted interpretation of contemporary data on biodiversity and on landscape processes that clearly contradict this narrative, and which require quite different ecological models in the explanation of landscape and species population change (Western and Gichohi, 1993; Brockington and Homewood, 1996; Leach and Mearns, 1996; Shackleton, 2000). The qualitative social and ecological character of drylands is inextricably intertwined with processes of continual disturbance through patchy and unpredictable rainfall, fire, grazing and browsing, as well as through a range of abiotic-biotic-anthropogenic relationships (e.g. Ellis and Swift, 1988; Dublin, 1995; Behnke and Scoones, 1993; Homewood and Brockington, 1999). Dryland biodiversity, for example, is based less on local endemism and more on the ability of dryland species to disperse, colonize and persist in a patchy, unpredictably fluctuating and continually 'disturbed' environment. In such landscapes habitat disturbance *per se*, therefore, is not necessarily detrimental to species survival (Davis *et al.*, 1994; Stattersfield *et al.*, 1998; Homewood and Brockington, 1999). For example, measures of dryland biodiversity increase with the extent of the landscape throughout which mobile species are able to disperse in pursuit of seasonal and annual fluctuations of productivity, such that it can become misleading to limit these measures to the formal administrative boundaries defining the spatial extent of a protected area (Western and Ssemakula, 1981; Western and Gichohi, 1993). In the East African contexts, large mammal density, frequency and abundance are at least as great in unfenced protected-area buffer zones (Maddox, 2002) and can be greater (Norton Griffiths, 1998). Local hunting of species with high reproductive rates (e.g. ungulates/rodents) appears sustainable across much dryland/cropland mosaic, reflected recently in an Australian legal precedent which ruled that Aboriginal hunting of protected species was deemed not to be 'poaching' but to be a legal and sustainable resource use activity (Davies *et al.*, 1994).

Nevertheless, conservation policy has sought first to protect as spatially extensive a set of areas as possible,[9] alongside the targeting of biodiversity hotspots for special protection (Myers *et al.*, 2000; Balmford *et al.*, 2001). Throughout 'Old World' drylands, protection has been based primarily on 'fortress conservation', i.e. requiring the exclusion of local users through fencing and legislation (Brockington, 2002); enforcement through paramilitary style ranger forces (Leader-Williams and Albon, 1988; Campbell and Borner, 1995; Campbell and Hofer, 1995; Clynes, 2002; Sullivan, 2002); and the retention of tourism/scientific research as appropriate uses within protected areas where consumptive use of natural resources for local livelihoods is banned.

A number of challenges to this overall policy of strict protection have now emerged, on the grounds of flawed theory (Bell, 1987), poor conservation outcomes (Western and Ssemakula, 1981) and problematic development implications (Bell, 1987; Escobar, 1996a). In its place, various forms of community-based conservation (CBC) have assumed ascendancy, based on the potentially conservation-compatible and positive role of local land uses, the growing urgency of a universal human right to improved livelihoods and welfare and the realization that state resources cannot maintain the levels of enforcement needed for fortress conservation. In Australia, there have been complex interactions (including some synergy) between CBC (or community wildlife management, CWM) and changing Aboriginal land rights (Davies *et al.*, 1999; Roe *et al.*, 2000), and some associated unease over the extent to which Aboriginal land management is and may continue to be conservation-compatible (Holmes and Mott, 1993). In Old World drylands, particularly sub-Saharan Africa, International

Monetary Fund-led structural adjustment policies (and the associated reduction in public expenditure) have made it necessary to enlist the support of reserve-adjacent dwellers, rather than simply excluding them (IIED, 1994; Homewood *et al.*, 2001). Community-based conservation and community-based natural resources management (CBNRM) have thus been conceived and marketed by development agencies and donors as a people-friendly alternative to fortress conservation (Sullivan, 2002). CBC benefit-sharing schemes seek to compensate local people for the resources they forgo to protected areas by distributing income, employment and other benefits from wildlife tourism. In other cases, communities are contracted to manage part of their land for conservation aims (Roe *et al.*, 2000; Hulme and Murphree, 2001; Davies *et al.*, 1999).

As with agriculture and livestock developments, community-based wildlife management initiatives frequently depend on the demarcation of landscape boundaries and the registering of community membership within these boundaries, and can require the setting aside of areas of communal land for conservation purposes. For local people this can mean curtailing through passage, land-use options and mobility, and may also further extend the arm of the state over rural (and otherwise 'peripheral') populations (e.g. Fairhead, 2000; Sullivan, 2002). An outcome can be less to involve local people in protected area conservation, than to extend conservation control from the centre over indigenous uses of resources occurring outside protected areas (e.g. as in the case of Wildlife Management Areas in Tanzania, and the retaining of conservation control over areas ceded to Aboriginals in Australia; Holmes and Mott, 1993; Davies *et al.*, 1999). Increasingly, private land title and/ or private sector access is proffered to entrepreneurs establishing eco-tourism and high-paid trophy-hunting ventures from which local people are considered to benefit via employment and other income opportunities (Wøien and Lama, 1999). In many cases, this process is managed by central government and bypasses control by the local rural population altogether. Within the co-operative 'community-based' wildlife associations established on Kenya's Group Ranches, for example, small subsets of well-placed individuals were able to identify and secure legal title to key areas of high tourist potential, and then moved rapidly to exclude other members from sharing the potential benefits (Galaty, 1999; Thompson and Homewood, 2002). In Zimbabwe's CAMPFIRE programme, initiatives hailed by donors and implementers as successful for the 'community' excluded gatherer-hunters who effectively became refugees and criminalized poachers on their own land (Marindo-Ranganai and Zaba, 1994). Nevertheless, there are recurrent institutional pressures to cast community-based wildlife management initiatives as *the* route for producing win/ win outcomes favourable to development and to wildlife conservation (e.g. LWAG, 2002). Despite claims of success, 'community-based conservation' is also critiqued as an extension of a 'northern' corporatism that requires commoditization of land and natural resources, normalizes particular socio-economic uses of, and relationships with, these resources, and acts to assert access and control by wealthy 'outsiders' via ecotourism, trophy-hunting and the globalizing of neoliberal economic (and conservation) agendas (Holmes and Mott, 1993; Escobar, 1996b; Brockington,

2002; Sullivan, 2002). A growing response has been the emergence of local protest to such initiatives in several contexts (Patel, 1998; Alexander and MacGregor, 2000; Sullivan 2003).

(En)gendering modernity in drylands

While cognisant of the problems of essentializing categories, a view is emerging that an expanding frontier of modernity in drylands has tended to have a particularly disempowering impact on women. Androcentric colonial and donor assumptions that men are heads of households, the holders of land title and the owners of livestock have created and exacerbated gender inequalities in drylands (e.g. Hodgson, 2000). Women's workloads, together with loss of control over their own labour, have been exacerbated by sedentarization and tenure changes, male labour migration, changes in livestock entitlements and by a compromised access to natural resources due to reductions in common land area and transformations of landscapes under commercial agriculture. Women's dependence in some cases has increased due to the common passing to men of formal title to land. Those whose husbands have mismanaged their land and their herd, or who are divorced or widowed, find themselves dispossessed and excluded in circumstances where previously their access, use of resources and livelihoods might have been safeguarded under customary forms of tenure and entitlement (e.g. Talle, 1988; Joekes and Pointing, 1991). Conversely, even where formal tenure allows women to own land, the clash between imposed national legal frameworks which state this right, and the realities of customary practice and local hierarchies of power within and between households, can mean that women do not in fact benefit from their supposed legal right (as recorded in Agarwal, 1999).

The growing ascendance of market pressures over social obligations also make it increasingly common for livestock to be disposed of by men without consulting their wives (Talle, 1988). This can extend to the production and management of milk, an item conventionally associated with pastoralist women as heads of houses. The social redistribution of milk among pastoralists is important not only for poorer individuals who benefit from the milk as food, but also in establishing those women who manage milk as centrally responsible for matters of importance to the household and therefore to the broader social grouping. As urban agglomerations grow in semi-arid and arid areas, however, and with the associated increase in sales of milk and other pastoral products, urban dairying activities by pastoralist women become increasingly common (Waters Bayer, 1985; Herren, 1990; Little, 1994). When this shift occurs men often gain control of the actual marketing and of the revenue, engendering a corresponding deterioration in women's autonomy and income that can have a knock-on effect on the food and health of dependants (Salih, 1985; Talle, 1990). Progressively greater diversion of milk to market outlets thus affects the fabric of social relations (e.g. Grandin, 1988; Ndagala, 1990, 1992), the commoditization and commercialization of milk precipitating a loss of control by women of both the management and the proceeds of milk sales. This is particularly likely to happen where there is the possibility of establishing larger scale dairying enterprises.

Sedentarization, codification and commercialization have affected another component of rural women's economic security, namely their use of gathered resources. The rapid increase of private and exclusive ownership throughout drylands, the spread of fencing, and increasing human and livestock populations around settlements, compromise women's access to wild plants for fuels, foods, fibres, medicines (e.g. Gadgil, 1993; Konstant *et al.*, 1995; Sullivan *et al.*, 1995; Schreckenberg, 1996).This affects women and their dependents at every level of income, workloads and food security (Anonymous, 1990). Spending more time seeking fuel or other plant resources, or having to find the money to purchase fuel, means restructuring domestic activities, for example, by spending more time on producing items that can be sold to finance alternative purchases. These activities and gendered areas of environmental knowledge may be further masked by a tendency to focus on a masculinized wildlife of large mammals in conservation initiatives (Sullivan, 2000a).

These problematic outcomes of development initiatives mean that statements regarding 'development' in drylands (whether oriented towards agriculture, livestock or wildlife) today are couched in explicitly participatory, inclusive and 'pro-poor' terms. The extent to which these translate into significant reorientation of action on the ground, however, is debatable. On the face of it there has been progressive recognition that an orientation towards local identification of problems and priorities, addressed through low-capital, low-tech and indigenous practices offers more chance of 'sustainable development' than expensive interventions transplanted from western systems. There also has been a general 'development policy shift' (at least rhetorically) towards strengthening livelihood security, health, education and political representation, as opposed to attempts to maximize income and monetary profit. Ostensibly, this participatory (and 'pro-poor') rhetoric, together with the low-impact inputs with which it is associated, might minimize opportunities for élites and middlemen to benefit from the development process at the expense of target groups. Nevertheless, those administering dryland areas, both Old and New World, have been quick to respond to changing official priorities and development fashions. In some cases, those constituting local, regional and national bureaucracies (and their inevitable alliances with politicians and commercial entrepreneurs) have indicated compliance with donor and structural adjustment agendas, restructuring themselves to attract and retain funding flows while ensuring limited implementation so that in practice little changes. In African drylands this is expressed in policy documents that are contradictory both internally and in their outcomes. They pay lip service to establishing and addressing local priorities with local means, while at the same time maintaining a hard line on replacing indigenous dryland production systems with imported western-style enterprises, supported by conventional equilibrium narratives of ecosystem processes. Comparable contradictions have been evident in 'New World' drylands (Young and Solbrig, 1993). Even where the aim is to devolve decision-making responsibility this, unsurprisingly, tends to be heavily circumscribed in practice; with élites seeking to protect their privileged positions, and recipients experiencing the socio-economic problematic of attempting to break from historical circumstances that locate them in prior positions of inferiority (*vis à vis* the centre) and marginalization (e.g. Little, 1985; IIED, 1994; Brockington, 2002).

To summarize then, development trends in drylands have emphasized interventions that are capital-intensive and frequently subsidized, amounting to hi-tech inputs that require reliance on exogenously produced petrochemicals, and emphasizing production for single product external markets. They have necessitated the rationalization of land tenure into static, fenced and privately owned landholdings, and they have supported conservation initiatives that fetishize a spectacular animal wildlife and 'wilderness' landscapes, and that affect control over landscapes and biodiversity by distantly located consumers. Their problematic outcomes have included:

- increasing wealth differentials, landlessness and the disruption of reciprocal welfare safety nets;
- severe transformation of landscapes through the establishing of capital-intensive agricultural land-use schemes;
- erosion and loss of local environmental knowledge;
- and erosion of rights to productive resources and decision-making arenas held by women.

But the key point is that these processes and their outcomes are understandable, even predictable, if they are considered as part and parcel of the suite of hegemonic rationalizing and ideological assumptions underscoring modernity (see introductory section), i.e. which emphasize regulation and management from the centre, the fixing of people to places and the purification of difference and apparent disorder.

Concluding remarks

In this chapter we have attempted to add to current debate regarding equilibrium and non-equilibrium dynamics, and the implications of these conceptual principles to drylands and their inhabitants, by asking a number of questions. In what relationship do these concepts exist with each other? Why have equilibrium concepts been so overwhelmingly naturalized within science and policy communities, to the detriment both of the understanding of drylands, and the possibilities for self-determination by the peoples who live in these environments? And why are non-equilibrial framings of dynamics apparently so threatening to states and experts? While clarification of different positions is important and necessary (cf. Illius and O'Connor, 1999; Sullivan and Rohde, 2002), we feel that it can become problematic if it entrenches positions and promotes defensive attitudes in relation to these. Thus we have tried here to move beyond our own positions to date, and to write with the intention of promoting conversations across dualisms.

In trying to think about how we think about things and why, however, we have not been able to avoid considering the devastating associations between equilibrium thinking, state science, the assumed superiority of the core and the corresponding justification of top-down policies of control over landscapes and people. Again we should ask who benefits: ecologists as purveyors of a higher understanding? Bureaucracies as regulators of land use? Enforcers as having their role and control legitimized? Men as gaining disproportionately from introduced systems of land title and the commodification of natural resources?

One element of the debate relates to an urgent need to shift from a formal science perspective that maintains that 'reality' can be satisfactorily measured and predicted through the separation and abstracting of parameters from the contexts in which they occur. With prescience of currently emerging complex systems theories, De Rosnay (1979, in Saner, 1999) argues that we need to take a macroscopic as opposed to a microscopic view of phenomena; an approach that is trans-disciplinary, accepts the hybrid nature of knowledge production (cf. Latour, 1987, 1993), and that responds to the need to integrate '[b]oth the science of parts and the science of the integration of parts' (Holling, 1998). Such a shift would underscore a rebalancing towards policies that facilitate opportunistic tracking of environmental dynamics, and a relinquishing of decision-making and administrative power by the centre to the periphery. Given the inherently conservative nature of states and institutions, however, it is perhaps wishful thinking that such a reorientation will occur in meaningful terms.

Further, we are unwilling to avoid what we feel are the broader historical and contemporary processes of purification with which an adherence to the linear, equilibrium thinking of state science is entwined. Thus we ask ourselves if the processes we describe in this chapter for drylands are qualitatively distinct from the spectacular and violent power driven by desire for purification of the dehumanized 'other' throughout the last two millennia? We think not. In attempts to bring pastoralists into the fold of the settled state; to constrain perceptions of drylands to the filter of a constructed dynamical norm; to demonize drylands as degraded through the equally demonic land-use practices of their dwellers; and to impose static boundaries over both landscapes and people, we feel that we can see the seeds of some of the worst excesses of purification occurring through history.

But where to now? How to feel optimistic or confident enough to make recommendations, other than to say that it is critical for all who place themselves in the position of writing about, acting on behalf of, or drawing up policy for others, to consider where ideas and views about environmental dynamics and best professional practice come from, what conceptions of reality they uphold, and what outcomes they are likely to support. And following Hardt and Negri (2000), perhaps to not be surprised by an increase in fragmented and dispersed forms of resistance to interventions that involve the further surveillance, codification, rationalization and control of peoples' lifestyles and landscapes.

Acknowledgments

Sian Sullivan gratefully acknowledges support from UK's Arts and Humanities Research Council (AH/K005871/2, http://www.futurepasts.net) in the final reworking of this paper.

Notes

1 This chapter was first commissioned in 2002 and a version has been available online as a working paper since 2003 (Sullivan and Homewood, 2003). In revisiting the piece for publication in the present collection we have decided against systematically updating our source material in the text. The chapter is already reference-heavy and a major integration of new material arising over the intervening 14 years would add too much length to an already long piece. We observe, however, that even with inclusion of new research it is likely that we would come to similar conclusions, since the issues we discuss here remain systemic. The possible exception is global climate change, for which concern has heightened considerably since we first wrote this chapter and which has particular implications for the already very variable environmental settings of the dryland contexts we explore here. For a more recent synthesis of pastoralist ecology in African contexts and the impacts of varied development interventions see Homewood (2008).

2 Nonetheless, a major initiative – the Great Green Wall for the Sahara and Sahel Initiative (GGWSSI) – was launched in 2007 by the African Union, in association with the United Nations Convention to Combat Desertification, to finance the establishment of a 'green wall' of sustainable land and forest management and restoration that creates a '15 km wide tree barrier linking Dakar to Djibouti in order to stop "desert encroachment"' (GGWSSI, 2014).

3 The panopticon is a circular building with an observation tower in the centre of an open space surrounded by an outer wall. This wall contains cells for occupants as part of a design that increases security by facilitating more effective surveillance. In his book *Discipline and Punish, The Birth of the Prison*, Michel Foucault (1977) describes the 'panopticon' as an experimental laboratory of power in which behaviour could be modified. Foucault viewed the panopticon as a symbol of the disciplinary society of surveillance.

4 This typology is somewhat problematic for the southern African context where livestock have been herded nomadically for some 2,000 years (Kinahan, 1991). We group southern Africa with the Americas and Australia, however, because of the shared experiences of these territories in terms of European settlement practices and the ensuing dislocation of indigenous peoples from the land via processes of genocide and proletarianisation. For dryland southern Africa see Bley (1996), Skotnes (1996), Gordon and Sholto Douglas (2000) and Suzman (2000).

5 This is not to deny that throughout history there have been long periods when settled peoples and places have lived under the hegemony of mobile, pastoralist groups, who have dominated and manipulated resources, production and social norms according to their own ideologics, whether religious, political, economic or military. For example, in the nineteenth century Tuareg and Fulani States dominated large areas of West Africa, with pastoralist nobles depending on the farm production and domestic labour of enslaved cultivating peoples. Maasai controlled much of East Africa and the Tutsi dominated nineteenth century Rwanda and Burundi. Similarly, herders may have a tradition of maintaining others in positions of subservience as labourers, as currently is the case with Herero in south-west Africa (Namibia and Botswana) in their hiring of 'Bushman' (i.e. Sān-speaking) workers (Suzman, 2000). Nonetheless, our focus here is on the ways in which variously nomadic peoples have met with, been incorporated within and been accommodated by the modern state, and our position is that this encounter has been systematically problematic for indigenous herders and nomads.

6 Tolkien, 1954: p. 260.

7 This subsection draws heavily on material developed for Sullivan and Homewood (2004).

8 Following Behnke (1983), it is intriguing to note that ranchers in these areas in many cases did not fence themselves in by choice as a means of enhancing production. If anything, fencing initially led to livestock losses. In North America, ranchers fenced the range so as to keep out land-hungry farmers and other ranchers. In Australia, they fenced

in response to a crisis in labour availability when the 1850s gold rush drew away their sheepherders.
9 Soulé and Sanjayan, 1998, for example, argue for 50% of the land surface area globally and nationally to be protected.

References

Abel, N. (1993) 'Carrying capacity, rangeland degradation and livestock development policy for the communal rangelands of Botswana', *Pastoral Development Network Paper 35c*, Overseas Development Institute, London.

Abramson, A. (2000) 'Mythical land, legal boundaries: Wondering about landscape and other tracts', In Abramson, A. and D. Theodossopoulos (Eds) *Land, Law and Environment: Mythical Land, Legal Boundaries*, Pluto Press, London.

Abramson, A. and D. Theodossopoulos (Eds) (2000) *Land, Law and Environment: Mythical Land, Legal Boundaries*, Pluto Press, London.

Adams, W. (1992) *Wasting the Rain: Rivers, People and Planning in Africa*, Earthscan, London.

Agarwal, B. (1999) *A Field of One's Own: Gender and Land Rights in South Asia*, Cambridge University Press, Cambridge.

Anderson, D. (1984) 'Depression, dust bowl, demography and drought: The colonial state and soil conservation in East Africa during the 1930s', *African Affairs* 83(332): 321–43.

Alexander, J. and J. MacGregor (2000) 'Wildlife and politics: CAMPFIRE in Zimbabwe', *Development and Change* 31(3): 605–62.

Anker, P. (2002) *Imperial Ecology: Environmental Order in the British Empire*, Harvard University Press, Cambridge.

Anonymous (1990) 'The impact of fuelwood scarcity on dietary patterns: Hypotheses for research', *Unasylva* 41(160): 29–34.

Arab World Geographer (2002) Special edition on Alternative Perceptions of Authority and Control, Vol. 5(2): 71–140.

Attwell, C. A. M. and F. P. D. Cotterill (2000) 'Postmodernism and African conservation science', *Biodiversity and Conservation* 9(5): 559–77.

Aubréville, A. (1938) *La Forêt Coloniale: Les Forêts de L'Afrique Occidentale Française*. Annales d'Academie des Sciences Coloniales IX, Societé d'Editions Géographiques, Maritimes et Coloniales, Paris.

Balmford, A., J. L. Moore, T. Brooks, N. Burgess, L. Hansen, P. Williams and C. Rahbek (2001) 'Conservation conflicts across Africa', *Science* 291(5513): 2617–19.

Barrow, E. (1988) 'Trees and pastoralists: The case of the Pokot and Turkana', *ODI Social Forestry Network Paper* 6b, Overseas Development Institute, London.

Barrow, E. (1990) 'Usufruct rights to trees: The role of Ekwar in dryland central Turkana, Kenya', *Human Ecology* 18(2): 163–76.

Baumann, M. (2000) 'On nature, models, and simplicity', *Conservation Ecology* 4(2): r4.

Behnke, R. H. (1983) 'Production rationales: The commercialisation of subsistence pastoralism', *Nomadic Peoples* 14: 3–34.

Behnke, R. H. and I. Scoones (1993) 'Rethinking range ecology: Implications for rangeland management in Africa', In Behnke, R. H., Scoones, I. and C. Kerven (Eds) *Range Ecology at Disequilibrium: New Models of Natural Variability and Pastoral Adaptation in African Savannas*, Overseas Development Institute, London.

Behnke, R. H., I. Scoones and C. Kerven (Eds) (1993) *Range Ecology at Disequilibrium: New Models of Natural Variability and Pastoral Adaptation in African Savannas*, Overseas Development Institute, London.

Beinart, W. (2000) 'African history and environmental history', *African Affairs* 99(395): 269–302.

Beinart, W. (2003) *The Rise of Conservation in South Africa: Settlers, Livestock, and the Environment 1770–1950*, Oxford University Press, Oxford.

Belenky, M. F., B. M. Clinchy, N. R. Goldberger and J. M. Tarule (1986) *Women's Ways of Knowing: The Development of Self, Voice and Mind*, Basic Books, USA.

Bell, R. (1987) 'Conservation with a human face', In Anderson, D. and R. Grove (Eds) *Conservation in Africa: People, Policies and Practice*, Cambridge University Press, Cambridge.

Belsky, A. J. (1989) 'Landscape patterns in a semi-arid ecosystem in East Africa', *Journal of Arid Environments* 17(2): 265–70.

Belsky, A. J. (1994) 'Influences of trees on savanna productivity: Tests of shade, nutrients and tree-grass competition', *Ecology* 75(4): 922–32.

Belsky, A. J. (1995) 'Spatial and temporal patterns in arid and semi-arid African savannas', In Hansson, L., Fahrig, L. and G. Merriam (Eds) *Mosaic Landscapes and Ecological Processes*, Chapman and Hall, London.

Belsky, A. J. and C. D. Canham (1994) 'Forest gaps and isolated savanna trees: An application of patch dynamics in two ecosystems', *Bioscience* 44(2): 77–84.

Belsky, A. J., R. G. Amundson, S. J. Riha, A. R. Ali and S. W. Mwonga (1989) 'The effect of trees on their physical, chemical and biological environment in a semi-arid savanna (Tsavo) Kenya', *Journal of Applied Ecology* 26(3): 1005–24.

Belsky, A. J., S. M. Mwonga, R. G. Amundson, J. M. Duxbury and A. R. Ali (1993a) 'Comparative effects of isolated trees on their understory environments in high and low rainfall savannas', *Journal of Applied Ecology* 30(1): 143–55.

Belsky, A. J., S. M. Mwonga and J. M. Duxbury (1993b) 'Effects of widely spaced trees and livestock grazing on the understory environments in tropical savannas', *Agroforestry Systems* 24(1): 1–20.

Bender, B. (ed.) (1993) *Landscape: Politics and Perspectives*, Berg, Oxford.

Berkes, F. (1989) *Common Property Resources: Ecological and Community Based Sustainable Development*, Belhaven Press, London.

Birch, T. (1996) '"A land so inviting and still without inhabitants": erasing koori culture from (post-) colonial landscapes', In Darian-Smith, K., Gunner, L. and S. Nuttall (Eds) *Text, Theory and Space: Land, Literature and History in South Africa and Australia*, Routledge, London.

Birley, M. H. (1982) 'Resource management in Sukumaland, Tanzania', *Africa* 52(2): 1–29.

Bley, H. (1996) *Namibia Under German Rule*, Germany LitVerlag, Hamburg.

Bollig, M. (1998a) 'The colonial encapsulation of the north-western Namibian pastoral economy', *Africa*, 68(4): 506–36.

Bollig, M. (1998b) 'Power & trade in precolonial & early colonial Northern Kaokoland 1860s–1940s', In Hayes, P., Silvester, J., Wallace, M. and W. Hartmann (Eds) *Namibia Under South Africa Rule: Mobility and Containment 1915–1946*, James Currey, London.

Botelle, A. and R. Rohde (1995) *Those Who Live on the Land: A Socio-economic Baseline Survey for Land Use Planning in the Communal Areas of Eastern Otjozondjupa*, Republic of Namibia, Ministry of Lands, Resettlement, and Rehabilitation, Windhoek.

Bourdieu, P. (1980) *The Logic of Practice*, Polity, Cambridge.

Briske, D. D., S. D. Fuhlendorf and F.E. Smeins (2003) 'Vegetation dynamics on rangelands: A critique of the current paradigms', *Journal of Applied Ecology* 40(4): 601–14.

Broch-Due, V. and D. M. Anderson (1999) 'Preface', In Anderson, D. M. and V. Broch-Due (Eds) *The Poor Are Not Us: Poverty and Pastoralism in Eastern Africa*, James Currey, Oxford, East African Educational Publishing, Nairobi, and Ohio University Press, Ohio.

Brockington, D. (2002) *Fortress Conservation: The Preservation of the Mkomazi Game Reserve, Tanzania*, James Currey, Oxford.

Brockington, D. and K. Homewood (1996) 'Wildlife, pastoralists and science: Debates concerning Mkomazi Game Reserve, Tanzania', In Leach, M. and R. Mearns (Eds) *The Lie of the Land: Challenging Received Wisdom on the African Environment*, The International African Institute, London, James Currey, Oxford and Heinemann, Portsmouth.

Brockington, D. and K. Homewood (1998) 'Pastoralism around Mkomazi Game Reserve: The interaction of conservation and development', In Coe, M., McWilliam, N., Stone, G. and M. Packer (Eds) *Mkomazi: The Ecology, Biodiversity and Conservation of a Tanzanian Savanna*, Royal Geographical Society (with the Institute of British Geographers), London.

Brockington, D. and K. Homewood (2001) 'Degradation debates and data deficiencies. The case of the Mkomazi Game Reserve, Tanzania', *Africa* 71(3): 179–227.

Bromley, D. and W. Cernea (1989) *The Management of Common Property Resources: Some Conceptual and Operational Fallacies*, World Bank, Washington, DC.

Campbell, B. M., R. Costanza and M. van den Belt (2000) 'Special section: Land use options in dry tropical woodland ecosystems in Zimbabwe: Introduction, overview and synthesis', *Ecological Economics* 33(3): 341–51.

Campbell, K. and H. Hofer (1995) 'People and wildlife: Spatial dynamics and zones of interaction', In Sinclair, A. R. E. and P. Arcese (Eds) *Serengeti II. Dynamics, Management and Conservation of an Ecosystem*, University of Chicago Press, Chicago.

Campbell, K. and M. Borner (1995) 'Population trends and distribution of Serengeti herbivores: Implications for management', In Sinclair, A. R. E and P. Arcese (Eds) *Serengeti II. Dynamics, Management and Conservation of an Ecosystem*, University of Chicago Press, Chicago.

Caro, T. (1999a) 'Densities of mammals in partially protected areas: The Katavi ecosystem of western Tanzania', *Journal of Applied Ecology* 36(2): 205–17.

Caro, T. (1999b) 'Demography and behaviour of African mammals subject to exploitation', *Biological Conservation* 91(1): 91–97.

Carswell, G. (2002) 'Farmers and fallowing: Agricultural change in Kigezi District, Uganda', *The Geographical Journal* 168(2): 130–40.

Carter, A. J. and T. G. O'Connor (1991) 'A two-phase mosaic in a savanna grassland', *Journal of Vegetation Science* 2(2): 231–36.

Caughley, G., N. Shepherd and J. Short (Eds) (1987) *Kangaroos: Their Ecology and Management in the Sheep Rangelands of Australia*, Cambridge University Press, Cambridge.

Charney, J., P. Stone and W. Quirk (1975) 'Drought in the Sahara: A biogeophysical feedback mechanism', *Science* 187(4175): 434–35.

Chatty, D. (2003) 'Environmentalism in the Syrian Badia: The assumptions of degradation, protection and bedouin misuse', In Anderson, D. G. and E. Berglund (Eds) *Ethnographies of Conservation: Environmentalism and the Distribution of Privilege*, Berghahn Books, Oxford.

Chatwin, B. (1987) *The Songlines*, Picador, London.

Cilliers, P. (1998) *Complexity and Postmodernism: Understanding Complex Systems*, Routledge, London.

Clements, F. E. (1916) *Plant Succession: An Analysis of the Development of Vegetation*, Carnegie Institute Publications 242, Washington, DC.

Clynes, T. (2002) 'They shoot poachers, don't they?' *The Observer*, 24 November 2002, *OM Dispatches* pp35–47.

Coombes, H. C., J. Dargavel, J. Kesteven, H. Ross, D. Smith and E. Young (1990) *The Promise of the Land: Sustainable Use by Aboriginal Communities*, Centre for Resource and Environmental Studies, Australian National University, Canberra.

Corbett, M. (2003) 'Sound organisation: A brief history of psychosonic management', *ephemera: critical dialogues on organization*, 3(4): 265–76.

Corbin, A. (1986) *The Foul and the Fragrant: Odour and the Social Imagination*, Harvard University Press.

Cossins, N. and M. Upton (1987) 'The Borana pastoral system of Southern Ethiopia', *Agricultural Systems* 25(3): 199–218.

Coughenour, M. B., D. L. Coppock and J. E. Ellis (1990) 'Herbaceous forage variability in an arid pastoral region of Kenya: Importance of topographic and rainfall gradients', *Journal of Arid Environments* 19(2): 147–59.

Coughenour, M. B., J. E. Ellis, D. M. Swift, D. L. Coppock, K. Galvin, J. T. McCabe and T.C. Hart (1985) 'Energy extraction and use in a nomadic pastoral ecosystem', *Science* 230(4726): 619–24.

Cowling, R. (2000) 'Challenges to the "new" rangeland science', *Trends in Ecology and Evolution (TREE)* 15(8): 303–04.

Dahl, G. (1987) 'Women in pastoral production: some theoretical notes on roles and resources', *Ethnos* 52(1–2): 246–79.

Davies, J., K. Higginbottom, D. Noack, H. Ross and E. Young (1999) 'Sustaining Eden: Indigenous community wildlife management in Australia', *Evaluating Eden Series* No.1, International Institute of Environment and Development, London.

Davis, S.,V. Heywood and A. Hamilton (Eds) (1994) *Centres of Plant Diversity*, WWF/IUCN, Gland, Switzerland.

Debaine, F. and R. Jaubert (2002) 'The degradation of the steppe, hypotheses and realities', *The Arab World Geographer* 5(2): 124–40.

De Certeau, M. (1984) *The Practice of Everyday Life*, University of California Press, Berkeley.

Deleuze, G. and F. Guattari (1980) *A Thousand Plateaus: Capitalism and Schizophrenia* (translation and foreword by Brain Massumi), The Athlone Press, London.

Dennis, C. (2003) 'Burning issues', *Nature* 421: 204–6.

De Rosnay, J. (1979) *The Macroscope: A New World Scientific System* (translated by Robert Edwards), Harper and Row, New York.

Dirk Moses, A. (2000) 'An antipodean genocide? The origins of the genocidal moment in the colonization of Australia', *Journal of Genocide Research* 2(1): 89–106.

Douglas, M. (1966) *Purity and Danger*, Routledge and Kegan Paul, London.

Dublin, H.T. (1991) 'Dynamics of the Serengeti-Mara woodlands: An historic perspective', *Forest and Conservation History* 35(4): 169–78.

Dublin, H. T. (1995) 'Vegetation dynamics in the Serengeti-Mara ecosystem: The role of elephants, fire and other factors', In Sinclair, A. R. E. and P. Arcese (Eds) *Serengeti II. Dynamics, Management and Conservation of an Ecosystem*, University of Chicago Press, Chicago.

Dublin, H.T., Sinclair, A. R. E. and J. McGlade (1990) 'Elephants and fire as causes of multiple states in the Serengeti-Mara woodlands', *Journal of Animal Ecology* 59(3): 1147–64.

Duffy, R. (2000) *Killing for Conservation: Wildlife Policy in Zimbabwe*, The International African Institute, London, James Currey, Oxford, and Indiana University Press, Bloomington.

Ellis, J. E. (1994) 'Climate variability and complex ecosystem dynamics: Implications for pastoral development', In Scoones, I. (ed.) *Living with Uncertainty*, Intermediate Technology Publications, London.

Ellis, J. E., M. B. Coughenour and D. M. Swift (1993) 'Climate variability, ecosystem stability, and the implications for range and livestock development', In Behnke, R. H, Scoones, I. and C. Kerven (Eds) *Range Ecology at Disequilibrium: New Models of Natural Variability and Pastoral Adaptation in African Savannas*, Overseas Development Institute, London.

Ellis, J. E. and D. M. Swift (1988) 'Stability of African pastoral ecosystems: alternative paradigms and implications for development', *Journal of Range Management* 41(6): 450–59.

Escobar, A. (1996a) *Encountering Development*, Princeton University Press, Princeton.

Escobar, A. (1996b) 'Constructing nature: Elements for a poststructural political ecology', In Peet, R. and M. Watts (Eds) *Liberation Ecologies: Environment, Development, Social Movements*, Routledge, London.

Evans-Pritchard, E. E. (1940) *The Nuer: A Description of the Modes of Livelihood and Political Institutions of a Nilotic People*, Oxford University Press, Oxford.

Fairhead, J. (2000) Book review article of Richards, P. W. (1996) *The Tropical Rain Forest* (2nd edition), Cambridge University Press, Cambridge, *Progress in Physical Geography* 24(4): 609–13.

Foucault, M. (1977) *Discipline and Punish, The Birth of the Prison*, Pantheon Books, New York.

Foucault, M. (1981) *Power/knowledge: Selected Interviews and Other Writings, 1972–1977* (edited by Gordon, C.), Harvester Wheatsheaf, Hemel Hempstead.

Fraser, S. (2003) *Parks, Pastoralists & Development Policy in the Nigerian Savanna: A Situated Study of the Politics of Land and Land-Use Reform in Relation to Migratory Pastoralists*, Ph.D. Thesis, University of London.

Fuller, B. B. Jnr (1993) *Institutional Appropriation and Social Change Among Agropastoralists in Central Namibia 1916–1988*, PhD Dissertation, Boston Graduate School, Boston.

Galaty, J. G. (1999) 'Double-voiced violence in Kenya', paper presented at the workshop on *Conflict's Fruit: Poverty, Violence and the Politics of Identity in African Arenas*, 21–24 October 1999, Sophienburg Castle, Denmark.

Galaty, J. G. and P. Bonte (1992) *Herders, Warriors and Traders: Pastoralism in Africa*, Westview Press, Boulder.

Gadgil, M. (1993) 'Restoring the productivity of Indian savannas', In Young, M. D. and O. Solbrig (Eds) *The World's Savannas: Economic Driving Forces, Ecological Constraints and Policy Options for Sustainable Land Use*, Vol. 12, Man and Biosphere Series, UNESCO and Parthenon Publishing Group, Carnforth and New York.

GGWSSI (2014) *Harmonised Regional Strategy for Implementation of the 'Great Green Wall Initiative of the Sahara and the Sahel'*, African Union Commission and Secretariat of the Panafrican Agency, www.greatgreenwallinitiative.org/sites/default/files/publications/harmonized_strategy_GGWSSI-EN_.pdf

Gordon, R. J. (1972) 'Towards an ethnography of Bergdama gossip', *Namib und Meer* 2: 45–7.

Gordon, R. J. (1991) 'Vernacular law and the future of human rights in Namibia', *NISER Discussion Paper* 11, Namibian Institute for Socio-economic Research, Windhoek, Namibia.

Gordon, R. J. (1998) 'Vagrancy, law and "shadow knowledge": Internal pacification 1915–1939', In Hayes, P., Silvester, J., Wallace, M. and W. Hartmann (Eds) *Namibia Under South African Rule: Mobility and Containment 1915–1946*, Ohio University Press, Ohio.

Gordon, R. J. and S. Sholto Douglas (2000) *The Bushman Myth: The Making of a Namibian Underclass* (2nd edition), Westview Press, Boulder.

Government of Tanzania (1997) 'Livestock and agriculture policy', *Policy Statements*, Government of Tanzania, Dar es Salaam.

Graham, O. (1988) 'Enclosure of the East African rangelands: Recent trends and their impact', *ODI Pastoral Development Network Paper* 25a, Overseas Development Institute, London.

Grainger, A. (1992) 'Characterization and assessment of desertification processes', In Chapman, G. P. (ed.) *Desertified Grasslands: Their Biology and Management*, Academic Press, London.

Grainger, A. (1999) 'Constraints on modelling the deforestation and degradation of tropical open woodlands', *Global Ecology and Biogeography* 8(3–4): 179–90.

Grandin, B. (1988) 'Wealth and pastoral dairy production: A case study from Maasailand', *Human Ecology* 16(1): 1–21.

Grandin, B. and P. Lembuya (1987) 'The impact of 1984 drought at Olkarkar Group Ranch, Kajiado, Kenya', *ODI Pastoral Development Network Paper* 23e, Overseas Development Institute, London.

Greer, G. (1984) *Sex and Destiny: The Politics of Human Fertility*, Secker and Warburg, London.

Haldermann, J. M. (1985) 'Problems of pastoral development in East Africa', *Agricultural Administration* 18(4): 199–216.

Hampshire, K. and S. Randall (1999) 'Seasonal labour migration strategies in the Sahel: Coping with poverty or optimizing security?' *International Journal of Population Geography* 5(5): 367–85.

Hampshire, K. and S. Randall (2005) 'People are a resource: Demography and livelihoods in Sahelian Fulbe of Burkina Faso', In Homewood. K (ed.) *Rural Resources and Local Livelihoods in Africa*, James Currey, Oxford.

Hardin, G. (1968) 'The tragedy of the commons', *Science* 162(3859): 1234–48.

Hardt, M. and A. Negri (2000) *Empire*, Harvard University Press, Cambridge, MA.

Hartmann, B. (2002) 'Degradation narratives: Oversimplifying the link between population, poverty and the environment', IHDP Update 4, 6–8, IHDP/IGBP Bonn.

Haugerud, A. (1989) 'Land Tenure and Agrarian Change in Kenya', *Journal of the International African Institute* 59(1): 61–90.

Herren, U. (1990) 'The commercial sale of camel milk from pastoral herds in the Mogadishu hinterland of Somalia', *Pastoral Development Network* 30a, Overseas Development Institute, London.

Hilhorst, T. and F. Muchena (2000) *Nutrients on the Move: Soil Fertility Dynamics in African Farming Systems*, International Institute of Environment and Development, London.

Hodgson, D. (1999) 'Images and interventions: The problems of pastoralist development', In Anderson, D. M. and Broch-Due, V. (Eds) (1999) *The Poor Are Not Us: Poverty and Pastoralism in Eastern Africa*, James Currey, Oxford, East African Educational Publishing, Nairobi, and Ohio University Press, Athens.

Hodgson, D. L. (ed.) (2000) *Rethinking Pastoralism in Africa: Gender, Culture and the Myth of the Patriarchal Pastoralist*, Ohio University Press, Athens.

Hodgson, D. L. and R. A. Schroeder (1999) 'Mapping the Maasai: Dilemmas of counter-mapping community resources in Tanzania', paper presented at conference on *African Environments – Past and Present*, July 5–8, 1999, St. Anthony's College, University of Oxford.

Hogg, R. (ed.) (1997) *Pastoralists, Ethnicity and the State in Ethiopia*, HAAN/Institute for African Alternatives, London.

Holland, J. H. (1992) *Adaptation in Natural and Artificial Systems*, MIT Press, Cambridge, MA.

Holland, J. H. (1998) *Emergence: From Chaos to Order*, Oxford University Press, Oxford.

Holland, J. H. (2000) *Hidden Order: How Adaption Builds Complexity*, Addison-Wesley Longman, Boston.

Holling, C. S. (1973) 'Resilience and stability of ecological systems', *Annual Review of Ecology and Systematics* 4: 1–23.

Holling, C. S. (1998) 'Two cultures of ecology', *Conservation Ecology* 2(2): 4.

Holmes, J. and J. Mott (1993) 'Towards the diversified use of Australia's savannas', In Young, M. D. and O. Solbrig (Eds) *The World's Savannas: Economic Driving Forces, Ecological Constraints and Policy Options for Sustainable Land Use*, Vol. 12, Man and Biosphere series, UNESCO and Parthenon Publishing Group, Carnforth and New York.

Homewood, K. (1994) 'Pastoralists, environment and development in East African rangelands', In Zaba, B. and J. Clarke (Eds) *Environment and Population Change*, Ordina Editions, Belgium IUSSP.

Homewood, K. (2008) *Ecology of African Pastoralist Societies*, Ohio University Press, Athens.

Homewood, K. and A. Rodgers (1987) 'Pastoralism, conservation and the overgrazing controversy', In Anderson, D. and R. Grove (Eds) *Conservation in Africa: People, Policies and Practice*, Cambridge University Press, Cambridge.

Homewood, K. and D. Brockington. (1999) 'Biodiversity, conservation and development in Mkomazi, Tanzania', *Global Ecology and Biogeography* 8(3–4): 301–13.

Homewood, K., E. Coast and M. Thompson (2004) 'In-migration and exclusion in East African rangelands: Access, tenure and conflict', *Africa* 64(4): 567–610.

Homewood, K., E. F. Lambin, E. Coast, A. Kariuki, I. Kikula, J. Kivelia, M. Said, S. Serneels and M. Thompson (2001) 'Long-term changes in Serengeti-Mara wildebeest and land cover: Pastoralism, population or policies?' *Proceedings of the National Academy of Sciences* 98(22): 12544–49.

Hooker, R. (1996) *Yin and Yang*, Online, http://www.wsu.edu:8080/~dee/CHPHIL/YINYANG.HTM (15 May 2003)

Horowitz, M. (1979) 'The sociology of pastoralism and African livestock projects', *AID Programme Evaluation Discussion Paper* 6, United States Department for International Development (USAID), Washington, DC.

Hughes, D. M. (1999) 'Mapping the Mozambican hinterlands: Land rights, timber, and territorial politics in Chief Gogoi's area', Paper presented at conference on *African Environments – Past and Present*, July 5–8, 1999, St. Anthony's College, University of Oxford.

Hulme, D. and M. Murphree (Eds) (2001) *African Wildlife and Livelihoods: The Promise and Performance of Community Conservation*, James Currey, Oxford.

Igoe, J. and D. Brockington (1999) 'Pastoral land tenure and community conservation in East African rangelands: A case study from northeastern Tanzania', *Pastoral Land Tenure Series*, 11, International Institute of Environment and Development, London.

IIED (1994) *Whose Eden? An Overview of Community Approaches to Wildlife Management*, International Institute of Environment and Development, London.

Illius, A. and T. O'Connor (1999) 'On the relevance of nonequilibrium concepts to arid and semiarid grazing systems', *Ecological Applications* 9(3): 798–813.

Illius, A. W. and T. G. O'Connor (2000) 'Resource heterogeneity and ungulate population dynamics', *Oikos* 89(2): 283–94.

Ingersell, A. E. (n.d.) *A Critical User's Guide to Ecosystem and Related Concepts in Ecology*, Institute for Cultural Landscape Studies, Harvard University, Harvard.

Ingold, T. (1992) 'Culture and the perception of the environment', In Croll, E. and D. Parkin (Eds) *Bush Base: Forest Farm*, Routledge, London.

Ingold, T. (2000) *The Perception of the Environment: Essays in Livelihood, Dwelling and Skill*, Routledge, London.

Irigaray, L. (1996) 'The other: Woman', In Kemp, S. and J. Squires (Eds) *Feminisms*, Oxford University Press, Oxford.

Jacobs, J. M. (1996) *Edge of Empire: Postcolonialism and the City*, Routledge, London.

Jantsch, E. (1980) *The Self-Organising Universe: Scientific and Human Implications of the Emerging Paradigm of Evolution*, Pergamon Press, Oxford.

Joekes, S. and Pointing, J. (1991) 'Women in pastoral societies in East and West Africa', *Drylands Programme Issues Paper* 28, International Institute of Environment and Development, London.

Jowkar, F., M. H. Horowitz, C. Naslund and S. Horowitz (1991) *Gender Relations of Pastoralist and Agropastoralist Production: A Bibliography with Annotations*, Prepared for UNIFEM and UNDP, Institute of Development Anthropology, New York.

Kinahan, J. (1991) *Pastoral Nomads of the Central Namib Desert: The People History Forgot*, Namibian Archaeological Trust and New Namibian Books, Windhoek, Namibia.

Klink, C., A. Moreira and O. Solbrig (1993) 'Ecological impact of agricultural development in the Brazilian Cerrados', In Young, M. D. and O. Solbrig (Eds) *The World's Savannas: Economic Driving Forces, Ecological Constraints and Policy Options for Sustainable Land Use*, Vol.12, Man and Biosphere Series, UNESCO and Parthenon Publishing Group, Carnforth and New York.

Knight, K. (1991) *Blood Relations: Menstruation and the Origins of Culture*, Yale University Press, New Haven and London.

Konstant, T. L., Sullivan, S. and A. B. Cunningham (1995) 'The effects of utilization by people and livestock on *Hyphaene petersiana* basketry resources in the palm savanna of north-central Namibia', *Economic Botany* 49(4): 345–56.

Kristjansen, P., M. Radeny, D. Ndekianye, R. Kruska, R. Reid, H. Gichohi, F. Atieno and R. Sanford (2002) *Valuing Alternative Land Use Options in the Kitengela Wildlife Dispersal Area of Kenya*, International Livestock Research Institute (ILRI), Nairobi.

Kuhn, T. (1962) *The Structure of Scientific Revolutions*, 2nd edition, University of Chicago Press, Chicago.

Kumar, S. (2002) *You Are Therefore I Am*, Green Books, Totnes.

Kurimoto, E. and S. Simonse (Eds) (1998) *Conflict, Age and Power in North East Africa*, James Currey, Oxford and Ohio University Press, Athens.

Lambin, E., B. Turner, H. Geist, S. Agbola, A. Angelsen, J. Bruce, O. Coomes, R., G. Fischer, C. Folke, P. George, K. Homewood, J. Imbernon, R. Leemans, X. Li, E. Moran, M. Mortimore, P. Ramakrishnan, J. Richards, H. Skånes, W. Steffen, G. Stone, U. Svedin, T. Veldkamp, C. Vogel and J. Xu (2001) 'Our emerging understanding of the causes of land-use and -cover change', *Global Environmental Change* 11(4): 261–9.

Lamprey, H. F. (1983) 'Pastoralism yesterday and today: The overgrazing problem', In Bourliere, F. (ed.) *Tropical Savannas*, Ecosystems of the World Vol.13, Elsevier, Amsterdam.

Lancaster, W. and F. Lancaster (1986) 'The concept of territory among the Rwala Bedouin', *Nomadic Peoples* 20: 41–48.

Lane, C. and J. N. Pretty (1990) 'Displaced pastoralists and transferred wheat technology in Tanzania', *Sustainable Agriculture Programme Gatekeeper Series* 20, International Institute of Environment and Development, London.

Latour, B. (1987) *Science in Action*, Open University Press, Milton Keynes.

Latour, B. (1993) *We Have Never Been Modern*, translated by Catherine Porter, Harvester Wheatsheaf, Hemel Hempstead.

Leach, M. and J. Fairhead (2000) 'Fashioned forest pasts and the occlusion of history: Landscape, conservation and politics in the historiography of West Africa', *Development and Change* 31: 35–59.

Leach, M. and R. Mearns (Eds) (1996) *The Lie of the Land: Challenging Received Wisdom in African Environmental Change and Policy*, the International African Institute, London, James Currey, Oxford and Heinemann, Portsmouth.

Leader-Williams, N. and S. Albon (1988) 'Allocation of resources for conservation', *Nature* 336(6199): 533–35.

Levins, R. and R. Lewontin (1985) *The Dialectical Biologist*, Harvard University Press, Cambridge, MA.

Levi-Strauss, C. (1966) *The Savage Mind*, University of Chicago Press, Chicago.

Lewis, I. (2001) 'Why the warlords won: How the United States and the United Nations misunderstood the clan politics of Somalia', *Times Literary Supplement* 5123: 3–5.

Lind, J. (2003) 'Adaptation, conflict and cooperation in pastoralist East Africa: A case study from South Turkana, Kenya', *Conflict, Security and Development* 3(3): 315–34.

Little, P. (1985) 'Absentee herd owners and part-time pastoralists: The political economy of resource use in Northern Kenya', *Human Ecology* 13(2): 131–51.

Little, P. (1994) 'Maidens and milk markets: The sociology of dairy marketing in Southern Somalia', In Fratkin, E., Galvin, K. and E. A. Roth (Eds) *African Pastoral Systems: An Integrated Approach*, Lynne Rienner Publishers, Boulder.

LWAG (2002) *Wildlife and Poverty Study*, Livestock and Wildlife Advisory Group, Rural Livelihoods Department, Department for International Development, London.

Mace, R. (1991) 'Overgrazing overstated', *Nature* 349(6307): 280–81.

Mackenzie, J. M. (1987) 'Chivalry, social Darwinism and ritualised killing: The hunting ethos in Central Africa up to 1914', In Anderson, D. and R. Grove (Eds) *Conservation in Africa: People, Policies and Practice*, Cambridge University Press, Cambridge.

Maddox, T. (2002) 'The ecology of cheetahs and other large carnivores in a pastoralist-dominated buffer zone', unpublished PhD Thesis, University of London, London.

Markakis, J. (1966) *Resource Conflicts in the Horn of Africa*, Sage, London and International Peace Research Institute, Oslo.

Markakis, J. (ed.) (1993) *Conflict and the Decline of Pastoralism in the Horn of Africa*, Macmillan, Basingstoke.

Marindo-Ranganai, R. and B. Zaba (1994) 'Animal conservation and human survival: A case study of the Tembomvura people of Chapoto Ward in the Zambezi Valley, Zimbabwe', Harare University Research Paper, Harare.

May, R. M. (1977) 'Thresholds and breakpoints in ecosystems with a multiplicity of stable states', *Nature* 269(5628): 471–77.

Merchant, C. (1980) *The Death of Nature: Women, Ecology and the Scientific Revolution*, Wildwood House Ltd, London.

Morphy, H. (1993) 'Colonialism, history and the construction of place: The politics of landscape in Northern Australia', In Bender, B. (ed.) *Landscape: Politics and Perspectives*, Berg Publishers, Oxford.

Mortimore, M. (1998) *Roots in the African Dust: Sustaining the Drylands*, Cambridge University Press, Cambridge.

Mukerji, C. (1997) *Territorial Ambitions and the Gardens of Versailles*, Cambridge University Press, Cambridge.

Murdoch, J. and J. Clark (1994) 'Sustainable knowledge', *Centre for Rural Economy Working Paper Series* 9, Newcastle University, Newcastle-upon-Tyne.

Myers, N., R. Mittermeier, C. Mittermeier, G. da Fonseca and J. Kent (2000) 'Biodiversity hotspots for conservation priorities', *Nature* 403(6772): 853–58.

Nachtergaele, F. (2002) 'Land degradation assessment in drylands', *LUCC Newsletter* 8, International Geosphere–Biosphere Program and Human Dimensions Program on Global Environmental Change, Louvain, Belgium.

Nader, L. (ed.) (1996a) *Naked Science: Anthropological Enquiry into Boundaries, Power, and Knowledge*, Routledge, London.

Nader, L. (1996b) 'Preface' to *Naked Science: Anthropological Enquiry into Boundaries, Power, and Knowledge*, Routledge, London.

Ndagala, D. K. (1990) 'Pastoral territoriality and land degradation in Tanzania', In Palsson, G. (ed.) *From Water to World-Making: African Models and Arid Lands*, Scandinavian Institute of African Studies, Uppsala.

Ndagala, D. K. (1992) 'Production diversification and community development in African pastoral areas', In Ornas, A. H. (ed.) *Security in African Drylands. Research, Development and Policy*. Reprocentralen HSC, Uppsala.

Negri, A. (2002) 'Approximations: Towards an ontological definition of the multitude', translated by Arianna Bove, *Multitudes* 9: 36–48.

Neumann, R. P. (1996) 'Dukes, earls and ersatz Edens: Aristocratic nature preservationists in colonial Africa', *Environment and Planning D: Society and Space* 14(1): 79–98.

Niamir-Fuller, M. (ed.) (1999a) *Managing Mobility in African Rangelands: The Legitimization of Transhumance*, IT Publications, London and FAO, Rome.

Niamir-Fuller, M. (1999b) 'Towards a synthesis of guidelines for legitimizing transhumance', In Niamir-Fuller, M. (ed.) *Managing Mobility in African Rangelands: The Legitimization of Transhumance*, IT Publications, London and FAO, Rome.

Nicholson, S., C. Tucker and M. B. Ba (1998) 'Desertification, drought and surface vegetation an example from the west African Sahel', *Bulletin of the American Meteorological Society* 79(5): 815–29.

Norton-Griffiths, M. (1998) 'The economics of wildlife conservation policy in Kenya', In Miner Gulland, E. J. and R. Mace (Eds) *Conservation of Biological Resources*, Blackwell Science, Oxford.

Noy-Meir, I. (1973) 'Desert ecosystems: Environment and producers', *Annual Review of Ecology and Systematics* 4: 25–51.

Noy-Meir, I. (1975) 'Stability of grazing systems: An application of predator-prey graphs', *Journal of Ecology* 63(2): 459–81.

Oba, G. (2000) '"Where the bulls fight, it is the grass that suffers": Impact of border administration on drought-coping strategies of the Obbu Booran during the 20th century', *Journal of Oromo Studies* 7 (1&2): 87–108.

Oba, G., N. C. Stenseth and W. J. Lusigi (2000) 'New perspectives on sustainable grazing management in arid zones of sub-Saharan Africa', *BioScience* 50(1): 35–51.

Oba, G., R. B. Weladji, W. Lusigi and N. C. Stenseth (2003) 'Scaling effects of grazing on rangeland degradation as a test of equilibrium and nonequilibrium hypotheses: A case study from northern Kenya', Paper presented at International Rangelands Congress on *Rangelands at Equilibrium and Non-equilibrium*, 26–27 July 2003, Durban.

Orestes, N., K. Shrader-Frechette and K. Belitz (1994) 'Verification, validation, and confirmation of numerical models in the earth sciences', *Science* 263(5147): 641–46.

Osbahr, H. (2001) *Livelihood Strategies and Soil Fertility in Southwestern Niger*, PhD Thesis, Geography, University College London.

Pagels, E. (1979) *The Gnostic Gospels*, Vintage, New York.

Pantuliano, S. (2002) 'Sustaining livelihoods across the rural–urban divide. Changes and challenges facing the Beja pastoralists of north eastern Sudan', *Pastoral Land Tenure Series* No.14, International Institute of Environment and Development, London.

Parmesan, C. and G. Yohe (2003) 'A globally coherent fingerprint of climate change impacts across natural systems', *Nature* 421(6918): 37–42.

Patel, H. (1998) Sustainable Utilization and African Wildlife Policy: The Case of Zimbabwe's Communal Areas Management Programme for Indigenous Resources (CAMPFIRE), Indigenous Environmental Policy Centre, Cambridge, Massachusetts.

Peluso, N. L. (1995) 'Whose woods are these? Counter-mapping forest territories in Kalimantan, Indonesia', *Antipode* 27(4): 383–406.

Platteau, J.-P. (2000) 'Does Africa need land reform?', In Toulmin, C. and J. Quan (Eds) *Evolving Land Rights, Policy and Tenure in Africa*, Department for International Development, International Institute of Environment and Development and Natural Resources Institute, London.

Pockley, P. (2002) 'Bushfires leave ecologists hot under the collar', *Nature* 415(6868): 105.

Potkanski, T. (1994) 'Property concepts, herding patterns and management of natural resources among the Ngorongoro and Salei Maasai of Tanzania', *IIED Drylands Programme Pastoral Land Tenure Series* No.6, International Institute of Environment and Development, London.

Prior, J. (1994) 'Pastoral development planning', *OXFAM Development Guidelines* No.9, OXFAM, Oxford.

Rae, J. (1999) *Tribe and State: Management of the Syrian Steppe*, PhD thesis, University of Oxford, Oxford.

Ramisch, J. (1999) 'In the balance? Evaluating soil nutrient budgets for an agro-pastoral village of southern Mali', *Managing Africa's Soils* 9, IIED Drylands Programme, International Institute of Environment and Development, London.

Rangan, H. (2000) *Of Myths and Movements: Rewriting Chipko into Himalayan History*, Verso, London.

Richards, P. (1985) *Indigenous Agricultural Revolution: Ecology and Food Production in West Africa*, Unwin Hyman, London.

Richards, P. W. (1996) *The Tropical Rain Forest* (2nd edition), Cambridge University Press, Cambridge.

Ritchie, C. (1987) *The Political Economy of Resource Tenure: San Survival in Namibia and Botswana*, MA thesis, Boston University, Boston.

Roe, D., J. Mayers, M. Grieg-Gran, A. Kothari, C. Fabricius and R. Hughes (2000) 'Evaluating Eden: Exploring the myths and realities of community based wildlife management', *Evaluating Eden Series* 8, International Institute of Environment and Development, London.

Roe, E., L. Huntsinger and K. Labnow (1998) 'High reliability pastoralism', *Journal of Arid Environments* 39(1): 39–55.

Rohde, R. F. (1994) 'Tinkering with chaos: Towards a communal land tenure policy in former Damaraland', *SSD Discussion Paper* 8, Social Sciences Division, MRC, University of Namibia, Windhoek.

Rohde, R. F. (1997a) *Nature, Cattle Thieves and Various Other Midnight Robbers: Images of People, Place and Landscape in Damaraland, Namibia*, PhD thesis, University of Edinburgh, Edinburgh.

Rohde, R. F. (1997b) 'Looking into the past: Interpretations of vegetation change in Western Namibia based on matched photography', *Dinteria* 25: 121–49.

Rohde, R. F., T. A. Benjaminsen and M. T. Hoffman (2001) 'Land reform in Namaqualand: Poverty alleviation, stepping stones and "economic units"', *PLAAS Land Reform and Agrarian Change in Southern Africa Occasional Paper* 16, University of the Western Cape, Cape Town.

Rosenberg, A. (1995) *Philosophy of Social Science* (2nd ed.), Westview Press, Boulder.

Salih, M. (1985) 'Pastoralists in town: Some recent trends in pastoralism in the north west of Omdurman District', *ODI Pastoral Development Network* 20b, Overseas Development Institute, London.

Sandford, S. (1983) *Management of Pastoral Development in the Third World*, Overseas Development Institute, London, with John Wiley and Sons, Chichester and New York.

Saner, M. A. (1999) 'Two cultures: Not unique to ecology', *Conservation Ecology* 3(1): r2.

Schreckenberg, K. (1996) *Forests, Fields and Markets: A Study of Indigenous Tree Products in the Woody Savannas of the Bassila Region, Benin*, PhD Thesis, University of London, London.

Scoones, I. (1991) 'Wetlands in drylands', *Ambio* 20(8): 366–71.

Scoones, I. (1992) 'Land degradation and livestock production in Zimbabwe's Communal Areas', *Land Degradation and Rehabilitation* 3(2): 99–114.

Scoones, I. (1993) 'Why are there so many animals? Cattle population dynamics in the Communal Areas of Zimbabwe', In Behnke, R. H. Jr, Scoones, I. and C. Kerven (Eds) *Range Ecology at Disequilibrium: New Models of Natural Variability and Pastoral Adaptation in African Savannas*, Overseas Development Institute, International Institute for Environment and Development, and Commonwealth Secretariat, London.

Scoones, I. (1995) 'Exploiting heterogeneity: Habitat use by cattle in dryland Zimbabwe', *Journal of Arid Environments* 29(2): 221–37.

Scoones, I. (1997) 'The dynamics of soil fertility change: Historical perspectives on environmental transformation from Zimbabwe', *The Geographical Journal* 163(2): 161–69.

Scoones, I. (1999) 'New ecology and the social sciences: What prospects for a fruitful engagement?' *Annual Review of Anthropology* 28: 479–507.

Scott, J. (1998) *Seeing Like a State: How Certain Schemes to Improve the Human Condition Have Failed*, Yale University Press, New Haven.

Seddon, G. (1997) *Landprints: Reflections on Place and Landscape*, Cambridge University Press, Cambridge.

Shackleton, C. (2000) 'Comparison of plant diversity in protected and communal lands in the Bushbuckridge lowveld savanna, South Africa', *Biological Conservation* 94(3): 273–85.

Silva, J. and A. Moreno (1993) 'Land use in Venezuela', In Young, M. D. and O. Solbrig (Eds) *The World's Savannas: Economic Driving Forces, Ecological Constraints and Policy Options for Sustainable Land Use*, Vol.12, Man and Biosphere Series, UNESCO and Parthenon Publishing Group, Carnforth and New York.

Simpson, J. and P. Evangelou (1984) *Livestock Development in Sub Saharan Africa: Constraints, Prospects, Policy*, Westview Press, Boulder.

Sinclair, A. R. E. and J. M. Fryxell (1985) 'The Sahel of Africa: Ecology of a disaster', *Canadian Journal of Zoology* 63(5): 987–94.

Skotnes, P. (ed.) (1996) *Miscast: Negotiating the Presence of the Bushmen*, University of Cape Town Press, Cape Town.

Smith, M. (2001) 'Repetition and difference: Lefebvre, Le Corbusier and modernity's (im)moral landscape', *Ethics, Place and Environment* 4(1): 31–44.

Sobania, N. (1991) 'Feasts, famine and friends: Nineteenth century exchange and ethnicity in the eastern Lake Turkana region', In Galaty, J. and P. Bonte (Eds) *Herders, Warriors and Traders: Pastoralism in Africa*, Westview Press, Boulder.

Soulé, M. E. and M. A. Sanjayan (1998) 'Conservation targets: Do they help?', *Science* 279(5359): 2060-1.

Stattersfield, A., M. Crosby, A. Long and D. Wedge (1998) 'Endemic bird areas of the world. Priorities for biodiversity conservation', *Birdlife Conservation Series* No.7, Birdlife International, Cambridge.

Stott, P. (1997) 'Dynamic tropical forestry in an unstable world', *Commonwealth Forestry Review*, 76(3): 207–09.

Stott, P. (1998) 'Biogeography and ecology in crisis: The urgent need for a new metalanguage', *Journal of Biogeography*, 25: 1–2.

Sullivan, S. (1996a) 'The "communalization" of former commercial farmland: Perspectives from Damaraland and implications for land reform', *Research Report 25*, Social Sciences Division of the Multidisciplinary Research Centre, University of Namibia, Windhoek.

Sullivan, S. (1996b) *People and Plants on Communal Land in Namibia: The Relevance of Indigenous Range and Forest Management Practices, and Land Tenure Systems, to In Situ Plant Genetic Resources Conservation in the Arid and Semi-arid Regions of Namibia*, International Plant Genetic Resources Institute (IPGRI), Nairobi.

Sullivan, S. (1996c) 'Towards a non-equilibrium ecology: Perspectives from an arid land', *Journal of Biogeography* 23: 1–5.

Sullivan, S. (1999a) 'Folk and formal, local and national: Damara cultural knowledge and community-based conservation in southern Kunene, Namibia', *Cimbebasia* 15: 1–28.

Sullivan, S. (1999b) 'The impacts of people and livestock on topographically diverse open wood- and shrub-lands in arid north-west Namibia', *Global Ecology and Biogeography* (Special Issue on Degradation of Open Woodlands) 8(3–4): 257–77.

Sullivan, S. (2000a) 'Gender, ethnographic myths and community-based conservation in a former Namibian "Homeland"', In Hodgson, D. (ed.) *Rethinking Pastoralism in Africa: Gender, Culture and the Myth of the Patriarchal Pastoralist*, James Currey, Oxford.

Sullivan, S. (2000b) 'Getting the science right, or introducing science in the first place? Local 'facts', global discourse – 'desertification' in north-west Namibia', in Stott, P. and S. Sullivan (Eds) *Political Ecology: Science, Myth and Power*, Edward Arnold, London.

Sullivan, S. (2002) 'How sustainable is the communalising discourse of "new" conservation? The masking of difference, inequality and aspiration in the fledgling "conservancies" of north-west Namibia', In Chatty, D. and M. Colchester (Eds) *Conservation and Mobile Indigenous People: Displacement, Forced Settlement and Sustainable Development*, Berghahn Press, Oxford.

Sullivan, S. (2003) 'Protest, conflict and litigation: Dissent or libel in resistance to a conservancy in North-west Namibia', In Berglund, E. and D. Anderson (Eds) *Ethnographies of Conservation: Environmentalism and the Distribution of Privilege*, Berghahn Press, Oxford.

Sullivan, S. (2005) 'Detail and dogma, data and discourse: Food-gathering by Damara herders and conservation in arid north-west Namibia', In Homewood, K. (ed.) *Rural Resources and Local Livelihoods in sub-Saharan Africa*, James Currey, Oxford.

Sullivan, S. and K. Homewood (2003) 'On non-equilibrium and nomadism: Knowledge, diversity and global modernity in drylands (and beyond …)', *CSGR Working Paper* 122/03, Centre for the Study of Globalisation and Regionalisation (CSGR), University of Warwick, Coventry.

Sullivan, S. and K. Homewood (2004) 'Natural resources: Use, access, tenure and management', In Potts, D. and T. Bowyer-Bower (Eds) *Eastern and Southern Africa*, Institute of British Geographers' Developing Areas Research Group, Addison Wesley Longman, London.

Sullivan, S. and R. Rohde (2002) 'On non-equilibrium in arid and semi-arid grazing systems: A critical comment on A. Illius and T.G. O'Connor (1999) On the relevance of nonequilibrium concepts to arid and semiarid grazing systems, *Ecological Applications*, 9, 798–813', *Journal of Biogeography* 29: 1595–1618.

Sullivan, S., T. Konstant and A. B. Cunningham (1995) 'The impact of utilization of palm products on the population structure of the Vegetable Ivory Palm (*Hyphaene petersiana*, Arecaceae) in north-central Namibia', *Economic Botany* 49(4): 357–70.

Suzman, J. (2000) '"Things from the bush": A contemporary history of the Omaheke bushmen', (with an introduction by Robert J. Gordon), *Basel Namibia Studies Series* 5, P. Schlettwein Publishing, Basel.

Talle, A. (1987) 'Women as heads of houses: The organisation of production and the role of women among the pastoral Maasai in Kenya', *Ethnos* 52(1–2): 50–80.

Talle, A. (1988) *Women at a Loss: Changes in Maasai Pastoralism and their Effects on Gender Relations*, Studies in Social Anthropology, University of Stockholm, Stockholm.

Talle, A. (1990) 'Ways of milk and meat among the Maasai: Gender identity and food resources in a pastoral economy', In Palsson, G. (ed.) *From Water to World-making: African Models and Arid lands*, Scandinavian Institute of African Studies, Uppsala.

Thompson, M. and K. Homewood (2002) 'Elites, entrepreneurs and exclusion in Maasailand', *Human Ecology* 30(1): 107–38.

Tilley, C. (1994) *A Phenomenology of Landscape: Places, Paths and Monuments*, Berg, Oxford.

Timberlake, L. (1988) *Africa in Crisis: The Causes, the Cures of Environmental Bankruptcy*, Earthscan, London.

Tolkien, J. R. (1954) *The Fellowship of the Ring* (the first part of *The Lord of the Rings*), George Allen and Unwin, London.

Toulmin, C. and J. Quan (Eds) (2000) *Evolving Land Rights, Policy and Tenure in Africa*, Department for International Development, International Institute of Environment and Development and Natural Resources Institute, London.

Trafzer, C. E. (2000) *As Long as the Grass Shall Grow and Rivers Flow: A History of Native Americans*, Harcourt College Publishers, London.

Turner, M. (1998a) 'Long-term effects of daily grazing orbits on nutrient availability in Sahelian West Africa I', *Journal of Biogeography* 25: 669–82.

Turner, M. (1998b) 'Long-term effects of daily grazing orbits on nutrient availability in Sahelian West Africa II', *Journal of Biogeography* 25: 683–94.

Turner, M. (1999) 'Spatial and temporal scaling of grazing impact on the species composition and productivity of Sahelian annual grasslands', *Journal of Arid Environments* 41: 277–97.

Virilio, P. (1975) *L'insecurite du Territoire*, Stock, Paris.

Waldrop, M. M. (1993) *Complexity: The Emerging Science at the Edge of Order and Chaos*, Viking, London.

Waller, R. (1985) 'Ecology, migration and expansion in east Africa', *African Affairs* 84(336): 347–70.

Ward, D., B. T. Ngairorue, J. Kathena, R. Samuels and Y. Ofran (1998) 'Land degradation is not a necessary outcome of communal pastoralism in arid Namibia', *Journal of Arid Environments* 40(4): 357–71.

Waters Bayer, A. (1985) 'Dairying by settled Fulani women in Central Nigeria and some implications for dairy development', *ODI Pastoral Development Network* 20c, Overseas Development Institute, London.

Western, D. and H. Gichohi (1993) 'Segregation effects and the impoverishment of savanna parks: The case for ecosystem viability analysis', *African Journal of Ecology* 31(4): 268–71.

Western, D. and J. Ssemakula (1981) 'The future of savanna ecosystems', *African Journal of Ecology* 19(1–2): 7–19.

Wiens, J. (1984) 'On understanding a non-equilibrium world: Myth and reality in community patterns and processes', In Strong, D. R., Simberloff, D., Abele, L. G. and A. B. Thistle (Eds) *Ecological Communities: Conceptual Issues and the Evidence*, Princeton University Press, Princeton.

Wøien, H. and L. Lama (1999) 'Market commerce as wildlife protector? Commercial initiatives in community conservation in Tanzania's northern rangelands', IIED Pastoral Land Tenure Series No.12, International Institute of Environment and Development, London.

Young, M. D. and O. Solbrig (Eds) (1993) *The World's Savannas: Economic Driving Forces, Ecological Constraints and Policy Options for Sustainable Land Use*, Vol.12, Man and Biosphere Series, UNESCO and Parthenon Publishing Group, Carnforth and New York.

Zaal, F. and T. Dietz (1999) 'Of markets, meat, maize and milk: Pastoral commoditization in Kenya', In Anderson, D. M. and V. Broch-Due (Eds) *The Poor Are Not Us: Poverty and Pastoralism in Eastern Africa*, James Currey, Oxford, East African Educational Publishing, Nairobi, and Ohio University Press, Athens.

5

SUB-SAHARAN AFRICA, KENYA AND THE MALTHUSIAN PARADIGM IN CONTEMPORARY DEVELOPMENT THINKING

Eric B. Ross

Introduction

In the two centuries since the first edition of Thomas Malthus's *Essay on the Principle of Population* in 1798, it has become a central argument of dominant Malthusian discourse that poverty, underdevelopment and associated patterns of mortality and environmental degradation could all be regarded chiefly as the products of human population pressure on the means of subsistence. This reflects the central argument of Malthus's work, which was conceived initially as a reaction against a radical belief in human progress associated with the French Revolution (Rothschild, 1996). Over the next half century, however, as the industrial revolution gave rise to new social and economic conflicts, the Malthusian perspective rapidly evolved into a general defence of capitalist economy (Ross, 1998b). As such, it appealed to ruling classes because it insisted that efforts to ameliorate the living conditions of the poor would only tend to make matters worse by encouraging them to have more children. In this way, by dismissing as illusory any alternative to the inequalities produced by capitalist relations of production, the ideas of Malthus and his adherents became an essential ideological weapon against systemic change.

More subtly, such ideas expressed themselves in the patterns of development thinking that emerged after the Second World War and have continued, in various forms, ever since (Ross, 1998b, 1998c; Hartmann, 1999; Daoud, 2010; Diamond, 2005). In particular, they have helped to legitimize an explanation of the origins of poverty and underdevelopment that only allows 'solutions' that are compatible with the advance of capitalist interests. These have changed little since the sixteenth century and are typified by the priority of private accumulation over the sustainable livelihoods of the poor (Box 5.1). The result, therefore, has been a continuous process of clearance or enclosure, of the displacement of subsistence producers, and of environmental degradation, usually as a result of the expansion of commercial agriculture (cf. Dobb, 1947; *The Ecologist*, 1992; Blum, 1978; Lis and Soly, 1979; Holt-Giménez and Altieri, 2012). But, this process was never more intense than during the years after the Second World War when it was associated with – but not exclusively confined to – what became known as the Green Revolution.

BOX 5.1 IRELAND: THE POTATO, POPULATION AND THE POLITICAL ECONOMY OF FAMINE

The Great Famine of 1846–49 in Ireland, which resulted in the death or emigration of some two million people, was widely regarded – and is still described – as a classic illustration of Malthus's principles. That is, rather than attributing it to a lack of access to land or to foods other than the potato as a result of centuries of English colonial rule – what Sen would regard as an 'entitlement failure' – advocates of Malthusian logic blamed the plight of the Irish peasantry on its own laziness and over-dependence on the potato, their 'favourite root' (Niven, 1846), while population growth had outstripped the means of production. It was, in this popular view, 'The fatal luxuriance with which this vegetable flourished in the soil of Ireland [that] caused population to run fearfully ahead of the requirements and capabilities of the country' (Anon, 1847).

In fact, Ireland was a land of extraordinary agricultural potential (Owen, 1823; Cobbett, 1984) and, at the time of the famine, it was producing and exporting agricultural commodities on a prodigious scale (Ross, 1998b). By then, however, Irish tenants had been reduced to reliance on the potato, which had been introduced from South America and whose cultivation had been encouraged by landlords because it grew so well on poor soils and allowed them to take over better land for commercial crops or pasture (Ross, 1998b).

Malthusians interpreted the famine as a manifestation of 'Supreme Wisdom' that would bring about 'the commencement of a salutary revolution in the habits of a nation long singularly unfortunate' (Trevelyan, 1848). But, in fact, the tragic years of the famine provided the English with an unprecedented opportunity to accelerate the process of land clearance which had been going on since the end of the Napoleonic Wars had ended a boom market in grain and spurred landlords to raise more cattle. Such clearances forced many tenants onto the roads, where visitors to Ireland saw what seemed dramatic evidence of its excess population (Ross, 1998b).

Before, during and after the famine, the poor of rural Ireland were encouraged to emigrate, not because Ireland was resource poor, but because its valuable resources were so important as a subsidy for the industrialization of the English economy. As early as the first third of the nineteenth century, Ireland set the pattern which characterizes so much of the Third World today, when capitalist agriculture, with its declining demand for human labour, forces millions of rural people to seek a living in dire conditions abroad.

(Anon, 1847; Cobbett, 1984 (orig. 1834); Niven, 1846; Owen, 1823; Ross, 1998a; Trevelyan, 1848)

The Green Revolution, which was publicly rationalized in Malthusian terms and widely described as a humanitarian response to impending famine in the Third World, was, in fact, predicated on the rejection of the yield-raising potential of land redistribution and of indigenous innovation. It expressed instead a commitment to an entrepreneurial mode of production that was oriented to the world market rather than to local subsistence needs and that viewed farms as business units rather than as a socially dynamic reality. Such development has adversely affected peasant communities throughout the world over the past half century. But, while it has been argued that this was merely an unfortunate and unanticipated consequence, I have suggested elsewhere (Ross, 1998b, 1998c, 2003) that one of the ultimate aims of creating such an agricultural regime in developing countries was actually to minimize the role of peasants as an economic and political force (cf. Cleaver, 1972; Feder, 1971) and, in the process, for local food production in developing countries actually to be *reduced*, as Third World agriculture was transformed into an export-producing sector and the United States profited as a supplier of agricultural inputs[1] and as the principal source of food grains for the Third World (cf. DiGiacomo, 1996).

In the following sections of the chapter, I will discuss how modernization theory, relying on certain Malthusian precepts, advanced the interests of Western capitalism during and after the Cold War; and how it rationalized the advance of commercial agriculture and therefore helped to sustain policies that exacerbated, rather than ameliorated, many of the structural inequalities that were inherited from the colonial past. In the end, it has been the advance of commercial agriculture – promoted by international development institutions and multinational capital – that has systematically marginalized the potential of smallholder agriculture as a source of local innovation, an agent of social and ecological diversity and a prerequisite for national food security.

Modernization in theory and practice

The rationale for the expansion of capital-intensive, market-oriented agriculture lay in the West's perception of what became known as 'modernization', a process that was described, on the one hand, as a desired goal for the developing world, yet, at the same time, filled the West with foreboding, as, in Robert McNamara's words, a 'sweeping surge of development ... turned traditionally listless areas of the world into seething caldrons of change' (cited in Shafer, 1988). Such change had long been associated with the advance of anti-capitalist struggles. In 1857, for example, the arch-conservative British historian, Thomas Macaulay, had written of how the combination of economic distress and population density 'made the labourer mutinous and discontented, and inclined him to listen with eagerness to agitators who tell him that it is a monstrous iniquity that one man should have a million while another cannot get a full meal' (in Schuster, 1940). A century later, against the backdrop of national liberation struggles that threatened the interests of Western capital, such ideas became increasingly common. Drawing on the Malthusian tradition, it became commonplace among Western policy-makers and development strategists to emphasize how the adverse (or unacceptable) effects of the modernization process seemed to have their origin primarily in population pressures, rather than in a real need to remedy economic or social injustice.

It was after Communist forces ousted the Kuomintang from mainland China in 1949 that Western policy makers – including the Ford and Rockefeller Foundations, which were becoming closely intertwined with the highest echelons of the US government (Ross, 1998b, 1998c) – turned their attention to the peasant wars that seemed to threaten their influence in much of Asia and most notably in India, which seemed particularly vulnerable to communist influence (Ross, 1998a). But, before that happened, on the very eve of the communist victory in China, the Rockefeller Foundation had sent Frank Notestein, one of the architects of modern demography (Ross, 1998b) to the Far East. He returned with the conclusion that what Asia needed – and, by inference, the rest of the Third World – was not land reform and social justice, but a cheap and effective method of contraception. 'We doubt', he wrote, 'that any other work offers a better opportunity for contributing to Asia's and the world's fundamental problems of human welfare' (cited in Caldwell and Caldwell, 1986).[2]

Notestein, who became the president of the Rockefeller- and Ford Foundation-funded Population Council in 1961 (Ross, 1998b), would soon add his influential voice to the advocates of commercial agriculture. In a paper to the Eighth International Conference of Agricultural Economists in 1952 he observed that, if modernization typically brought disorder, increased food production might be one way to ameliorate it. But, echoing the views that the Rockefeller Foundation had developed over many decades of support for agricultural missionary work in pre-revolutionary China (Stross 1986; Ross, 1998b), he argued that this should be achieved through moderate change (Notestein, 1953). As with fertility, this meant relying primarily on technological innovation.

But, it had further implications: if the prevailing view was that 'The social organization of a peasant society is ill-adapted to the achievement of high technological proficiency' (Notestein, 1953); and if, as the proponents of 'modernization theory' maintained,[3] peasants were too backward and unimaginative to be significant agents of increased food production in the context of a modern market economy, then this was an argument for relying on – and even enhancing – the role of large landowners and commercial farmers who reflected the 'rationalist and positivist spirit' (Lerner, 1958) that was conventionally regarded as having been the catalyst for Western capitalism.

In this way, the expansion of commercial agriculture was promoted as a necessary and benign response to presumed population pressures in the Third World. This dominant view of the Green Revolution was enshrined by the award of the Nobel peace prize in 1970 to the Rockefeller Foundation geneticist, Norman Borlaug. A quarter of a century later, despite growing attention to the adverse consequences of the Green Revolution (Pearse, 1980; Bayliss-Smith and Wanmali, 1984; Feder, 1983), prominent Neo-Malthusians such as Paul Ehrlich still described Borlaug as 'a founder of the green revolution' (Ehrlich *et al.*, 1993) as if the latter had never been more than a package of politically neutral technologies. Viewed in this way, the Green Revolution has been represented as the major factor preventing the famines regularly predicted for the Third World since the 1960s (cf. Paddock and Paddock, 1967). But it also helped to ensure that policy debate over food security shifted from one about agrarian relations of production – in particular, access to cultivable land – to one about technical innovation, a focus that not only remains deeply influential as a conceptual tool of policy makers and of the agribusiness sector, but which continues to marginalize peasants as a source of creative agricultural change.

This was a tendency that gained special currency during the decade after the Chinese Revolution when peasants came to be viewed as a perilous factor in the development process. The renowned agricultural economist, Wolf Ladejinsky, summed up the implications when he observed:

> We must realize how serious a threat an agrarian revolution could be at this point of history ... The only way to thwart Communist designs on Asia is to preclude such revolutionary outbursts through timely reforms, peacefully before the peasants take the law into their own hands and set the countryside ablaze.
>
> *(Walinsky, 1977)*

Few writers commented on the paradox that peasants were being described as too conservative to be the agents of agricultural innovation at the same time that their political aspirations were deemed too radical. But, logical consistency was not the point. It was the ideological parameters of the Cold War that defined the West's response to global poverty and nutritional deprivation in narrow technical terms that would take 'the wind out of the Communist sails in a peasant ocean' (Walinsky, 1977) and spread the Green Revolution, in its various forms. This would have far-reaching consequences for billions of people in developing countries who, in the face of the relentless commercialization of lands once devoted to subsistence cultivation, would be steadily deprived of secure and sustainable livelihoods. Lacking the capital resources on which such agriculture depended, and forced to adapt to conditions in which local labour demand was often dramatically diminished, they have been forced into a problematic and migrant existence which has led to the last decades of the twentieth century being characterized as an 'age of mass migration' (Castles and Miller, 1993; see Box 5.2).

BOX 5.2 THE GREEN REVOLUTION AND THE FATE OF THE MEXICAN PEASANTRY

After the consolidation of the Mexican Revolution, the land aspirations of the peasantry languished until the 1930s, when President Lazaro Cardenas initiated a programme of land reform, which involved the breaking up of large estates into communal units called *ejidos* (Perelman, 1977). This programme has been described as 'the most far reaching in Latin America before Cuba and one which really did incorporate the peasants into national life' (Frank, 1969).

By 1940, just under half of all cultivable land had been redistributed to *ejidos* and state support for traditional agriculture had been sufficiently developed that productivity on *ejidos* surpassed that of private holdings, accounting for just over half of the value of all Mexican farm production (Hewitt de Alcantara, 1976; DeWalt and Barkin, 1991; Marchesi, 2016). As a result, not only did the number of landless labourers in Mexico fall from 68% to 36% of the rural workforce (Hewitt de Alcantara, 1976), but migration out of the countryside was at its lowest rate in half a century (Unikel, 1975).

By then, however, the United States was exerting tremendous pressure on Mexico to reverse its progressive policies. Even before the end of his term, the Cardenas administration had begun to step back from its commitment to the peasantry; land distribution declined and what land continued to be distributed was of poor quality. Then, under Cardenas's conservative successor, Avila Camacho, Mexican national development policy returned to a course based on 'private initiative', with the agricultural sector subsidising industrial growth (Hewitt de Alcantara, 1976). It was in this new context that the Rockefeller Foundation established an agricultural research project in Mexico that was the effective cradle of what became known as the Green Revolution. It focused on developing 'the very highest yielding genetic material, capable of great productivity under optimum conditions available only in the richest agricultural areas of the country' (Hewitt de Alcantara, 1976). The key element was seeds that depended on a general package of external inputs and were ideally suited for well-endowed market-oriented farmers.

Over the next 40 years, *ejido* agriculture either stagnated or collapsed (Barkin, 1990). By 1970, at least one half of the agricultural labour force consisted of landless workers who depended on migration and seasonal work for what was increasingly a tenuous existence, not least because the increasing mechanization of large farms meant that the demand for labour was on the wane (Stavenhagen, 1970). Widespread and growing rural unemployment therefore produced migration to Mexican cities and the United States on a scale that was 'without precedent in the demographic development of the country' (Unikel, 1975). Thanks to the Green Revolution, rural Mexico was largely reduced to the role of a labour reserve for capitalist interests within and outside the country (Gledhill, 1995; Ross, 1998b).

(Barkin, 1990; DeWalt and Barkin, 1991; Frank, 1969; Gledhill, 1995; Hewitt de Alcantara, 1976; Marchesi, 2016; Perelman, 1977; Ross, 1998a; Stavenhagen, 1970; Unikel, 1975)

Colonialism and the Malthusian myth in sub-Saharan Africa

A widespread distrust of Third World peasants has its roots in the very processes that gave birth to them as a product of Western colonialism. As European colonialists rationalized their activities in terms of their cultural, racial and economic superiority, they defined the peoples they colonized as 'non-adult races' (Lugard, 1922) who needed Europeans to bring 'to the dark places of the earth, the abode of barbarism and cruelty, the torch of culture and progress' (Lugard, 1922). To such colonialists, Asians and Africans, Latin Americans and others seemingly had much to gain and relatively little to lose when Europeans took over their vast resources, which, in the commonly held view, they had neglected and which the Europeans, with their burgeoning populations and rising standards of living, urgently required. As Lord Frederick John Lugard – who brought Uganda under British authority in 1890 and was the first British governor of Nigeria – wrote, from the vantage point of his experience in Africa:

> Who can deny the right of the hungry people of Europe to utilise the wasted bounties of nature, or that the task of developing these resources was … a 'trust for civilisation' and for the benefit of mankind? Europe benefited by the wonderful increase in the amenities of life for the mass of her people which followed the opening up of Africa at the end of the nineteenth century. Africa benefited by the influx of manufactured goods, and the substitution of law and order for the methods of barbarism.
>
> *(Lugard, 1922)*

It was a view that found great favour among Europeans who wanted, on the one hand, to account for underdevelopment in terms of indigenous cultural habits and, on the other, to explain European development in terms of some specific genius. It implied, moreover, that developing countries – whose 'backwardness' was regarded as a function of primitive and irrational traditions – could only advance as a result of the moral and political guidance of the advanced capitalist countries. This became the generic and popular rationale for colonialism and, in time, in a new discourse, for the post-colonial order.

As a result of such thinking, while the populations and resources of sub-Saharan Africa were systematically transformed in the interests of Western profits, European and American writers were busily attributing Africa's ills – pre-colonial, colonial and post-colonial – to local cultural patterns that were loosely described as 'traditional'. Foremost among these were sexual practices that readily lent themselves to Malthusian interpretations of everything from rural-urban migration to the contemporary problem of AIDS. We will return to these issues shortly.

Africa, for the most part, was late to be colonized in a systematic way. By the time that European capitalism began to penetrate the continent, which took place largely after 1870 (Hobson, 1902; Barratt Brown, 1974), most of Latin America had already been liberated from Spain and Portugal for more than half a century (Lambert, 1967). But, when European enterprise began to take a deliberate interest in the interior of Africa in the late nineteenth century, it methodically crafted the region to Western needs. All colonial economies, whether in the Dutch East Indies or Spanish America, were labour-demanding (cf. Alexander, 1986), but the various European powers transformed African societies into what Amin has aptly called 'Africa of the labour reserves' (1976; cf. Harris, 1959), creating a complex constellation of labour-generating enclaves, particularly in Southern and East Africa, which remained locked into this role long after independence and which goes far towards explaining why the subsequent demographic and economic history of that vast region is so different from that of other continents.

Where the Europeans created systems of direct or indirect forced labour in Africa, local communities were compelled to enter a new market economy. The forms of male labour migration that emerged had a tremendous effect on subsistence agriculture, especially in regard to the adaptations that women were compelled to make. A general process of what Whitehead calls the 'feminization of food production' (Whitehead, 1990) took place – still widespread in sub-Saharan Africa – as a result of the complex survival strategies that colonial regimes demanded of African households. It was not a uniform process, nor, above all, did it mean that women were not also being drawn into the market economy. On the contrary, colonial regimes forced African families to produce for the market even as they had to continue as units of subsistence. This, in turn, was a major factor in the determination of patterns of marriage and fertility, in a situation where children were required to help maintain the family economy during the father's absence.

Thus, the nature of agrarian society in sub-Saharan Africa was transformed. But, in part for the same reasons, much of it survived precisely because of the complex interdependence that colonial economies required of men and women, of rural livelihoods and migrant labour. As a result (in contrast, for example, to Latin America), some 70% of the region's population remains rural (Bryceson, 1995).

The important points, then, are these: first, though they remain largely rural, the economies of the sub-Saharan countries are not therefore 'traditional'. Secondly, the on-going impact of the systems that the Europeans imposed upon post-colonial Africa can hardly be exaggerated. And, finally, it was the relationship between labour reserves and the European market, not African procreative impulses, that created the semblance of 'population pressure'.

The case of Kenya

Africans continue to migrate to seek work and one of the most compelling factors behind such migration continues to be a scarcity of arable land, a problem that has persisted since the end of the colonial era, largely because of the dominance of a system of commercial agriculture. This is certainly the case in Kenya where in 1990 large farms controlled a large proportion of prime arable lands, even though they significantly underused them (Foeken and Tellegen, 1994). At the same time, they produced most of the country's marketed maize, the chief food crop in the country, and employed a large proportion of rural wage labour (Foeken and Tellegen, 1994).

From early in the century, when it was established as the East African Protectorate, Kenya was administered primarily in the commercial interests of a small European settler community who had secured for themselves the most productive lands of the temperate highlands, where they established plantations of tea and coffee (Mkangi, 1983). In contrast, Africans generally were denied the right to cultivate such cash crops and were forced instead to adapt their household economy to the labour demands of the European commercial sector.

Indigenous groups such as the Maasai, Nandi and Kikuyu were not only pressured into giving up their lands, but, in order to ensure that white settlers had a sufficient supply of cheap labour, were confined to highly circumscribed areas which were wholly inadequate for their own subsistence needs, let alone for the paying of colonial taxes. Such policies deliberately created a Malthusian scenario – characterized by overcrowding and environmental degradation (Leys, 1975) – that was wholly independent of the pace of reproduction. As Mkangi observes:

> by confining Africans to reserves whose land could support a limited number of people, it was envisaged that, with normal population growth, the excess population would be released into the labour market. The more the reserves were reduced in size, the more labour was squeezed out, for [according to Ross] 'the existence of unnecessarily extensive reserves is directly antagonistic to an adequate labour supply'.
>
> *(Mkangi, 1983)*

The multitude of factors that systematically jeopardized the security and potential of indigenous agriculture contrasted sharply with the many benefits enjoyed by white settlers. These included rich land, cheap labour, a monopoly over the most lucrative crops (such as coffee) and access to markets through rail lines. Such a contrast in the relative availability of strategic resources was consistently overlooked when arguments were made that communal African land tenure was inherently inferior to private landholding. Thus, many of the so-called adverse consequences of African agriculture, including soil erosion and exhaustion, were directly or indirectly attributed to over-population, when they were, in fact, the result of constraints imposed on African subsistence by colonial policies. Over the following decades, such explanations of the causes of environmental stress lent support to the argument, which itself reflected a general view in international circles, that communal tenure was a major impediment to productive agricultural development and rural food security.

This reflected a pattern that was established in the years after the Second World War, when Kenya became a major centre for foreign capital investment. The Kenyatta government assumed a pivotal role in the Cold War political strategies of Britain and the United States, facilitating the steady advance of private capitalist interests at the expense of the African rural poor. As a result, by the 1970s multinational corporations dominated most of the country's manufacturing and played a pre-eminent role in agriculture as well, with profoundly detrimental consequences for domestic food production. Coffee and tea remained the main cash crops. One English company alone, Brooke Bond, produced almost a quarter of Kenyan tea on its estates and controlled almost 38% of the country's tea exports. The estate sector also produced well over half of Kenya's coffee (Langdon, 1978).

Meanwhile, landlessness or land scarcity continued to be – and remains – the predominant structural problem in Kenya's rural economy, one of the chief causes of high unemployment and of so many people living in absolute poverty, and the main reason that Hunt, among others, underscored the signal importance of redistributive land reform as the best strategy for absorbing labour and alleviating poverty (Hunt, 1984).

A 'population problem' or political economy?

If the land issue underlies Kenya's so-called population 'problem', it is not because land is scarce *per se*, but because relative scarcity results from structural conditions. These have habitually compelled adult males to seek work on large plantations or in urban areas, forcing their families in turn to try to secure an adequate subsistence in their absence. This has been a major influence on rural population growth, since it has created

> an incentive among smallholder families to have more children in order to increase the labour power available to manage the small farms. This is due to the fact that the male is often absent from the land at the very time terrace construction is needed to prevent soil erosion. The decision to have more children means more potential labourers to manage the duties of the farm.
>
> *(Hunt, 1984)*

Despite this, many writers continue to seek the root cause of Kenyan underdevelopment in a 'population problem'. Thus, a 1991 report by the United Nations Population Fund noted that, while Kenya's economy since independence had been among the best performers in sub-Saharan Africa, 'the benefits of a strongly growing GNP have been markedly reduced by population growth' (UNFPA, 1991). In fact, as we have already seen and discuss further below, the distribution of those benefits has been severely circumscribed, while a large proportion of the population has been excluded from the resources from which they are derived.

The report also attributes declining agricultural production to the pressure of population on the environment, which it argues is driving people into the cities where unemployment is rife (ibid.). As elsewhere throughout the Third World, however, such migration is primarily a product of structural conditions that were initially imposed by colonial rule. The apparent excess population in rural areas and the migration of Africans out of rural communities that characterize Kenya and other parts of East Africa both have a long history which is closely tied to the way the colonial economies confined indigenous Africans to inadequately resourced areas, while the benefits of their labour and of the country's natural resources largely accrued to the European settler class or to foreign firms.

This has been too readily ignored. The preoccupations of the West – particularly of the United States – with other regions, such as Latin America and Southeast Asia, has unfortunately tended to reinforce a view that sub-Saharan Africa was (and is) marginal to the concerns of foreign capital. Thus, in looking at the relationship between insurgent peasantries and anti-colonialist struggles – which so alarmed Western policy makers in the years after the Second World War – even Eric Wolf's classic work, *Peasant Wars of the Twentieth Century* (1969), dismissed Africa below the Sahara. Such intellectual marginalization contributed to the impression that the social and environmental problems in the region were largely endogenous in origin. But, this is extraordinarily deceptive, as Amin has observed (Amin, 2002).

In fact, in the years between the end of the Second World War and independence, Kenya, for example, was not only a particularly prosperous colony, but one whose eventual fate was closely connected, by the British at least, to the course of political developments elsewhere in the world. Thus, Furedi observes:

> It is worth recalling that on the eve of the post-Second World War era, Britain had no intention of abandoning its empire in Africa. Tropical Africa in general and Kenya in particular were central to the calculation of the British Chiefs of Staff and with the decline of British power globally, Africa emerged as an important economic asset.
>
> *(Furedi, 1989)*

This accounts, to a very great degree, for the lengths to which the British went to control the process that eventually led to Kenya's independence and to do so in a way that ensured, as it had done elsewhere, that power was handed over to moderates who would have a stake in curbing a more radical form of nationalism (Furedi, 1989). This became especially important after the Mau Mau revolt of the early 1950s.

The Mau Mau rebellion was essentially a peasant uprising. It was primarily confined to the Kikuyu, because they were the ethnic group whose livelihoods had been most severely disrupted by English settlement (Furedi, 1989). This dated from the beginning of the nineteenth century, when the English, having acquired ample supplies of African land, needed to ensure an adequate and compliant supply of labour. In Kenya generally, this was provided by the system of 'squatting', or tenancy. According to Rosberg Jr and Nottingham, the colonial government …

> rapidly came to regard the growth of a profitably based white settlement as a primary goal, and the role of Africans in this economy as principally that of wage earners …
>
> One early and constant source of labor for European farms was the African squatter, who either was made landless as a result of alienation, or was unable to meet his needs in crowded African rural areas. These people were encouraged to settle on European farms, cultivate crops and pasture cattle, in return for varying periods of service each year. Squatters were always inadequate to the labor needs of Europeans, and from the beginning of settlement the creation of an expanding African labor force was a central feature of the economy history of the country.
>
> *(Rosberg, Jr and Nottingham, 1966)*

It was so effective a system that, 'in the mid-1940s, Kikuyu squatters and their families numbered around a quarter of a million and constituted around a quarter of the Kikuyu population' (Furedi, 1989). By then, however, under pressure from landlords, the British administration and the Colonial Office, such tenants had been transformed, practically and legally, into wage labourers (ibid.). It was the growing resistance of the Kikuyu to the resultant deterioration in their lives – and the relative indifference of European settlers (Edgerton, 1989) – that eventually gave rise to the Mau Mau revolt (ibid.).

Once suppressed, it was the subject of a thorough mystification by the British colonial authorities and moderate Africans who 'aimed at presenting Mau Mau as a criminal organization. The colonial government went to great lengths to portray Mau Mau as an irrational force of evil, dominated by bestial impulses and influenced by world communism' (ibid.).

It was an interpretation that ignored the economic and social sources of African dissent and grew out of the kind of sentiments Lugard had promoted just a few decades earlier and which modernization theory contained at its core. This presumed that, for the sake of economic progress, traditional African culture needed to be transformed through Africans' cooperative appreciation of the merits of colonial development. Mau Mau was made to seem like the most brazen illustration of the alternative, when Africans resisted the European road to modernity (Rosberg, Jr and Nottingham, 1966).

But the process was even more complex, because the impact of English colonization on the Kikuyu had not only created some of the system's strongest opponents, but also, perhaps paradoxically, its allies. The Kikuyu were among the most educated and market-oriented ethnic groups. So, in the end, it was one of the leaders of Mau Mau, Jomo Kenyatta, with his PhD in anthropology from the London School of Economics, who was released from detainment to become the central figure in the British strategy to forge a moderate and manageable path to independent statehood. His party, KANU (the Kenya African National Union), became for the British, the 'acceptable vehicle for the realization of national independence' (Furedi, 1989; cf. Edgerton, 1989).

During the 1960s, both before and after independence, Kenya's average annual GDP growth was an impressive 6% (Langdon, 1981). This owed much to foreign aid and especially to multinational investment, which reflected the logistical importance of the country and the fact that 'Nairobi had become something of a regional centre for all of Eastern and Central Africa' (ibid.). With this in mind, the British, the United States and international financial agencies such as the World Bank had done everything they could to ensure that Kenyan independence was accompanied by an orderly transfer of power to an African bourgeoisie that would benefit from and be constrained by an on-going relationship with international capital. They had also ensured that this privileged class of Kenyans continued to profit from the perpetuation of such relations (ibid.).

Thus, the decades after independence were characterized by a remarkable continuity with the colonial past. Despite the early promise by KANU, that 'all privileges and vestiges of colonialism will be swept away' (quoted in Cheche Kenya, 1983), the years after 1968 saw 'the wholesale adoption by the KANU government of the inequality underlying the colonial system' (Cheche Kenya, 1983). Above all, no serious effort was made to question the prevailing (European) view that the private ownership of land was sacrosanct. Thus, the pattern of land distribution in Kenya remained largely unchanged. In the mid-1970s, Lele noted:

> Except for removing white hegemony in the former scheduled areas, the present Kenyan government, which inherited this inequitable system of land distribution, has not made a significant effort to correct the imbalance through a systematic land distribution policy. Indeed, the government's policies in the agricultural sector may be exacerbating the rate of landlessness. In achieving rural development ... Kenya has explicitly opted for the Western model of private capitalism and individual entrepreneurship.
>
> *(Lele, 1975)*

As a result, by the early 1980s, only 2.4% of landholdings still accounted for 32% of arable land (Woldesmiate and Cox, 1987). But, by then, the problem was not simply that government policies, including land registration and schemes for 'progressive' farmers, had eroded communal property and that a new rural elite had emerged (Lele, 1975). Many of the large holdings were also foreign-owned. Indeed, Kenya was perhaps the main focus of multinational corporate investment in sub-Saharan Africa.

There was nothing new in this. As Kaplinsky points out, the fact that Kenya had the greatest concentration of direct foreign investment of any country in Africa has led many writers 'to see the role of Kenya as an emerging sub-metropole of global imperialism' (Kaplinsky, 1978). This situation continues, and the principal beneficiaries of Kenyan capital accumulation remain the countries in which such companies are based, with subsidiary benefits accruing to the privileged African bourgeoisie who ensure a favourable climate for international investment. The majority of the population, however, has been excluded from this process. Today, when unemployment rates have reached 50%, the agricultural sector – which, as Hunt has aptly noted, is the only one which potentially could absorb sufficient numbers of workers – is shrinking. Rather than being encouraged to achieve food sufficiency, the country is described by the US Department of Agriculture as 'an East African oasis of opportunity' for food producers and processors in the United States (USDA, 1999, 2015), a situation that is ensured by the international environment of free trade. But, while Washington, taking a broad view of the country's investment opportunities, characterizes Kenya as having a healthy economy, one would hardly know this from the current condition of the rural and urban poor.

This is because the dominant interpretation of Kenya's ills reflects the general tendency of Cold War modernization theory, blaming them on factors that have little to do with the country's actual historical experience. Among these was an apparent Malthusian trend that, as noted, was built into the nature of the colonial reserve system. This view explicitly came to the fore, in Kenya as elsewhere, in the 1950s and is exemplified by such works as *Defeating Mau Mau*, by the famous white Kenyan anthropologist, Louis Leakey, the son of colonial missionaries. In it he attributed the revolt primarily to 'pressure on the land' that arose, in his view, from the 'alarmingly great' population increase among the Kikuyu (Leakey, 1954). For Leakey, as much as for the demographer Frank Notestein, one of the most strategic preventatives for such conflicts was therefore, advantageously, not to provide more land for a Kenyan peasantry that had been dispossessed by European colonialism, but, rather, 'finding satisfactory methods of birth control for peasant communities' (Leakey, 1954).

Such Malthusian views carried over into the independence era when the Kenyan government demographer, John Blacker, located in the Ministry of Planning and Development (MPD), argued that the slowing of Kenya's population growth rate was essential to ensure the new country's economic development (Watkins and Hodgson, 1998). According to Ajayi and Kekovole, such concerns 'have been a recurring theme in national development plans from the first (1966–70) to the current 8th Development Plan (1997–2001)' (Ajayi and Kekovole, 1998). Most importantly, they were given the imprimatur of the New York-based Population Council by its famous mission to Kenya, at the request of the MPD, in 1968 (Bondestam, 1980). The Population Council's report unambiguously asserted that any effort to narrow the gap between people's aspirations and economic development depended on 'a programme directed toward a decrease in the rate of population growth' (quoted in Bondestam, 1980). But, as Watkins and Hodgson have noted (1998), the report reflected Kenyan realities far less than it did the dominant neo-Malthusian thinking in the field of 'population control', where the council, with its links to influential forces in US society (e.g., the Ford and Rockefeller Foundations) was in the vanguard (Ross, 1998b).

However, as Watkins and Hodgson also make clear, when the government endorsed the report the following year, making Kenya the first country in sub-Saharan Africa officially to adopt a population policy, it was not so much because of a fervent conviction that such a policy should actually be implemented as the signal of a general commitment to the modern market economy (Watkins and Hodgson, 1998). Certainly, as the World Bank entered the field of family planning and often made it a condition for development, such initiatives placed Kenya in an enviable position as a potential recipient of Bank funds (ibid.). By the late seventies, moreover, the appeal of a Neo-Malthusian approach to the Kenyan political elite was even more opportunistic. As Watkins and Hodgson observe: 'as a reading of the 1979–1983 Development Plan illustrates, neo-Malthusianism allows rapid population growth to be used as a scapegoat for economic stagnation' (Watkins and Hodgson, 1998).

The Machakos interlude

Given this role, it is perhaps strange that Kenya also provides an important exception to the Malthusian scenario, on a local scale. But, so effectively did modernization theory distort the perceptions of rural realities that it seemed to come as a surprise when Mary Tiffen and Mike Mortimer demonstrated how population growth in the semi-arid Machakos region of Kenya actually seemed to have been a catalyst for agricultural innovation. The story of the so-called Machakos 'miracle' (Tiffen, Mortimore and Gichuki, 1994; Rocheleau, 2002; Greiner and Sakdapolrak, 2013) seemed to confirm the ideas of the Swedish economist, Ester Boserup, whose book, *The Conditions of Agricultural Growth* (1965), had proposed that population pressure was the trigger for innovation and increased productivity.

Thus, in one possible reading of the evidence, Machakos appeared to offer hope that the Malthusian trap was not as bad or inexorable as it once seemed. Yet, in itself, it provides no new formula for resolving local poverty in Kenya. Evidence that peasant farmers, in situations of population growth, can innovate and enhance the productivity of their environment is not in itself new (Altieri *et al.*, 2012; Johnson, 1971; Netting, 1993). But, the evidence from Machakos and elsewhere is that it is not enough. Even with local innovation, the process of environmental regeneration still depends, as Leach and Fairhead (2000) note, on economic and institutional factors that are not typically developed in the Boserup perspective. In Machakos, it was contingent on access to non-farm income, usually in the form of remittances by migrants to Nairobi, which borders on Machakos District (Wellard and Mortimore, 2000; cf. Tiffen and Mortimore, 1994; Greiner and Sakdapolrak, 2013). Thus, a UN report concludes that: 'the future of agricultural innovation and land productivity in Machakos also depends in no small part on the larger economy in which the district operates' (UNDP *et al.*, 2002).

It is notable that, in a country with an area larger than France and a population density generally far below that of the European Community as a whole (Townsend, 1993), the poor have suffered such limited access to land that Machakos should even seem to offer a solution. The problem, however, has never been a lack of peasant ingenuity, even in the absence of sufficient resources. It is instead that having appropriated land from the Africans, the European settler economy then set about to create a system that constrained their economic initiative. As Mosley notes, whatever notions of capitalist rationality the Europeans brought with them and by which they measured and judged the behaviour of Africans, the fact is that they never actually permitted the latter to exercise it (Mosley, 1983). In this sense, the example of Machakos best serves to demonstrate how much potential that system suppressed. The long-term solution to poverty, in the Kenyan countryside at least, continues to lie in the systematic alleviation of resource scarcity and in the radical transformation of an economic and political system that continues to limit the optimal use of natural and social resources by the majority of people.

That kind of change is constrained by the continuing relationship of Kenya to global capital. Thus it was that just before independence in 1962, the Leavey Report, which was intended to help the new government in its economic development planning, urged it to avoid 'prejudicing the production of those areas upon which the economy so much depends for maintaining export earnings' (quoted in Townsend, 1993). The same recommendation was repeated throughout most of British-dominated Africa and, in the case of Kenya, it meant effectively endorsing a situation in which only 3,600 farms and plantations owned by Europeans or Asians accounted for over 75% of the country's agricultural exports (Townsend, 1993). Despite the fact that many estates subsequently were taken over by Africans, and that occasional efforts were made to accommodate the aspirations of the land-poor, the general pattern and structure of inequality remained largely unchanged, even before multinational agri-businesses came to dominate the picture.

Fertility, land and labour

Malthusian thinking has tended to discourage any substantive critique of Kenyan development policy by separating demographic from socio-economic factors, making it seem that fertility (in the view of outside agencies and many Kenyan policy makers) could be regarded chiefly as a function of the rate of contraceptive use (Frank and McNicoll, 1987; cf. Hartmann, 1987), rather than of the consequences of underdevelopment for the demand for children.

In this way, the emphasis on fertility in the African context reflects a widespread and historical tendency to regard poverty generally as a function of pro-natalist cultural patterns. This was certainly the case throughout the nineteenth century in Ireland where Malthusian theory was first applied (Ross, 1998b) and it has broadly prevailed in regard to sub-Saharan Africa to the present day. The dominant view of Kenya's situation was more or less defined by the World Bank in the early 1980s, when it said that rapid population growth was 'the single most important obstacle to sustaining rising living standards' (in Hartmann, 1987).

The factual basis of such a facile generalization is highly ambiguous at best, but, more to the point, the rate of population growth in countries such as Kenya has not been static and actually only rose to what were considered alarming proportions during the 1960s and 1970s, precisely when the country was undergoing rapid socio-economic change. This is not to say that Kenya was 'traditional' in 1946. Quite to the contrary, local life, as we have already noted, had already been significantly affected by colonial rule. But, at least in the 1950s and 1960s, fertility was 'roughly similar to the rest of Africa at between six and seven children per woman' (Frank and McNicoll, 1987). This would change, however, in the face of on-going land privatization.

As Hunt has observed, the 'land reform' which began in the 1950s in order to head off political tensions – and which was continued in the post-independence period – had the effect of reducing traditional claims to land and increasing land inequality (Hunt, 1984). In particular, it was associated with a process of land trans-fer to an African bourgeoisie, including politicians and members of the armed forces (Hunt, 1984), which ensured continuity between the pre- and post-colonial state in Kenya and therefore helped to secure the relationship between Kenyan economic development and international capital.

During the following years, a policy of support for large farms meant that, by the early eighties, there was a continuing subdivision of smallholdings (Foeken and Tellegen, 1994), which impelled an on-going process of rural proletarianiza-tion. As a result, African workers had little access to any productive resources other than their own labour (Foeken and Tellegen, 1994), which inevitably affected their reproductive output.

As the Kenyan economy declined in the late seventies and eighties, for reasons linked to outside events that we will explore shortly, the living conditions of many rural workers, which were already poor, deteriorated further. This was especially so in the tea and coffee sectors, where declining prices impelled employers to main-tain their profits at the expense of workers' wages, and by feminizing their work-force (Foeken and Tellegen, 1994).

During the serious economic decline of the eighties, the annual population growth rate for Kenya as a whole rose. Having been around 2.5% just after the war, it reached about 3.8% in the 1980s (Kelley and Nobbe, 1990). Such a trend, of course, cannot be ascribed to a sudden decline in knowledge of or access to modern contraception and may have resulted from a general 'fertility-enhancing effect of shortened breast-feeding and postpartum abstinence' (ibid.), trends that reflected significant changes in the rural economy over that period which altered the economic and social role of women. While their access to land and other property was severely circumscribed both by customary law and state policy, their role as farm workers has increased to the point where, according to a recent report by Human Rights Watch, 'Women account for only 5 percent of registered landholders nationally ... [but] constitute over 80 percent of the agricultural labor force' (Human Rights Watch, 2003).[4]

While the definitive impact of such developments on rural fertility is hard to demonstrate, Frank and McNicoll reasonably suggest that:

> the lessened economic security of women that has resulted from land pri-vatisation may be a factor in the fertility increase recorded since the 1950s:

reflecting their worsened negotiating position within marriage, women have had to give up virtually their last measure of control over fertility that had been represented by postpartum abstinence.

(Frank and McNicoll, 1987)

A decline in breast-feeding would also have reflected increasing demands on female labour in the more insecure rural economy that resulted from agricultural commercialization. This, in turn, depended on the role of international capital in Kenya's development, and the consequences for rural labour use and access to food resources. Such connections need to be further explored. But, there is comparatively little insight to be gained by viewing demographic trends primarily in the conventional framework of transition theory or in the narrow terms of contraceptive prevalence.

Most importantly, it is pointless to acknowledge, as do the neo-Malthusians, Paul and Anne Ehrlich, that 'much of the best land is used for producing cash crops such as coffee and tea', if one then ignores the implications of such land use (and of the forces that underlie it) by asserting that it was an 'explosive growth of the population [that] led to the increasing subdivision of Kenya's farms' (Ehrlich and Ehrlich, 1990). What is necessary, of course, is to understand the causal *linkages* between the forces that have created land monopoly for a few and land scarcity for many, and the way in which poor Kenyans have had to accommodate to such scarcity. Yet, the Malthusian view, as exemplified by the Ehrlichs, simply continues to dismiss sub-Saharan African countries such as Kenya as 'a demographic basket case' (Ehrlich and Erlich, 1990) as if population growth was only the product of endogenous factors and virtually the sole cause of underdevelopment, nutritional insecurity and environmental distress.

Deeply influenced by this view, outside development agencies and international institutions have exerted little pressure on Kenya to institute a programme of land redistribution, while, in contrast, they have required it to conform to global market priorities as defined by the West. As a result, having previously been the subject of a World Bank mission and study shortly after independence – which had the effect of retarding effective land redistribution (Harbeson, 1973) – less than two decades later Kenya became 'the first African country to receive a structural adjustment loan from the World Bank in 1980, and has been engaged in Bank, IMF, and donor supported adjustment programs in most of the past twenty years' (O'Brien and Ryan, 1999).

The current situation: AIDS and poverty in sub-Saharan Africa

Until the early seventies, the Kenyan economy was regarded as fairly successful, with an annual GDP growth rate of over 5% (O'Brien and Ryan, 1999), despite the fact that the benefits of economic growth were not equitably distributed (Hunt, 1984) and that the incidence of poverty was high. But, it 'worsened considerably

during the following decades. A recent study shows that the level of absolute poverty in 1994 was around 44 per cent; for the urban areas it was 29 per cent, while
for the rural areas it was nearly 47 per cent' (EAMAT/ILO, 1999).

As a result, according to a recent paper by the Central Organisation of Trade
Unions (COTU) in Kenya, the country 'entered the 21st century facing what
is arguably the worst economic crisis since the country gained independence'
(Chanyisa Chune, 2002).

The relative prosperity of countries such as Kenya began to come to an end as
a result of changes in the world economy, in particular when oil prices increased
five-fold in 1973 (O'Brien and Ryan, 1999). The dramatic effect of this was a direct
result of the high degree of dependence on imported petroleum products, which
itself was linked to the 'Green Revolution' in agricultural production. Oil prices
rose again in 1979, by which time the United States had driven up interest rates to
unprecedented levels, causing debts that had originally been contracted with float-
ing interest rates to spiral upward (Geda, 1997).

At the same time, as in many other parts of the Third World, most countries
of sub-Saharan Africa faced a serious decline in the prices of their major exports
(tea and coffee in the case of Kenya) (Geda, 1997). By the early 1980s, such trends
had forced most of these economies, which were experiencing a severe rise in
their balance of payments current account deficit, to increase their dependence on
multilateral loans and to assume unprecedented levels of external debt. In Kenya,
for example, there was 'a rise in the debt service ratio from 2.6 percent of export
revenues in 1977 to 14 percent in 1983' (O'Brien and Ryan, 1999). Structural
adjustment, as a requirement for debt rescheduling, came soon after. But, it scarcely
ameliorated the problem of the rising debt owed by poor countries to international
financial institutions, which had the general effect of reversing the international
flow of capital. While in 1982 US$18.2 billion still went *to* developing countries,
by 1990 US$52 billion a year was actually being transferred *out*. As a result, many
developing countries had to reduce their investment in social services, particularly
health and education, while intensifying export production (Lugalla, 1995) at the
expense of their natural resource base and subsistence cultivation.

In spite of this, by 1998 the International Monetary Fund (IMF) was taking
credit for an 'African economic renaissance', claiming, along with the World Bank,
that structural adjustment had paid off. This claim is, of course, widely disputed
(cf. Naiman and Watkins, 1999; Sanyal, 2014), not least by former World Bank senior economist and Nobel laureate, Joseph Stiglitz (Stiglitz, 2000). But, the Bretton
Woods institutions tend to define the economic well-being of a developing country in ways that often do not directly correspond to the welfare of the majority of
its people.

While the idea that African countries are economically on the upswing is highly
contentious (de Brie, 1997; Jerven, 2013), it reflects the prevailing framework
within which the Bank and OECD countries typically address a wide variety of
policy issues facing the Third World. Among these is AIDS.[5] Because the IMF and
the Bank are now optimistic about the economies of countries such as Kenya,

they are disinclined to view AIDS in sub-Saharan Africa as a function of poverty. On the contrary, while the World Bank was saying in 1983 that rapid population growth was 'the single most important obstacle to sustaining rising living standards in Kenya over the long term' (in Hartmann, 1987), today, when population growth is lower and the Kenyan economy is in crisis – though the Bank would argue that point – the Bank actually tends to regard AIDS as the main impediment to sustained economic prosperity in sub-Saharan Africa.

A 1999 draft report by the Human Development Sector of the Africa Region of the World Bank, *Why Strategies to Reduce Poverty in Africa Should Focus on HIV/ AIDS and Sexual and Reproductive Health* exemplifies this position (World Bank, 1999). The thrust of this paper, as the title suggests, is that AIDS is the primary threat to economic development in the region. Because it attributes AIDS exclusively to HIV and HIV transmission to sexual behaviour, the paper effectively reduces underdevelopment in Africa to a reproductive health issue, which hinges on individual responsibility.

The first cases of AIDS were documented in Africa in 1982 (Turshen, 1984). It was first reported in Kenya in 1983 (Nduru, 1997). Given the history and heritage of colonial assumptions about African culture (cf. Harrison-Chirimuuta and Chirimuuta, 1997), the immediate response in the West was that AIDS would pose a serious threat precisely because of African sexual behaviour (cf. Caldwell, Caldwell and Orubuloye, 1992, 1999) – that is, attributed to the same cause as its putative 'population problem'. For the present, I will merely add that such immune deficiency has many possible causes and that the so-called AIDS diseases – which now include tuberculosis and sometimes seem to embrace malaria as well – have widely varied etiologies.[6] Nevertheless, AIDS has increasingly come to be spoken about as if it were a single disease with a unique (and definitive) retroviral cause that can indisputably be linked to sexual practices.

Initially, only a handful of writers, among them Meredith Turshen, seriously questioned whether we should not also be looking at political economic factors and the vulnerability of people whose immune systems were compromised by poverty and malnutrition (Turshen, 1984). A similar position was taken by the World Development Movement in a 1999 report, *Deadly Conditions? Examining the Relationship Between Debt Relief Policies and HIV/AIDS*, which notes that, 'While not restricted solely to poor people, AIDS is a disease of poverty, marginalization and social and economic injustice' (WDM, 1999). This was certainly likely to be the case in a country such as Kenya where, according to the World Bank in 1995, 'half of the population was … unable to consume a minimum requirement of food and essential non-food commodities' (World Bank, 1995).

Turshen, however, still subscribed to the conventional view that the sexual transmission of HIV caused AIDS. But University of California retrovirus specialist, Peter Duesberg, has made a powerful argument over the past 15 years that this is very unlikely to be the case (Duesberg, 1987, 1996; Duesberg and Rasnick, 1998). While not disavowing the idea of HIV causation, Gisselquist *et al.* nonetheless have recently lent significant support to Duesberg's position by noting that there is little correspondence between patterns of HIV prevalence (not to mention AIDS) in sub-Saharan Africa and those of sexual behaviour (Gisselquist *et al.*, 2002; Sawers and Stillwaggon, 2010). This strongly suggests that we must seriously reconsider the epidemiology of AIDS, especially in this region. At the heart of such a rethink should be a focus on the economic, social and environmental consequences of debt and structural adjustment and their cumulative impact on the human immune system and its capacity to respond to the resurgence of many of the diseases that are currently subsumed under the rubric of AIDS.

In 1999 the World Development Movement noted that:

> One third of children in Africa are significantly malnourished. UNICEF suggest that SAPs[7] have contributed to the widespread deterioration in the nutritional status of children, pregnant women and lactating mothers in rural and urban areas undergoing structural adjustment programmes.
>
> *(WDM, 1999)*

The implications for the spread of AIDS diseases are clear, for, as *The Merck Manual* notes:

> Malnutrition can also seriously impair the immune system. The malnutrition may involve a deficiency of all nutrients, or it may involve primarily proteins and certain vitamins and minerals … When malnutrition results in a weight that is less than 80 percent of the ideal weight, the immune system is usually somewhat impaired. When the weight is reduced to less than 70 percent of ideal weight, the immune system is usually severely impaired.
>
> *(Merck, 1995–2000)*

Thus, in 1998, in its *State of the World's Children Report*, UNICEF notes 'that malnutrition impairs the immune system of at least 100 million young children and several million pregnant women – none of them infected by HIV' (UNICEF, 1998).

Among other causes of immuno-deficiency are such diseases as malaria and tuberculosis – both of which are on the rise globally, for reasons related to economic development policies since the early eighties. Thus, malaria is on the increase in many parts of the world because anopheles mosquitoes (the malaria vector) are spreading into new niches being created by deforestation (cf. Chivian and Bernstein, 2010; Malakooti, Biomndo and Shanks, 1998; Johns Hopkins, 1999; Walsh *et al.*, 1993).

Deforestation, in turn, is rarely the result of solely local actions or even of national policies, but owes much to the 'rigid insistence on certain kinds of monetary, fiscal, trade and privatization policies by the international financial institutions in the name of stabilization and structural adjustment' (Barraclough, 1992). These have led to an increase in commercial agriculture, logging, loss of fuel subsidies or simply to the financial inability of the state to manage and protect forest resources (cf. Ahmed and Lipton, 1997; Tockman, 2001). In general, the unrelieved burden of Third World debt has forced poor countries with significant forest reserves to generate export earnings by cashing in on their natural resources (George, 1992).

In some places, World Bank-promoted expansion of non-traditional export crops has also meant a 'continued assault of pesticides' (Shiva, 1991; Murray, 1994) that has led to widespread mosquito resistance, at the same time that malarial parasites have also become resistant to conventional drugs. This is a special problem for developing countries, whose shrinking public health budgets cannot assume the costs of newer, more effective treatments. Meanwhile, many of the agro-chemicals used in commercial agriculture in such countries themselves pose a serious threat to the integrity of the human immune system (Schafer, Kegley and Patton, 2001).

There is no doubt that there have been sporadic but important efforts to highlight the impact of structural adjustment on the rise of ill-health in sub-Saharan Africa. Some writers have explicitly suggested that, 'The Sub-Saharan debt crisis and IMF-mandated structural adjustment policies have helped to spread AIDS throughout the region' (Global Exchange, 2001). Yet, in general, even this argument, while understanding that malnutrition itself weakens the immune system, assumes that this, in turn, simply makes people 'more vulnerable to viruses such as HIV' (Global Exchange, 2001), when the deterioration of the immune system is precisely what AIDS is.

The failure to recognize that the cumulative impact of the structural determinants of malnutrition and adverse environmental change on the immune function of the global poor is enough to account for the rise of AIDS in developing countries has meant that the World Bank has been able to remain committed to the conventional view that AIDS is the result of patterns of sexual behaviour. The Bank even suggests that the key to the future economic well-being of sub-Saharan Africa is a change in the sexual practices of its people. This view has been echoed in a report for the World Health Organization by the Commission on Macroeconomics and Health, chaired by Jeffrey Sachs, which ascribes the high prevalence of 'HIV/AIDS' in sub-Saharan Africa to particular sexual patterns and practices (Commission on Macroeconomics and Health, 2001). While there is no doubt that the social and economic costs of disease-caused mortality for African countries have been severe, the fact remains that neither these costs nor the prevalence of AIDS and other diseases are the main *cause* of underdevelopment in the region.

Reflections and conclusions

The implication that AIDS – or a particular retrovirus – is one of the principal determinants of African poverty seriously undermines the possibility of an alternative development strategy that would effectively liberate the human potential of Africa or of other poor regions from the influence of the institutions that dominate the contemporary global economy. It also reflects the enduring influence of Malthusian thinking on development policy. Such thinking has influenced the nature of the debate over the causes of poverty for two centuries and has scarcely abated. Its role in mystifying the origins of global inequalities during the era of the Cold War – and its central place in the dominant development ideology – was specific to certain historical conditions. But, just as the end of the Cold War has not meant the end of the longer struggle between disparate visions of how human society should be productively organized, so Malthusian thinking remains a powerful defence of the systemic conditions that underlie the contemporary global distribution of poverty and privilege.

Far from abating, these conditions have worsened over the last decade as policies of 'free trade', unregulated foreign investment and open capital markets have deepened the contradictions of capitalist development. The world economy has grown more unstable, environmental degradation has accelerated by the self-interested predations of multinational corporations and, as a consequence of growing inequalities throughout the Third World (Chossudovsky, 2003) and in many parts of the West, the poor are being increasingly displaced and forced to join the ranks of an international labour reserve army (Castles and Miller, 1993; Chomsky, 2017).

While rural unemployment linked to the advance of commercial agriculture can be shown to be perhaps the major cause of migration in the developing world, such movements are nonetheless still widely attributed to Malthusian pressures. Just as the fertility of the poor is claimed to be one of the principal causes of the decline of the natural capital of developing countries, so Third World migrants are increasingly being portrayed as 'environmental refugees' (Piquet, 2013; Saunders, 2000) whose movements threaten the sustainability of the natural resources and the well-being of the West to which they are emigrating (Pimental, 1991). Such a view stimulates and legitimizes efforts to construct increasingly restrictive immigration policies by richer countries (Nowak, 1997), policies that would compel the Third World poor to remain locked within their local systems of structurally determined poverty.

In such circumstances, a compelling view of alternative sustainable development has been proposed by David Barkin, who argues in favour of what he calls 'autonomous local production', which would rely on the knowledge systems and labour of local communities in a way that reflected a greater level of self-determination than is currently permitted by the dominant structures of globalization (Barkin, 2000). Writing on similar lines half a century ago, Yates observed that the principal objective of such development could be to foster 'self-contained systems which do not require excessive supplies of scarce raw materials or external sources of power' (Yates, 1951).

What this requires, above all, is for peasants and other subsistence producers to secure the opportunity to innovate in ways that would enlarge and secure the productivity of their own environments.

While this flies in the face of half a century of conventional theory about the prerequisites of economic growth, the fact remains that, even as the advocates of modernization theory were advancing the cause of commercial agriculture, there was ample evidence that large landowners (in Kenya and elsewhere) typically under-used lands that peasants could have brought creatively into food production; and that increased productivity did not depend upon capital investment in new agro-technologies *per se*, but could be achieved by releasing the enormous potential for peasant innovation and intensification that regularly persists in spite of resource scarcity. Thus, the UN Economic Commission on Latin America pointed out over 30 years ago that:

> the skill with which the small producers work their often inadequate farms is evidence of their ability to use land efficiently, and it can be assumed that, if the difficulties which now curtail activities were removed, their contribution to agricultural output would be even larger.
>
> *(UNECLA, 1968)*

Part of the answer, as Ernest Feder and others realized long ago (cf. Ross, 2011), has always been to offer smallholders more equitable access to land. But, as the case of Kenya amply demonstrates, what the Third World got was something very different: an agricultural regime that not only reduced the overall efficiency of food production, but wasted potential local resources, increasing dependence on costly imported inputs that have injured both the environment and human health (Viswanathan, 1991; Wheat, 1996) and further marginalized the peasant sector. In the process, local practices – including a greater diversity of local adaptations – became increasingly improbable even though they would have harnessed the creative energies of rural labour, enabling people to lead their lives without the need to emigrate. This, above all, would have dispelled the illusion that 'over-population' was only remediable through fertility control.

The main force behind this was the spread of commercial agriculture, driven by the strategies of international capitalism during the years of the Cold War and sustained by the arguments of contemporary Malthusianism (with its linkage to policy makers), which ruled out an effective role for land redistribution as a means for creating sustainable rural livelihoods. Thus, programmes of land equalization have been exceedingly rare (cf. Thiesenhusen, 1995) and generally have occurred only in the context of more general revolutionary change (cf. Bello and Guzman, 2000). Even then, they have often been thwarted by external pressure, usually from the United States (cf. Jonas, 1983), which has tended to characterize popular demands for land redistribution, not as a response to the structural determinants of land poverty, but as a symptom of 'over-population'.

The Malthusian scenarios that have helped to sustain this interpretation of world events have, as a result of the apocalyptic terms in which they have been drawn, effectively denied any real opportunity to consider alternative explanations – or solutions – for Third World deprivation. Yet, the need for such debate is now greater than ever, as such prominent organizations as the International Food Policy Research Institute (IFPRI), the Bill and Melinda Gates Foundation (BMGF) and the Council on Foreign Relations (CFR),[8] among others, call for a revival of the Green Revolution, arguing that developing countries have squandered past opportunities by failing to curb their fertility (Mathews, 1994; Pinstrup-Anderson, 1993; https://agra.org; http://www.gatesfoundation.org). Such calls chiefly address the needs of global capitalism, rather than the interests of land-hungry and undernourished people in developing countries. Their advocacy of increasing dependence on bio-technology as a means of forestalling Malthusian crisis is likely to bring about both the ultimate demise of what is left of viable peasant communities and the consolidation of a world food-production system dominated by a few giant agro-industrial corporations (ETC, 2001, 2015; Magdoff *et al.*, 2000; Ross, 2003). In such a world of diminished biodiversity and increased food insecurity, and of increasing privatization of communal resources, more and more people in Third World countries will be denied or deprived of secure access to the fundamental means of production on which creative livelihoods truly depend.

For such people, the only answer lies in real systemic change – the same answer that the rural poor have sought for generations. Only in a society where environmental resources are more or less equitably apportioned will their real reproductive and productive needs be able to be met or will gender equality, in real material terms, be possible.[9] But, so long as Malthusian thinking continues to perpetuate the illusion that the economic and reproductive behaviour of the poor is largely irrational and that it is the main source of most of their misery, such conditions will never emerge and capitalism and private resource use will be presented as their only source of hope.

The way that AIDS has been discussed in the context of sub-Saharan Africa provides one more illustration of the on-going nature of the perennial Malthusian bias of the dominant development discourse and the way in which it has been allied to a set of intractable European colonial stereotypes that continue to define the prerequisites of economic progress by locating the source of human poverty in the irrational economic and reproductive tendencies of the poor. It distracts attention from the structural causes of Third World poverty and minimizes the role of interventions that seek systemic change.

This bias, which has also regularly undermined the legitimacy of the struggles of the dispossessed against the forces of multinational capitalism, has benefited from the absence of any widespread critique within the general field of development studies. Thus, with rare exceptions, the efforts of the rural poor to reclaim control over the means of their own development (including their own knowledge base) from agencies and institutions that promote Malthusian thinking have received modest support, at best, from the Western intellectual community. In the interest of a more diverse and productive future for the global majority, that surely must change.

Notes

1 From its inception, this was one of the main consequences of the Green Revolution. The sale of fertilizers to developing countries became an important part of the solution to the US non-agricultural trade deficit (Doyle, 1986; Ross 1998a). One of the most important inputs for Green Revolution crops – water – is currently the great frontier for Western capital investment, dominated by such multinationals as Vivendi Universal, Suez Lyonnaise and, until recently, Enron (Grusky, 2001; Yaron, 2000).

2 On that basis, in 1952, when John D. Rockefeller III convened a special 'crisis conference' on population in Williamsburg, Virginia (Caldwell and Caldwell, 1986), it was Notestein who formulated the agenda (Ryder, 1984).

3 Cf. Lerner, 1958; Hagen, 1962. It is worth noting that many of the most prominent contributors to modernization theory were associated with the Center for International Studies (CENIS) at MIT, which had very close links to the CIA (Cumings, 1997; Ross, 1998a).

4 By 2014 some estimates indicated 'that as little as 1% of land is titled in the names of women and 5–6% is titled jointly by women and men' (Landesa, 2014).

5 A syndrome in which the breakdown of the immune system exposes an individual to many types of infections, one or more of which may eventuate in death.

6 Although the tendency to associate malaria and tuberculosis with AIDS varies among writers, it is very real. In a recent editorial in the *British Medical Journal*, Colebunders and Lambert (2002) clearly move in this direction when they comment that, 'About a third of the 36 million people living with HIV worldwide are co-infected with mycobacterium tuberculosis ... Tuberculosis is the leading cause of death among people with HIV infection'. Grant and de Cock (2001) also describe tuberculosis as 'the most important opportunistic infection complicating HIV infection in developing countries' (see also Harries *et al.*, 2002; Onipede *et al.*, 1999). For purposes of AIDS surveillance, the WHO defines someone as having AIDS if they test positive for HIV antibodies and have at least one of a number of conditions, including pulmonary or extra-pulmonary tuberculosis (Grant and de Cock, 2001).

7 Structural adjustment programmes.

8 IFPRI is part of the Consultative Group on International Agricultural Research (CGIAR), which was established in the early 1970s with the financial sponsorship of the Ford and Rockefeller Foundations. The headquarters of the CGIAR have always been at the World Bank, which provides its chair from among its senior staff (Clausen, 1986). Founded in 1921, the Council on Foreign Relations (CFR) is a major fixture of the so-called US establishment. Financed by the Ford and Rockefeller Foundations, among others, and by large corporations, the CFR is a crucial link between such interests and the highest levels of government, facilitating a regular exchange of personnel and ideas between the two sectors (cf. Domhoff, 1970).

9 Thus, although I recognize the important role of feminist writers in challenging Malthusianism, here and in Ross (1998a) my point is essentially that, in the absence of an equitable system of access to resources and of distribution of the benefits they confer, most discussions about fertility are distorted into a focus on the role of women in a way that actually constrains or distorts our analysis of the relationship between poverty and population (cf. Sexton, 2000).

References

Ahmed, I. with Lipton, M. (1997) 'Impact of structural adjustment on sustainable rural livelihoods: A review of the literature', Institute of Development Studies *Working Paper* 62, IDS, University of Sussex, Brighton, UK.

Ajayi, A. and Kikovole, J. (1998) 'Kenya's population policy: From apathy to effectiveness', in Jain, A. (ed), *Do Population Policies Matter? Fertility and Politics in Egypt, India, Kenya, and Mexico*, Population Council, New York.

Alexander, P. (1986) 'Labour expropriation and fertility: Population growth in nineteenth century Java', in Handwerker, W.P. (ed), *Culture and Reproduction: An Anthropological Critique of Demographic Transition Theory*, Westview Press, Boulder, Colorado.

Altieri, M.A., Funes-Monzote, F.R. and Petersen, P. (2012) Agroecologically efficient agricultural systems for smallholder farmers: Contributions to food sovereignty. *Agronomy for Sustainable Development* 32(1): 1–13.

Amin, S. (1976) *Unequal Development: An Essay on the Social Formation of Peripheral Capitalism*, Monthly Review Press, New York.

Amin, S. (2002) 'Africa: Living on the fringe', *Monthly Review* 53(10): 41–49.

Anon (1847) *Thoughts on Ireland*, James Ridgway, London.

Barkin, D. (1990) *Distorted Development: Mexico in the World Economy*, Westview Press, Boulder, Colorado.

Barkin, D. (2000) 'Overcoming the neoliberal paradigm: Sustainable popular development', *Journal of Developing Societies* 16(1): 163–180.

Barraclough, S. (1992) 'The struggle for land in the social dynamics of deforestation', paper presented at the Conference on Social Dimensions of Environment and Sustainable Development, Valletta, Malta, 22–25 April.

Barratt Brown, M. (1974) *The Economics of Imperialism*, Penguin, Harmondsworth.

Bayliss-Smith, T. and Wanmali, S. (eds) (1984) *Understanding Green Revolutions: Agrarian Change and Development Planning in South Asia*, Cambridge University Press, Cambridge.

Bello, W. and de Guzman, M. (2000) 'Perspective: Land reform no longer possible without a revolution', *Businessworld (Philippines)*, 19–20 June.

Blum, J. (1978) *The End of the Old Order in Rural Europe*, Princeton University Press, Princeton.

Bondestam, L. (1980) 'The foreign control of Kenyan population', in Bondestam, L. and Bergstrom, S. (eds) *Poverty and Population Control*, Academic Press, London.

Boserup, E. (1965) *The Conditions of Agricultural Growth: The Economics of Agrarian Change under Population Pressure*, Aldine Publishing Company, Chicago.

Bryceson, D.F. (1995) 'African women hoe cultivators: Speculative origins and current enigmas', in Bryceson, D.F. (ed.) *Women Wielding the Hoe: Lessons from Rural Africa for Feminist Theory and Development Practice*, Berg Publishers, Oxford.

Caldwell, J. and Caldwell, P. (1986) *Limiting Population Growth and the Ford Foundation Contribution*, Frances Pinter, London.

Caldwell, J., Caldwell, P. and Orubuloye, I.O. (1992) 'The family and sexual networking in sub-Saharan Africa: Historical regional differentials and present-day implications', *Population Studies* 46(3): 385–410.

Caldwell, J., Orubuloye, I.O. and Caldwell, P. (1999) 'Obstacles to behavioural change to lessen the risk of HIV infection in the African AIDS epidemic: Nigerian research', in Caldwell, J., Caldwell, P. Anarfi, J., Awusabo-Asare, K., Ntozi, J., Orubuloye, I.O., Marck, J., Cosford, W., Colombo, R. and Hollings, E. (eds) *Resistances to Behavioural Change to Reduce HIV/AIDS Infection in Predominantly Heterosexual Epidemics in Third World Countries*, Australian National University, Canberra, 113–24.

Castles, S. and Miller, M. (1993) *The Age of Migration: International Population Movements in the Modern World*, Macmillan, Basingstoke.

Chanyisa Chune, N. (2002) 'Highlights of current labor market conditions in Kenya. Central organisation of trade unions (Kenya)'. Unpublished paper.

Cheche, K. (1983) Independent Kenya, *Journal of African Marxists*, Zed Press, London.

Chivian, E. and Bernstein, A. (2010) *How Our Health Depends on Biodiversity*. Center for Health and the Global Environment at Harvard Medical School, Harvard. See: http://www.chgeharvard.org/sites/default/files/resources/182945%20HMS%20Biodiversit%20booklet.pdf.

Chomsky, N. (2017) *Requiem for the American Dream. The Principles of Concentrated Wealth and Power*, Seven Stories Press, New York.

Chossudovsky, M. (2003) *The Globalization of Poverty*, 2 Edition, Global Research, Pincourt, Canada.

Clausen, A.W. (1986) 'Foreword to Warren Baum', *Partners Against Hunger*, The Consultative Group on International Agricultural Research, Washington, DC.

Cleaver, H. (1972) 'The contradictions of the Green Revolution', *Monthly Review* 24(2): 80–111, www.eco.utexas.edu/faculty/Cleaver/cleavercontradictions.pdf.

Cobbett, W. (1984) (original 1834) *Cobbett on Ireland: A Warning to England*, Knight, D. (ed.) Lawrence and Wishart, London.

Colebunders, A. and Lambert, M.L. (2002) 'Management of co-infection with HIV and TB', *British Medical Journal* 324(7341): 802–3.

Commission on Macroeconomics and Health (2001) *Macroeconomics and Health: Investing in Health for Economic Development*. Report of the commission chaired by J. Sachs, World Health Organization, Geneva.

Cumings, B. (1997) 'Boundary displacement: Area studies and international studies during and after the Cold War', *Bulletin of Concerned Asian Scholars* 29(1): 6–26.

Daoud, A. (2010) 'Robbins and Malthus on Scarcity, Abundance, and Sufficiency: The Missing Sociocultural Element', *American Journal of Economics and Sociology* 69(4): 1206–29.

De Brie, C. (1997) 'The art of manipulating statistics: The virtual development of Africa', *Le Monde Diplomatique*, October, http://mondediplo.com/1997/10/africa.

DeWalt, B. and Barkin, D. (1991) 'Mexico's two Green Revolutions: Feed for food', in McMillan, D. (ed.) *Anthropology and Food Policy: Human Dimensions of Food Policy in Africa and Latin America*, University of Georgia Press, Athens, GA.

Diamond, J. (2005) *Collapse: How Societies Choose to Succeed or Fail*, Viking, New York.

DiGiacomo, G. (1996) 'US foreign agricultural policy', *Foreign Policy in Focus* 1(10), November, http://fpif.org/us_foreign_agricultural_policy.

Dobb, M. (1947) *Studies in the Development of Capitalism*, International Publishers, New York.

Domhoff, G.W. (1970) *The Higher Circles: The Governing Class in America*, Vintage Books, New York.

Doyle, J. (1986) *Altered Harvest: Agriculture, Genetics and the Fate of the World's Food Supply*, Penguin Books, London.

Duesberg, P. (1987) 'Retroviruses as carcinogens and pathogens: Expectations and reality', *Cancer Research* 47(5): 1199–220.

Duesberg, P. (1996) *Inventing the AIDS Virus*, Regnery Publishing, Washington, DC.

Duesberg, P. and Rasnick, D. (1998) 'The AIDS dilemma: Drug diseases blamed on a passenger virus', *Genetica* 104(2): 85–132.

EAMAT/ILO (1999) *Kenya: Meeting the Employment Challenges of the 21st Century*, Eastern Africa Multidisciplinary Advisory Team and ILO, Addis Ababa.

Edelman, M. (1992) *The Logic of the Latifundio: The Large Estates of Northwestern Costa Rica Since the Late Nineteenth Century*, Stanford University Press, Stanford.

Edgerton, R. (1989) *Mau Mau: An African Crucible*, Ballantine Books, New York.

Ehrlich, P. and Ehrlich, A. (1990) *The Population Explosion*, Simon and Schuster, New York.

Erlich, P., Erlich, A. and Daily, G. (1993) 'Food security, population and environment', *Population and Development Review* 19(1): 1–32.

ETC (2001) *Globalization, Inc. Concentration in corporate power: The unmentioned agenda*, Communique No.71, Action Group on Erosion, Technology and Concentration, ETC (formerly RAFI), Winnipeg.

ETC (2015) *Breaking Bad: Big Ag Mega-Mergers in Play*, ETC Communiqué No.115, Action Group on Erosion, Technology and Concentration, ETC (formerly RAFI), Winnipeg.

Feder, E. (1971) *The Rape of the Peasantry: Latin America's Landholding System*, Anchor Books, Garden City, New York.

Feder, E. (1983) *Perverse Development*, Foundation for Nationalist Studies, Quezon City, Philippines.

Foeken, D. and Tellegen, N. (1994) *Tied to the Land: Household Resources and Living Conditions of Labourers on Large Farms in Trans Nzoia District, Kenya*, African Studies Centre Research Series, Avebury.

Frank, A.G. (1969) *Latin America: Underdevelopment or Revolution. Essays on the Development of Underdevelopment and the Immediate Enemy*, Monthly Review Press, New York.

Frank, O. and McNicoll, G. (1987) 'Fertility and population policy in Kenya', *Population and Development Review* 13(2): 209–43.

Furedi, F. (1989) *The Mau Mau War in Perspective*, James Currey, London.

Geda, A. (1997) 'The historical origin of African debt and external finance problems', unpublished paper, Economic Research Seminar, Institute of Social Studies, The Hague.

George, S. (1992) *The Debt Boomerang: How Third World Debt Harms Us All*, Pluto Press with the Transnational Institute, London.

Gisselquist, D., Rothenberg, R., Potterat, J. and Drucker, E. (2002) 'HIV infections in sub-Saharan Africa not explained by sexual or vertical transmission', Editorial Review, *International Journal of STD & AIDS* 13(10): 657–66.

Gledhill, J. (1995) *Neoliberalism, Gransnationalization and Rural Poverty: A Case Study of Michoacan, Mexico*, Westview Press, Boulder.

Global Exchange (2001) 'How the International Monetary Fund and the World Bank undermine democracy and erode human rights: Five case studies', webpage, www.globalexchange.org/resources/wbimf/report.

Grant, A. and De Cock, K. (2001) 'HIV infection and AIDS in the developing world', *British Medical Journal* 322(7300): 1475–8.

Greiner, C. and Sakdapolrak, P. (2013) 'Rural-urban migration, agrarian change, and the environment in Kenya: A critical review of the literature', *Population and Environment* 34(4): 524–553.

Grusky, S. (2001) 'Privatization tidal wave: IMF/World Bank water policies and the price paid by the poor', *Multinational Monitor* 2(9): 14–19.

Hagen, E. (1962) *On the Theory of Social Change: How Economic Growth Begins*, The Dorsey Press, Homewood, Illinois.

Harbeson, J. (1973) *Nation-Building in Kenya: The Role of Land Reform*, North-Western European Press, Evanston.

Harries, A., Hargreaves, N., Chimzizi, R. and Salaniponi, F. (2002) 'Highly active antiretroviral therapy and tuberculosis control in Africa: Synergies and potential', *Bulletin of the World Health Organization* 80(6): 464–9.

Harris, M. (1959) 'Labor emigration among the Mozambique Thonga: Cultural and political factors', *Africa* 29(1): 50–65.

Harrison-Chirimuuta, R. and Chirimuuta, R. (1997) 'AIDS and Africa: A case of racism vs science', in Bond, G., Kreniske, J., Susser, I. and Vincent, J. (eds) *Aids in Africa and the Caribbean*, Westview Press, Boulder, Colorado.

Hartmann, B. (1987) *Reproductive Rights and Wrongs: The Global Politics of Population Control and Contraceptive Choice*, Harper & Row, New York.

Hartmann, B. (1999) 'Population, environment and security: A new trinity', in Silliman, J. and King, Y. (eds), *Dangerous Intersections: Feminism, Population and the Environment*, Zed Books, London.

Hewitt de Alcantara, C. (1976) *Modernizing Mexican Agriculture: Socioeconomic Implications of Technological Change, 1940–1970*, UN Research Institute for Social Development, Geneva.

Hobson, J. (1902) *Imperialism: A Study*, George Allen and Unwin, London.

Holt-Gimenez, E. and M.A. Altieri (2012) 'Agroecology, Food Sovereignty and the New Green Revolution'. *Agroecology and Sustainable Food Systems* 37(1): 90–102.

Human Rights Watch (2003) *Double Standards: Women's Property Rights Violations in Kenya*, www.hrw.org/reports/2003/kenya0303/kenya0303.pdf.

Hunt, D. (1984) *The Impending Crisis in Kenya: The Case for Land Reform*, Gower, London.

Jerven, M. (2013) *Poor Numbers: How We Are Misled by African Development Statistics and What to Do about It,* Cornell University Press, New York.

Johns Hopkins (1999) 'Deforestation effects on malaria in the Amazon', Johns Hopkins School of Public Health, website, http://iws1.jhsph.edu/globalchange/deforestation.html.

Johnson, A. (1971) *Sharecroppers of the Sertao: Economics and Dependence on a Brazilian Plantation,* Stanford University Press, Stanford.

Jonas, S. (1983) 'Contradictions of revolution and intervention in Central America in the transnational era: The case of Guatemala', in Dixon, M. and Jonas, S. (eds) *Revolution and Intervention in Central America,* Synthesis Publications, San Francisco.

Kaplinsky, R. (1978) 'Introduction: The role of the multinational corporation in the Kenyan economy', in Kaplinsky, R. (ed.) *Readings on the Multinational Corporation in Kenya,* Oxford University Press, Nairobi, 1–21.

Kelley, A. and Nobbe, C. (1990) 'Kenya at the demographic turning point? Hypotheses and a proposed research agenda', *World Bank Discussion Papers* 107, World Bank, Washington, DC.

Lambert, J. (1967) *Latin America: Social Structures and Political Institutions,* Helen Katel (trans), University of California Press, Berkeley.

Landesa (2014) *Women's Land and Property Rights in Kenya.* Accessed on 4 March 2017 at https://www.landesa.org/wp-content/uploads/LandWise-Guide-Womens-land-and-property-rights-in-Kenya.pdf.

Langdon, S. (1978) 'The multinational corporation in the political economy of Kenya', in Kaplinsky, R. (ed.) *Readings on the Multinational Corporation in Kenya,* Oxford University Press, Nairobi.

Langdon, S. (1981) *Multinational Corporations in the Political Economy of Kenya,* Macmillan, London.

Leach, M. and Fairhead, J. (2000) 'Challenging neo-Malthusian deforestation analysis in West Africa's dynamic forest landscapes', *Population and Development Review* 26(1): 17–43.

Leakey, L. (1954) *Defeating Mau Mau,* Methuen, London.

Lele, U. (1975) *The Design of Rural Development: Lessons from Africa,* The World Bank, Washington, DC.

Lerner, D. (1958) *The Passing of Traditional Society: Modernizing the Middle East,* The Free Press, Glencoe, NJ.

Leys, C. (1975) *Underdevelopment in Kenya: The Political Economy of Neo-Colonialism, 1964–1971,* Heinemann, London.

Lis, C. and Soly, H. (1979) *Poverty and Capitalism in Pre-industrial Europe,* Humanities Press, Atlantic Highlands.

Lugalla, J. (1995) 'Economic reforms and health conditions of the urban poor in Tanzania', paper presented at the American Anthropological Association meeting, 15–19 November, Washington, DC.

Lugard, Lord F. (1922) *The Dual Mandate in British Tropical Africa,* Blackwood and Sons, London.

Magdoff, F., Foster, J.B. and Buttel, F. (eds) (2000) *Hungry for Profit: The Agribusiness Threat to Farmers, Food and the Environment,* Monthly Review Press, New York.

Malakooti, M.K., Biomndo, K. and Shanks, G. (1998) 'Re-emergence of epidemic highland Malaria in western Kenya', *Journal of Emerging Infectious Diseases* 4(4): 671–76.

Marchesi, G. (2016) The Other Green Revolution: Land Epistemologies and the Mexican Revolutionary State. *Antipode,* 4 February 2016, 1–19. DOI:10.1111/anti.12215

Mathews, J. (1994) 'A small price to pay for proving Malthus wrong', *International Herald Tribune,* 9 June.

Merck (1995–2000) *The Merck Manual Home Edition*, website: www.merck.com/pubs/mmanual_home/sec16/168.htm.

Mkangi, G. (1983) *The Social Cost of Small Families and Land Reform: A Case Study of the Wataita of Kenya*, Pergamon, London.

Mosley, P. (1983) *The Settler Economies: Studies in the Economic History of Kenya and Southern Rhodesia 1900–1963*, Cambridge University Press, Cambridge.

Murray, D. (1994) *Cultivating Crisis: The Human Cost of Pesticides in Latin America*, University of Texas Press, Austin.

Naiman, R. and Watkins, N. (1999) *A Survey of the Impacts of IMF Structural Adjustment in Africa: Growth, Social Spending*, and *Debt Relief*, Center for Economic and Policy Research, Washington, DC, http://cepr.net/documents/publications/debt_1999_04.htm.

Nduru, M. (1997) 'Kenya–population: Heading into AIDS storm', InterPress News Service, 29 September, www.aegis.com/news/ips/1997/IP970909.html.

Netting, R.M. (1993) *Smallholders, Householders: Farm Households and the Ecology of Intensive Sustainable Agriculture*, Stanford University Press, Stanford.

Niven, N. (1846) *The Potato Epidemic and Its Probable Consequences*, James McGlashan, Dublin.

Notestein, F. (1953) 'Economic problems of population change', in Cumberlege, G. (ed.) *Proceedings of the Eighth International Conference of Agricultural Economists*, Oxford University Press, London.

Nowak, M. (1997) *Immigration and U.S. Population Growth: An Environmental Perspective*, Negative Population Growth, Washington, DC.

O'Brien, F. and Ryan, T. (1999) 'Aid and reform in Africa: Kenya case study', Working Paper 35727, World Bank Development Research Group, The World Bank, Washington, DC, http://documents.worldbank.org/curated/en/1999/08/6705624/aid-reform-africa-kenya-case-study.

Onipede, A.O., Idigbe, O., Ako-Nai, A.K., Omojola, O., Oyelese, A.O., Aboderin, A.O., Akinisho, A.O., Komolafe, A.O. and Wemambu, S.N. (1999) 'Sero-prevalence of HIV antibodies in tuberculosis patients in Ile-Ife, Nigeria', *East Africa Medical Journal* 76(3): 127–32.

Owen, R. (1823) *Statements Showing the Power that Ireland Possesses to Create Wealth Beyond the Most Ample Supply of the Wants of Its Inhabitants*, A. Applegath, London.

Paddock, W. and Paddock, P. (1967) *Famine – 1975!* Weidenfeld and Nicolson, London.

Pearse, A. (1980) *Seeds of Plenty, Seeds of Want: Social and Economic Implications of the Green Revolution*, Clarendon Press, Oxford.

Perelman, M. (1977) *Farming for Profit in a Hungry World: Capital and the Crisis in Agriculture*, Allenheld, Osmun, Montclair, NJ.

Piquet, E. (2013) 'From "primitive migration" to "climate refugees": The curious fate of the natural environment in migration studies', *Annals of the Association of American Geographers* 103(1): 148–162.

Pimental, D. (1991) 'Global warming, population growth and natural resources for food production', *Society and Natural Resources* 4(4): 347–63.

Pinstrup-Anderson, P. (1993) *World Food Trends and How They May Be Modified*. Report to CGIAR International Centres' Week, Washington, DC, 25–29 October, International Food Policy Research Institute, Washington, DC.

Rocheleau, D. (2002) 'The invisible ecologies of Machakos: Landscapes and life stories, 1900–2000', Radcliffe Institute Fellowships, website: www.radcliffe.edu/rrp/projects/2003/rocheleau.html

Rosberg, C. Jr, and Nottingham, J. (1966) *The Myth of 'Mau Mau': Nationalism in Kenya*, Praeger, New York.

Ross, E.B. (1998a) 'Cold warriors without weapons', *Identities* 4(3–4): 475–506.

Ross, E.B. (1998b) *The Malthus Factor: Poverty, Politics and Population in Capitalist Development*, Zed Books, London.

Ross, E.B. (1998c) 'Malthusianism, counterrevolution and the Green Revolution', *Organization & Environment* 11(4): 446–50.

Ross, E.B. (2003) 'Malthusianism, capitalist agriculture and the fate of peasants in the making of the modern world food system', *Review of Radical Political Economics* 35(4): 437–461.

Ross, E.B. (2011) 'A critic unfettered: The legacy of Ernest Feder', *Development and Change* 42(1): 33–48.

Rothschild, E. (1996) 'The debate on economic and social security in the late eighteenth century: Lessons of a road not taken', *Development and Change* 27(2): 331–51.

Ryder, N. (1984) 'Frank Wallace Notestein (1902–83)', *Population Studies* 38(1): 5–20.

Sanyal, K. (2014) *Rethinking Capitalist Development: Primitive Accumulation, Governmentality & Post-Colonial Capitalism*, Routledge, London.

Saunders, P. (2000) 'Environmental refugees: The origins of a construct', in Stott, P. and Sullivan, S. (eds) *Political Ecology: Science, Myth and Power*, Arnold, London.

Sawers, L. and Stillwaggon, E. (2010) 'Concurrent sexual partnerships do not explain the HIV epidemics in Africa: A systematic review of the evidence', *Journal of the International AIDS Society* 13(1): 34–44.

Schafer, K., Kegley, S. and Patton, S. (2001) *Nowhere to Hide: Persistent Toxic Chemicals in the US Food Supply*, Pesticide Action Network of North America, San Francisco.

Schuster, M.L. (1940) *A Treasury of the World's Great Letters*, Simon and Schuster, New York.

Sexton, S. (2000) 'Population talk: A review of The Malthus Factor', *Monthly Review* 52(2).

Shafer, D.M. (1988) *Deadly Paradigms: The Failure of U.S. Counterinsurgency Policy*, Princeton University Press, Princeton.

Shiva, V. (1991) 'The Green Revolution in Punjab', *The Ecologist* 21(2): 57–60.

Stavenhagen, R. (1970) 'Collective agriculture and capitalism in Mexico: A way out or a dead end?', *Latin American Perspectives* 2(2): 146–63.

Stiglitz, J. (2000) 'The insider: What I learned at the world economic crisis', *The New Republic*, 17 April.

Stross, R. (1986) *The Stubborn Earth: American Agriculturalists on Chinese Soil 1898–1937*, University of California Press, Berkeley.

The Ecologist (1992) 'Development as enclosure: The establishment of the global economy', *The Ecologist* 22(4): 131–47.

Thiesenhusen, W. (1995) *Broken Promises: Agrarian Reform and the Latin American Campesino*, Westview Press, Boulder.

Tiffen, M. and Mortimore, M. (1994) 'Malthus controverted: The role of capital and technology in growth and environmental recovery in Kenya', *World Development* 22(7): 997–1010.

Tiffen, M., Mortimer, M. and Gichuki, F. (1994) *More People, Less Erosion: Environmental Recovery in Kenya*, John Wiley, Chichester.

Tockman, J. (2001) *The IMF – Funding Deforestation*, American Lands Alliance, Washington, DC.

Townsend, P. (1993) *The International Analysis of Poverty*, Harvester Wheatsheaf, New York.

Trevelyan, C. (1848) *The Irish Crisis*, Longman, Brown, Green and Longmans, London.

Turshen, M. (1984) *The Political Ecology of Disease*, Rutgers University Press, New Brunswick.

UNDP, UNEP, World Bank and World Resources Institute (2002) *World Resources 2000–2001 – People and Ecosystems: The Fraying Web of Life*, World Resources Institute, Washington, DC.

UNECLA (1968) *Economic Survey of Latin America 1966*, UN Economic Commission for Latin America, New York.

UNFPA (1991) *Population, Resources and the Environment: The Critical Challenges*, United Nations Population Fund, New York.

UNICEF (1998) *The State of the World's Children*, Summary, website: www.unicef.org/sowc98/sum04.htm.

Unikel, L. (1975) 'Urbanization in Mexico: Process, implications, policies, and prospects', in Goldstein, S. and Sly, D. (eds), *Patterns of Urbanization: Comparative Country Studies*, Ordina Editions, Dolhain, Belgium, 465–568.

US Department of Agriculture (1999) 'Kenya's the market for US consumer-oriented foods', website: www.fas.usda,gov/info/agexporter/1999/kenyas.html.

US Department of Agriculture (2015) 'A Turning Point for Agricultural Exports to Sub-Saharan Africa'. Website: https://www.fas.usda.gov/data/turning-point-agricultural-exports-sub-saharan-africa.

Viswanathan, A. (1991) 'Pesticides: From silent spring to Indian summer', *Economic and Political Weekly* 26(44): 2039–40.

Walinsky, L. (ed.) (1977) *The Selected Papers of Wolf Ladejinsky: Agrarian Reform as Unfinished Business*, World Bank, Washington, DC.

Walsh J.F., Molyneux, D. and Birley, M. (1993) 'Deforestation: Effects on vector-borne disease', *Parasitology* 106, Supplement: S55–S75.

Watkins, S.C. and Hodgson, D. (1998) 'From mercantilists to neo-Malthusians: The international population movement and the transformation of population ideology in Kenya', Paper presented at the Workshop on Social Processes Underlying Fertility Change in Developing Countries, 29–30 January 1998. Committee on Population, National Academy of Sciences, Washington, DC.

WDM (1999) 'Deadly conditions? Examining the relationship between debt relief policies and HIV/AIDS', World Development Movement (now 'Global Justice Now'), London.

Wellard, K. and Mortimore, M. (2000) 'Farmer-led adoption of ox weeding in Machakos District, Kenya', in Starkey, P. and Simalenga, T. (eds) *Animal Power for Weed Control*, Technical Center for Agricultural and Rural Cooperation, Wageningen, The Netherlands.

Wheat, A. (1996) 'Toxic bananas', *Multinational Monitor* 17(9): 9–15, www.multinationalmonitor.org/hyper/mm0996.04.html.

Whitehead, A. (1990) 'Food crisis and gender conflict in the African countryside', in Bernstein, H., Crow, B., MacKintosh, M. and Martin, C. (eds) *The Food Question: Profits Versus People?* Monthly Review Press, New York.

Woldsmiate, T. and Cox, R. (1987) 'The food crisis in Kenya', in Yesilada, B. and Brockett, C. (eds) *Agrarian Reform in Reverse: The Food Crisis in the Third World*, Westview Press, Boulder, Colorado.

Wolf, E. (1969) *Peasant Wars of the Twentieth Century*, Harper and Row, New York.

World Bank (1995) *Kenya Poverty Assessment*, The World Bank, Washington, DC, http://documents.worldbank.org/curated/en/1995/03/697538/kenya-poverty-assessment.

World Bank (1999) 'Why strategies to reduce poverty in Africa should focus on HIV/AIDS and sexual & reproductive health', Human Development Sector, Africa Region, Draft discussion paper, The World Bank, Washington, DC.

Yaron, G. (2000) *The Final Frontier: A Working Paper on the Big 10 Global Water Corporations and the Privatization and Corporatization of the World's Last Public Resource*. Citizens' Council on Corporate Issues, USA. Accessed on 6 March 2017 at http://iatp.org/files/Final_Frontier_A_Working_Paper_on_the_Big_10_G.htm.

Yates, F. (1951) 'Manuring for higher yields', in Le Gros Clark, F. and Pirie, N.W. (eds) *Four Thousand Million Mouths: Scientific Humanism and the Shadow of World Hunger*, Oxford University Press, London.

6

PLANTS THAT SPEAK AND INSTITUTIONS THAT DON'T LISTEN

Notes on the protection of traditional knowledge

Nina Isabella Moeller

Wizards and fighter jets[1]

'How do you protect your knowledge?' I asked a middle-aged *yachak*,[2] a traditional Kichwa healer, wizard and community adviser, as we were preparing a large amount of *ayahuasca* brew, the hallucinogenic drink 'that makes you see', and ultimately, 'know and heal'.

'You need to be strong to protect yourself', he answered, pressing the vine and leaves deeper into the boiling water with a wooden stick. 'You need a lot of energy, *sinzhi*,[3] to protect yourself from attacks. Your enemies will always try to attack, make you ill or eradicate you completely. It is dangerous to be a *yachak*. That is why many are secret. But only a very powerful *brujo*[4] can get past my defences. I have many secrets, including a whole fleet of fighter jets, spiritually, that protect me. Sometimes I just use a mirror', he laughed 'and return the misdeed back to the one who sent it'.

'So, by protecting yourself from spiritual attacks, you protect your knowledge?' Domingo looked at me with the indulgent pity reserved for the stupid. I tried again: 'I mean, what happens to your knowledge when you get attacked? Does it disappear?'.

'Your power disappears. When you get attacked and you cannot protect yourself, you become weak. Maybe you get ill, maybe you die'.

'But if you get ill, and then recover, you will still have your knowledge?' I insisted, starting to be unsure about whether I was making any sense. What was this thing I called knowledge? 'Will you still know which plants to use to heal someone, for example, or will you forget such things?'

'It's not enough to know which plants heal. You need to have the knowledge to *make them* heal. That's why we diet.[5] It gives us *sinzhi*'. He paused. 'When they attacked my uncle, a very good *yachak*, and he got very ill, when he then recovered, he could not understand the [*ayahuasca*] visions. He could see, but he could not interpret them. For a long time he was no use as a healer. And he could not see the future very well. Not even the tobacco helped him. They took his power'.

It was through conversations such as this one that I realised that for many of the people with whom I met and worked in the Amazon, spiritual power and valuable knowledge were two sides of the same coin. Such power/knowledge is understood to be in danger of attack and even destruction from the negative energies of certain people, places and spirits that intentionally or unintentionally affect its holder. A 'powerful' *yachak* 'knows' not only in the sense of having access to a vast internal repertoire of information about such phenomena in the world as plants, animals, landscapes, diseases, spiritual energies and the ways these relate to one another, but also in the sense of (what we might call) her or his power of intuition being highly accurate. ('You've had a bad dream' said Ana to me unfailingly when I had indeed had one, and I never met her early in the morning when my tensed body could have still betrayed a nightmare.) This is not so much *knowledge held*, as an *ability to know*. It is a particular form of perceptiveness, which, I was told, is a skill not unlike 'a skill to play the piano'. You can learn it, but 'you will probably learn it better if you have the talent and the desire'.

This ability to 'know' things that were seemingly imperceivable became more and more a feature of ordinary reality the longer I spent with traditional healers from the forest. Some people are able to know surprising details about others in an (for want of a better word) intuitive way, *as if* they had been told, or *as if* there were conclusive clues in someone's body language, or gaze or particular scent. A visiting colleague, who had been suffering from recurring lower back problems for five years, was told by two different *yachakuna* on different occasions that she had blue light pouring out of her kidney, a spiritual injury that must have been provoked by a desert spirit in a far-away country. Neither of these *yachakuna* had ever been in a desert, nor had they been told either about my colleague's back pain, or its sudden origin during a trek through Botswana's drylands. The healer's knowledge in this context is thus more like the capacity to see or hear (in the very moment of ocular or auricular perception) than the capacity to recall or remember (memorised past experience and 'stored' information). It is a kind of knowledge that is based on real-time perception (like vision or sound), that might of course be mistaken, but not necessarily more often than people misinterpret what they 'normally' see or hear.

For the purposes of this chapter, in which I will question the idea of 'protecting traditional knowledge', it suffices to note that what exactly knowledge is, what it does, what it means, and what people value about it, does not become clear simply by invoking the term. What it might be threatened by, and what protecting it would involve, is hence even less obvious. For Domingo, at least in the context of our conversation above, protecting his medicinal knowledge meant summoning spiritual fighter jets, practising his diets, and generally taking care of his different powers and energies. The discussions in international policy-making fora revolve around very different ideas of knowledge and its protection. Indeed, through the policies made in these fora, the protection of traditional knowledge has become a vehicle for the reinforcement of intellectual property rights. In this way, indigenous peoples' lives and practices are being used in order to advance the process of capital expansion and market consolidation. Domingo's understanding of protection is unlikely to ever make it into international level discussions. While it is the concerns of people like Domingo that these discussions purport to address, they sideline and indeed silence those understandings of traditional knowledge which, if taken seriously, raise uncomfortable, critical questions about our current socio-economic order.

The hegemonic construction of traditional knowledge protection

The protection of traditional knowledge is by now undeniably a 'global' endeavour. Defined as the protection of 'knowledge, innovations and practices of indigenous and local communities embodying traditional lifestyles' by the Convention of Biological Diversity (CBD),[6] it is more or less directly addressed by the World Intellectual Property Organisation (WIPO),[7] the World Trade Organisation's (WTO) Doha Development Agenda,[8] the International Treaty on Plant Genetic Resources of the Food and Agriculture Organisation (FAO),[9] the United Nations Declaration on the Rights of Indigenous Peoples,[10] as well as by a host of ethical guidelines and codes of conduct of professional societies, such as the Natural Resources Stewardship Circle for the Beauty, Cosmetics, Fragrance, and Flavor Industries.[11] The creation of a legally binding international regime is being debated in several fora. Moreover, various countries have enacted special laws, or established regulatory frameworks for the protection of traditional knowledge at the national level, while indigenous peoples and subsistence farmers' organisations continue to fight for the recognition of their rights in this regard at all scales. Large amounts of resources continue to be mobilised for conferences, ad hoc meetings, fact-finding missions, capacity building and report writing in order to facilitate decision making about protective mechanisms and their implementation.

While the need to protect traditional knowledge is sometimes presented as arising from the erosion of traditional ways of life, its internationally dominant expression is in terms of illegitimate appropriation. This is to say that the protection of traditional knowledge has been generally seen as required due to the threat of 'biopiracy'.[12] Biopiracy is cast as the undue appropriation of, and exclusive commercial gain from plant and animal resources of traditional use in indigenous and farming communities, and its most infamous perpetrators are pharmaceutical and biotechnology companies by means of their bioprospecting endeavours[13] (Shiva, 1997; Mooney, 2000).

Talk of legitimate and illegitimate appropriation construes traditional knowledge as a kind of (intellectual) property of indigenous and farming peoples, implying the latter's rights to control access to and to benefit economically from their traditional knowledge.[14] It is in this way that traditional knowledge is treated as a commodity and understood as in need of the same kind of protection as other forms of (private, intellectual) property. Legislation and soft law guidelines regarding access and benefit sharing (ABS) are currently wielded as the main mechanisms for the protection of traditional knowledge.[15] ABS agreements – no matter how fair and equitable – enact or perform the protection of traditional knowledge as pseudo-intellectual property protection. By doing so, they hide from view, and indeed erode other possible understandings of traditional knowledge and its need for protection.

On the dominant account, knowledge is reduced to a tradeable object. The predicament of traditional knowledge is presented as an economic problem and the sharing of benefits (in whichever particular shape) as its fundamental solution. The present chapter aims to contribute to the destabilisation of this hegemonic construction. Elsewhere, I have argued that the dominant discourse of protection – the one developed and employed in national and international policy-making settings – perpetuates background assumptions ultimately instrumental to the continued expansion of capital and concomitant destruction of autonomous subsistence.[16] Here I illustrate, by way of ethnographic notes, how this dominant view gets perpetuated and taken up, how it suppresses other understandings of the value of traditional knowledge and its need for protection, and what some of these other understandings and their implications are. To this end, I present (1) a number of interactions which took place during the capacity-building course of the ABS project ProBenefit (see next section)[17] and which highlight the ways in which dominant understandings of traditional knowledge, the issue of protection, and its difference to scientific knowledge, were perpetuated, while alternative understandings of what was at stake were disregarded and subdued. (2) I present a series of conversations and encounters which I was party to during my work in the Ecuadorian Amazon and which make even clearer that projects such as ProBenefit, and the discourses which they introduce and perpetuate, veil the plural understandings and valuations of knowledge and people's concerns in this regard, and consequently undermine what they purport to protect.

Borrowing from Joan Martinez-Alier (2002), I conclude that the struggle surrounding the protection of traditional knowledge is not only a struggle regarding access over resources, but also a struggle over meanings and values – the meaning and value of knowledge embodied in traditional lifeways. I urge that the idioms in which these struggles are carried out continue to (or begin to) contest the dominance of market valuation, in order to keep alive the plurality of values through which people make sense of and give meaning to their worlds.

Introducing ProBenefit

ProBenefit ('PROcess-oriented development of a model for equitable BENEFIT-sharing for the use of biological resources in the Amazon Lowlands of Ecuador'),[18] was a €1.04 million project funded by the German Ministry for Education and Research. Its aim was to investigate the feasibility of fair and equitable access and benefit sharing as outlined by the CBD, and more particularly its Bonn Guidelines.[19] ProBenefit was to involve indigenous communities in the project realisation and execution. In this way, ProBenefit would respect and promote indigenous rights as well as contribute to capacity building of indigenous organisations. A partnership with a private company was a requirement on part of the funders, and so ProBenefit entered the Ecuadorian Amazon in 2003 with the proposal of facilitating a participatory process for the negotiation of a fair and equitable ABS agreement with the German pharmaceutical company Dr. Willmar Schwabe Ltd.

Schwabe Pharmaceuticals is a medium-sized enterprise with 727 employees in its German headquarters, and about 3,500 employees worldwide as part of the Schwabe Group, comprising subsidiaries and joint ventures in 18 countries. Schwabe has produced phytomedicines – i.e. plant-based medicines and health products – since 1866, relying on a high-tech manufacturing process. In 2011 their turnover was €590 million; they spent €27 million on research and development in the same year. Many of their products and manufacturing processes, such as special extraction methods, are protected by patents. Schwabe agreed to be part of ProBenefit not merely as a way to research new plants, but also in order to develop what could be marketed as 'fair trade' health products.[20]

ProBenefit was set to run for five years until the end of 2007 and was made up of two consecutive project phases:

- Phase 1: Entry into a model agreement with all actors representing relevant interests in the spirit of the CBD on access to natural resources in a part of the Ecuadorian Amazon region.
- Phase 2: Ethno-botanical and pharmacological investigations for the possible production of a plant extract with documented medicinal effect.

It was made very clear in all of ProBenefit's publications that without the successful completion of phase one, the activities planned for phase two would not begin. The project aimed to 'develop a suitable procedure for equitable benefit-sharing for the use of biological resources and the associated indigenous knowledge',[21] and not (or not chiefly) to develop the use itself, as other bioprospecting projects have done (such as the various incarnations of the International Cooperative Biodiversity Group[22] (cf. Berlin *et al.*, 1999; Berlin and Berlin, 2004; Greene, 2002; Hayden, 2003, 2005; Rosenthal and Katz, 2004; Rosenthal, 2006) or the InBio-Merck agreement[23] (cf. Martinez-Alier, 2002). The outcome of ProBenefit's endeavours was hoped to be a model ABS procedure, 'maybe the most ethical one world-wide', as I was told by a ProBenefit team member.

Given that there are an estimated 60,000–100,000 Kichwa living in the Ecuadorian Amazon region (plus some more in the adjacent Peruvian territory), full consultation was impossible within the parameters of the project. It was hence proposed to form an indigenous working group that could develop an 'indigenous framework' for basic access conditions. Extensive capacity building for such a working group was going to be provided by independent, and ideally indigenous professionals with expertise in the subject area. The consultation based on the conditions framed by the indigenous working group would then proceed via the mechanisms of the indigenous community organisations, federations and confederations to ensure the greatest possible coverage.

In May 2005, the Kichwa federation FONAKIN (*Federación de Organizaciones de la Nacionalidad Kichwa del Napo*) became the official indigenous counterpart of ProBenefit with contractual obligations to oversee the co-ordination of a delegation of indigenous representatives from various organisations (not all affiliated to FONAKIN). This delegation was to participate in a capacity-building workshop series (six four-day modules over three months), after which they would form an independent working group that would design and perform the actual activities constituting public consultation.[24]

However, ProBenefit failed to successfully complete its first phase. Indigenous participation stalled after the capacity-building workshop and made timely negotiation of an ABS proposal impossible; neither consent to, nor a clear rejection of, bioprospecting in Napo was therefore obtained.

ProBenefit frictions

As a project, ProBenefit was based on the belief that (sustainable) income-generating use of biodiversity would lead to its increased conservation, as long as local people partook in the income generated. These assumptions underlie the discourses that inform and draw upon such international frameworks as the CBD, and are also explicitly espoused by the main driver behind ProBenefit, the Institute of Biodiversity Network. Traditional knowledge was understood to be threatened by potentially unfair appropriation through private interests, as well as by increasing loss within communities as these underwent rapid changes in lifestyle. A fair and equitable access and benefit sharing contract with a pharmaceutical company was promoted as an ideal solution: any appropriation by outsiders would occur under strict conditions, consented to by the legitimate owners of the knowledge in question – the threat of *mis*appropriation would hence be averted through *ethical* appropriation. Moreover, the economic benefits gained from such an ABS contract would give value to traditional knowledge within the communities, and especially amongst the younger generation, leading to a renewed interest in maintaining and transmitting it. In this way of course, the value of traditional knowledge is cast solely in market terms, at the same time as human motivation is reduced to the function of an economic cost-benefit analysis.[25] Other ways of understanding what is at stake are suppressed by this hegemonic construction of knowledge, its value, threats and means of protection.

ProBenefit's inability to revise, or even reflect upon, some of these assumptions in the light of its work with representatives of indigenous community organisations contributed to conflicts (some of which I recount below) and led to the eclipsing of alternative visions and understandings that were raised by its indigenous participants. In this way, ProBenefit constituted an insidious imposition of a particular system of values. This imposition was not planned or intended, but rather an inevitable side effect of the project's set up and constraints. Crucially, for instance, ProBenefit team members were accountable to their funders, to whom they had of course certain contractual obligations, such as reports on expenses and progress.

This responsibility impeded a more flexible approach to working with their indigenous partners, and hence contributed to the project's premature ending.

After the capacity-building workshop, once the indigenous working group was officially formed and the German team had returned to Germany to await a proposal for continuation and plan for consultation, progress rapidly stalled. Communications between the two parties broke down for almost six months; they then picked up but were mired by a series of misunderstandings. The project ended without any wider public consultation, nor any agreement being reached.

From ProBenefit's point of view, the results of indigenous participation were disappointing: no dialogue with Schwabe Pharmaceuticals was entered into; the benefits that were offered by the company (capacity building, working group formation, travel possibility) were neither recognised as such, nor made sufficient use of. Moreover, the indigenous counterpart never made any proposals for expected benefits, nor were any conditions or contractual guidelines articulated, despite the support available from 'native experts' (ProBenefit, 2007). This outcome ran counter to ProBenefit's expectations. The project had assumed that participation would work due to the strong political organisation of indigenous communities in the Ecuadorian Amazon, in which the structures for consultation and negotiation were in place. Aware of the scandalous consultation carried out by oil companies in the Napo province in 2003,[26] ProBenefit assumed that as long as the planning and realisation of a public consultation was participatory, acceptance of the process would be high and co-responsibility of all project partners would be assured (ibid.). What exactly went wrong?

I identify four key areas of conflict that led to the premature ending of the project. One, the relevance of equitable access and benefit sharing to the lives of indigenous people was assumed and, ultimately, imposed rather than discovered as an actual priority for people. Two, ideas of participation were fixed and based on unexamined beliefs about rationality and the public sphere: the messiness of real indigenous participation conflicted with what was required for project legitimacy vis-à-vis funders and the overall global public. Three, the approach was based on the myth that the project was taking place on a level playing field, and what we might call the historical 'naivety' of ProBenefit team members complicated an already precarious 'partnership'. Four, other visions of what the protection of traditional knowledge might mean and ought to aim at were eclipsed during project activities: despite their best intentions ProBenefit thereby imposed a value system and world view on its indigenous participants. I will briefly address each of these conflict fields before moving on to the more detailed description of the ways in which alternative visions of traditional knowledge were ignored, silenced and disregarded in practice.

Contriving relevance

While the construal of indigenous peoples as affected by bioprospecting (due to their rights to their knowledge) offers itself as a useful tool to frame certain economic injustices, it insidiously supports the view that people's interests are primarily defined economically, and in terms of property. It downplays the possibility that people might actually not care about a pharmaceutical corporation elsewhere holding a patent on an active ingredient of a plant of ancient use. Yet this attitude might be more widespread than expected. Indeed, I have found that the relevance of the protection of traditional knowledge as protection from misappropriation had to be actively 'created' for people to conceive of it as a threat relevant to their lives.[27] Disinterest in the matter was generally dismissed as indigenous ignorance rather than as an indication of a valid, alternative perspective requiring exploration.

For example, the first few sessions of the capacity-building course were characterised by a lot of mobile phone use, joking and flirting on the part of the indigenous participants, who seemed to make use of the setting for what the Germans thought of as 'disturbing' sociability. In the evenings, several of the male participants, together with the people living in the community in the vicinity of the workshop venue, would indulge in alcoholic beverages to the point of getting severely intoxicated, which led to a series of absences during the morning sessions. It was hence reiterated again and again that this course was 'important', and that the participants had tasks to fulfil 'on behalf of all their communities', and, in fact, 'their whole people'. It was also during these first sessions in particular, that it was repeated how unique ProBenefit was, and what a great opportunity it would be for the Kichwa people, all indigenous nationalities, and Ecuador as a whole, if the participants made the best of this course. Certain 'ground rules' were then participatively defined, mobile phone use banned and greater attention pleaded for. The indigenous representatives themselves came up with these rules when the task was presented, and over the three months a particular project ethos came to characterise interactions, with participants disciplining each other if necessary to pay attention, participate and turn up on time for the morning sessions. The relevance (of ProBenefit, of commercialisation and protection of traditional knowledge) to Napo Runa lives was assumed and continuously performed (through reiterations by the German team in particular, but also by myself, as well as increasingly working group members themselves). The performance of *irrelevance*, however, was never taken as a legitimate expression of opinion or standpoint. In this way, forms of non-participation – such as when participants chatted and giggled amongst themselves about their private lives during the capacity-building course, or when nobody turned up to working group meetings to design a consultation process – were not interpreted as pointing towards potential flaws in the project, or its lack of relevance, but as indications of the incapacity of the participating indigenous organisations and the undisciplined nature of their members.

Participation power

Participative methods almost invariably increase group cohesion, and can trigger a feeling of co-ownership of the group process. Yet, such niceties can be deceptive: while authority is being decentralised through participative methods, certain unquestioned norms, values and power structures are easily internalised. The 'identification with' and 'co-ownership of' projects that use participative methods are often effective in producing successful outcomes in terms of project implementation. However, participation does not in and of itself lead to emancipatory or empowering results (cf. Cooke and Kothari, 2001). Indeed, an emphasis on the micro-level of intervention (participatory, decentralised, horizontal project activities and decision-making processes) can obscure and sustain broader, macro-level inequalities and injustices (geo-political asymmetries, institutional racism, gender inequalities, global colonial relations). From a Foucauldian view, participation can be a technique through which existing power relations express themselves in new ways – through the now self-disciplining participants.

Despite the insistence on an (ultra) transparent and (highly) participatory approach, ultimately the most vital aspects of the process were still defined by ProBenefit. The process still unfolded on their terms (partly to counter corrupt tendencies of indigenous organisations, in itself a problematic perspective), which were basically terms of a particular understanding of legitimation (one infused with images of openness, dialogue, transparency, rationality, etc., all to be found in the conventional ideas about the public sphere; see, e.g. Fraser, 1997). For example, it was possible for indigenous participants to insist that a neutral working group be formed (which fits with the imaginary of ethical legitimation of participatory approaches, and also would overcome the logistical difficulties of consulting directly with 60,000–100,000 people living in more or less remote rainforest locations); however, other key aspects of the process remained immutable. It was for instance impossible to extend the project timeframe[28] or to redefine criteria of transparency and representative participation.[29]

Neglect, wilful delay and sabotage are all 'weapons of the weak' (Scott, 1985), often used by the subaltern to exert a form of power over the processes affecting their lives. The question is, of course, why this should not be seen as a 'legitimate' form of expressing one's attitude or (unarticulated) opinion. Given that the primary concern of indigenous communities is their struggle for self-determination over their territories, neglecting or even sabotaging projects such as ProBenefit might well be the 'best' or most easily available way to exert some power in this regard. Is this not a form of participation, too?

The myth of a level playing field

The niceties of participation hid the deeper conflicts at the heart of ProBenefit – conflicts which harked back to colonial relationships with a history of 500 years, and which manifested themselves as seemingly unrelated frictions or complications throughout the project duration. It was difficult for the German team to understand and accept the suspicions with which they were faced, despite the transparent and participatory process which they had worked hard to achieve. As one team member remarked: 'It is always the same. Every time we go over and over the same issues: that we are not here to steal anyone's knowledge. That if we were, we could have long done so! That after all, FONAKIN and other organisations themselves decided on this particular process. It is quite exhausting'. For the indigenous delegates their worries were legitimate. As one participant put it: 'They [the ProBenefit team] have not come here for charity! This is a business proposal. They think they will make some money. But how are we to understand what is going on?!'. 'First the white foreigners came to steal outright, now they come to make business. What is the difference?'

It is interesting how readily it was assumed that a partnership could be constructed simply through a transparent and participatory process. That the expenses were paid by ProBenefit (or directly by the company) seemed to be taken as the (only) necessary levelling of the playing field. Trust was then assumed to be only a matter of transparent dialogue. However, the power asymmetry of the whole endeavour could not simply be readjusted through a participatory consultative process. The economic injustice which ethical bioprospecting professes to redress has, as we all know, a formidably bloody and brutal history of over 500 years. To leave this fact completely unaddressed is bound to fail to build trust or create partnerships with indigenous Amazonians. This approach is not specific to ProBenefit, but extends to other experiences of working with indigenous peoples: the historical context is rarely taken properly into account – indeed there is a general denial of the past, when white people come to do 'good'. A familiarity with the situation 'on the ground', especially in terms of people's perspectives being informed by often brutal historical realities, is most often lacking.[30] The repeated comments by more than one ProBenefit team member referring to the 'unbelievable patience' and 'goodwill' of, and 'great risks taken' by, the participating pharmaceutical company Schwabe indicate the belief on the part of the project team that the interaction was occurring on a relatively level playing field. However, an expense of about US$25,000 in 2006 (which Schwabe Pharmaceuticals Ltd. paid for the capacity-building course) does not necessarily back this view, given net sales of €490 million in 2007 (and research and development expenses of €27 million worldwide in the same year).

Eclipsing other visions: notes from a capacity-building course

The problems ProBenefit had to face were rooted in its structural inability to question some of its own fundamental assumptions regarding the value of traditional knowledge, the threats it faces and the most adequate strategies of protection. Its 'CBD assumptions' eclipsed other possible ways of understanding what was at stake. In many ways, this might have been a problem of 'late' participation: communication with indigenous organisations only began once the project had been conceived and was under way. Despite its willingness to delegate authority and of course responsibility regarding the consultation and negotiation process, the project was never meant to be a project primarily *for* indigenous peoples. Neither was it a project for Schwabe Pharmaceuticals; rather it was a knowledge-producing initiative informing the processes of the CBD, and the wider access and benefit sharing 'community'. In this way, the relevance of the aims and objectives of ProBenefit

FIGURE 6.1 AMUPAKIN cosmetics laboratory, Archidona 2008
Source: Nina Moeller

to the lives of indigenous people remained unexamined, and it remained a classic case of 'them' participating in 'our' project (Cooke and Kothari, 2001). ProBenefit, despite best intentions, imposed a value system and world view on its indigenous participants. The ProBenefit team, and most of the teachers and facilitators it hired for the course, were unable to see or consider the alternative understandings which were repeatedly voiced, a point which contributed to the strong sense of asymmetry felt by the Kichwa participants, underlining their historical sense of injustice. I use the remainder of this chapter to illustrate the way in which the eclipsing of other visions occurred in practice during the capacity-building course.

The premises of AMUPAKIN (*Asociación de Mujeres Parteras Kichwas del Alto Napo;*[31] Figure 6.1) are located on the outskirts of the community of Sábata, a typical near-urban indigenous settlement of wooden shacks and houses circling a football field. A big, yellow concrete arch and iron gate mark the entrance. Behind it appear several new-looking concrete buildings: the main health centre, a conference venue, a laboratory for the production of shampoos and natural medicines and three *cabañas*, the mosquito-netted accommodation for visitors. Tucked away out of view is a wooden ramshackle hut with tin roof, an open fireplace and a gas stove: the kitchen. All is set amongst overgrown flower beds, herb and vegetable gardens and surrounded by what are, by Amazonian standards, small trees. The construction of the 'House for Life' (*casa para la vida*), as AMUPAKIN's premises are known, began in 2001 with the financial support of the Spanish Red Cross that has left its mark in the form of a metal plaque on a concrete rock-imitating mound which everyone who enters passes.

On a hot and sunny Thursday morning in March 2006, a group of people started to gather in the conference building. The room was bright, the ceiling high and the windows big. The floor had been swept, and heavy, light-coloured, lacquered tables and chairs form a U-shape, opening onto a whiteboard and flipchart. A few white people, whom I knew to be German, were busy with papers and boxes, and a very European-looking Ecuadorian woman was talking to one of the midwives. Everyone else, about 20 people (all Kichwa except me), stood or sat quietly about. Soon, a desk was set up and topped with papers and a laptop. One by one the course participants were called up to the desk. Each one received a schedule, a pen and a notebook, and it was clearly explained that they were to stay on site for the full four days of each course module, and that they were from this point on accountable to their organisations as delegates and that they could not be replaced by anyone else at any point. Everyone signed their names on a register, then took a seat along the U-shape, and the introductory session of the first module began.

My own presence was warranted as a volunteer and independent adviser to FONAKIN. Rosa Alvarado, FONAKIN's President at the time, had welcomed me warmly into the organisation just a few weeks earlier. There had been foreign PhD student collaborators before. Everyone seemed generally happy to have me hang around their concrete office building in which Amazonian mould is winning its battle with industrial wall paint.[32] 'None of us knows anything about the protection of traditional knowledge and intellectual property. It's a new issue and politically very controversial. We have been severely criticised by other [indigenous] organisations just for signing the contract with ProBenefit. It is good that you are here, I want you to follow the whole process, and make sure that nothing goes wrong', Rosa said to me shortly after I had arrived. *'I wonder what "wrong" means in this context'*, I wrote in my notebook that day. As I familiarised myself with the complex and volatile politics of indigenous Ecuador, unfolding under enormous pressures from above (market) and below (grassroots), it became clear that FONAKIN was involved in a precarious balancing act of negotiating its (various) roles with regard to its members, the Ecuadorian indigenous movement as a whole, the Ecuadorian state, national and international funders and project partners. Things could hence 'go wrong' in a variety of ways, a damaged reputation within the indigenous movement and discontent amongst its members being amongst the *very* wrong. As a primarily representative organisation, FONAKIN's legitimate authority depended on good relations with its base communities, as well as other federations.

The first evening, after dinner, a party was organised to celebrate the start of the course. Several women briefly danced to some contemporary Kichwa music, one of the German facilitators got most people involved in some Bavarian yodel exercises and dancing and one of the older men crudely dramatised a shamanic healing ceremony which the Bavarian then had to imitate. Everyone seemed thoroughly amused. Florinda, one of the oldest midwives, ended the evening with a song about her grandfather's life and a call to all indigenous organisations that they may not forget that Napo Runa life really is in the forest. The song struck my European ears as more of a weeping. It was made up on the spot, which is the sign of a competent Kichwa singer: the ability to perform there and then moving, melodic poetry full of 'old words that our grandparents used'. 'Do you hear?' I was asked by a young man next to me. 'She knows a lot of traditional knowledge'. 'Yes' said another, 'she gives advice of how to live well, she reminds us what is important, she knows a lot'. This was my first direct encounter with a perspective on traditional knowledge as ethical rather than more purely empirical in kind. All day long we had been talking about traditional medicines, and how valuable this information about biological resources was for the whole of humankind, and how the elderly were like libraries, full of such important information. Yet Florinda was singing about being a bird, a toucan woman, about being full of yearning for her people to return to the forest. Her performance was proof of her 'traditional knowledge', which in this context meant a connection to and understanding of particular *values* rather than *data sets*. Such knowledge is of course uninteresting as far as pharmacological research is concerned – it might even be antithetical. Yet it was of obvious concern to Florinda and others around her. Protection of such knowledge would look very different to fair and equitable ABS arrangements.

The following two days were spent learning about cells, genes and biodiversity. At one point, the husband of one of the midwives commented: 'But the properties of plants can change! Their medicinal powers can become stronger or weaker when they get relocated or cultivated or tended to. Also, different properties of the same plant are more or less prevalent at different times. That's why we time the harvest. Sometimes it is better to harvest at night or during full moon, sometimes not'. 'Yes' said somebody else 'and also plants don't heal if you do not have a spiritual connection with them'. Others nod. 'Aha' said the facilitator, and continued to explain genetic inheritance while ignoring this traditional understanding of medicinal properties. An opportunity to gain actual insight into traditional knowledge (on changing medicinal properties) thus was ignored. A few PowerPoint slides later, the difference between biological and genetic resources was being defined. 'This is a *very* important distinction' emphasised the facilitator. A *genetic resource* is the genetic information contained in any part of a living organism, however small; while a *biological resource* is the whole of a living organism, or at least a significant part of it. The CBD deliberately refers to genetic resources only. 'You need to understand that access to genetic resources is not the same as access to biological resources. If the genetic information contained within a living organism is being scientifically or commercially used, we have to talk about access to genetic resources. However, if it is a whole plant, or a part of it, such as its sap, that is being used scientifically or commercially we are talking about access to biological resources. The company Schwabe that would like to do some bioprospecting in the area is seeking access to the biological and not the genetic resources'. 'Indeed, we are not interested in patenting genes', agreed the German representative of the pharmaceutical company who was present. 'So, we can do away with myths now' explained the facilitator, 'bioprospecting is not always bad! As long as it is done legally and with the consent of the communities, it could be a good thing. Bioprospecting is not biopiracy'. 'Shamans have also always done types of bioprospecting' added one of the German team members, 'in fact, they are like little companies, for you also have to pay them when they provide their services'.

I am recounting this particular exchange about biopiracy to illustrate how simple answers often foreclosed serious discussion about contested issues during the capacity-building course. Time, of course, was limited, and since a lot of subjects were supposed to be covered, lengthy discussion often needed to be cut short. In this case, however, one of the most crucial questions of the whole endeavour – when is bioprospecting legitimate? – was being brushed aside with simplistic explanations. This meant that participants often failed to receive the kind of information that is necessary in order to form an opinion about complex matters. Jodie Chapell (2011) argues, for example, that there are many 'biopiracies', and that the patenting of genetic materials only constitutes one such piracy. Moreover, patents on entire plants can be held in the United States under the U.S. Plant Patent Act of 1930, and indeed such a patent was granted to Loren Miller in the highly controversial *ayahuasca* patent case, which involved a protest by the Cofán people of the Ecuadorian Amazon (Fecteau, 2001; Moghaddam and Guinsburg, 2003; Dorsey, 2004; Schuler, 2004; Shiva, 2007). What is more, Schwabe Pharmaceuticals patents all its products. While these patents are not for actual plant varieties, they are usually for plant extracts (as well as their extraction methods) based on biological materials and not genetic information.

A similar incident concerned trade in plants and knowledge. While the facilitator explained the concept of agrobiodiversity, an inflatable globe was being passed around. She asked: 'Did you know that plantains and bananas originally come from Africa?'. 'No! They come from here. They are our *comida típica* [traditional food]' was the united response. 'No, no. They are from Africa. You see, different cultures have always exchanged and traded things and knowledge'. Based on this information, we then stuck pictures of different plants and foodstuffs on the globe corresponding to their place of origin. Again this example illustrates how complex issues were being obfuscated by simple answers. While it is undeniable that different social groups have always exchanged material objects and knowledge, the modes of such exchange vary widely. The plantains and bananas which actually originated in South Asia and not in Africa (Simmonds and Shepherd, 1955; Harlan, 1971; Zeller, 2005), for example, reached South America as part of the colonial trade system which moved slaves and exotic products in various directions across the Atlantic, and decimated indigenous populations.[33] The rhetoric that 'people have always exchanged things, why stop now?' does not take account of the historical and political context in which such exchange is taking place and by which it is determined. Instead of more in-depth discussion of such issues, we engaged in a little trust-building exercise to activate the mind through a little bit of movement: everyone formed a circle, and one blindfolded person stepped in the middle. He or she was pushed around and caught as she fell and stumbled from one side to the other. Such *dinamicas,* as they are called, were used often during the course. A useful method to enhance concentration, learning effectiveness and group cohesion, the deployment of such exercises ultimately serves those in whose interest the course content is.

Later the same day, the delegate from Schwabe Pharmaceuticals passed around little sealed plastic bags. The first one contained whole dried gingko biloba leaves, the second one powdered gingko biloba leaves, the third one a gingko biloba leaf extract – a very fine, yellow-brown powder – and the fourth one a handful of coated tablets, red-brown in colour. He also passed around the very same tablets in their shiny product packaging, including the package insert. The package read TEBONIN®. The 27-step manufacturing extraction process is patented internationally, and so is the extract EGb 761® itself. Nobody mentioned that at the time – I found it out later on the internet. Dazzled by the sparkling products that can be made out of some leaves, the course participants asked many questions: 'What is it for?' 'Where do the leaves grow?' 'How do you make the extract?' The German delegate explained how a difficult extraction process is required, involving a lot of the state-of-the-art technology owned by his company. 'Would the extraction process happen here in Napo, or would it all be in the labs in Germany?' asked someone. 'This is not clear yet' answered the German delegate. 'If a trustworthy, reliable counterpart can be found here that has the relevant capability for extraction, then yes, it's a possibility that it could happen here'. 'Mixing and exchanging our knowledge with Western science is fine, but my worry is that the company will have all of the lucrative benefits and the local organisations are left with nothing, no money and no knowledge, especially for the future and for the children. One hope is that the company would move all of the production process over here', said one of the participants. 'Well, yes, there are unclarities about the laws and potential partners, but it is not impossible. There always is the possibility of creating a multinational company if we find the right counterpart', explained the German delegate. I am later told by a few course participants that this incident made them feel uncomfortable: 'He could not answer our questions'… 'They are making empty promises! And who will eat the pills they make? White people'. … 'When he talks of making a company here, I don't believe it's any of us that he will employ. They will get people from Quito'.

'So where does traditional knowledge come from? How is it established?' asked the delegate from the German pharmaceutical company, sweating visibly, his naked legs covered in insect bites. His PowerPoint slides were in English, and hardly visible on the wall of the bright workshop centre, so he waited patiently for the translation of his question and the ensuing discussion in Spanish and Kichwa to end, wiping his brow. This question was more engaging than the previous ones. Everyone started speaking at once: 'The plants tell us' … 'Yes, the plant spirits talk' … 'When somebody in the family is ill, it's the plants that will tell us how to prepare them and make medicines from them'. … 'The *yachakuna* [traditional healers, shamans] speak regularly to the plants, so they know'. … 'When I was a little boy, and my mother was very ill, one day this plant – it grows here outside in the garden, I can show you – this plant came and it was laughing and dancing around in the house and we were a little scared but it told us to boil it and prepare a tea and then it left, so we found it again and made a tea and soon my mother was feeling better'. The translator hesitated at first, then explained the answers of the Kichwa workshop participants to the pharmaceutical delegate. 'Well', said the delegate after a confused little pause, 'okay, yes, but traditional knowledge comes from … ' he paused again as he flicked the remote control to populate his slide with prepared answers that fly across the screen in swoops before settling down as bullet points. 'Well', he commented the slide, 'it comes from accidents and coincidences, from one's own experience and self-testing, from hearsay, from knowledge exchange and from literature'. The translator translated and everyone remained quiet. Shortly after, the elderly midwife sitting next to me started to whisper angrily with a young man who nodded back at her. The pharmaceutical delegate, however, continued his PowerPoint presentation.

The inability on the German side to acknowledge or even register this very different understanding of the origin of traditional knowledge, which constitutes a central aspect of Napo Runa cosmovision,[34] was lamentable. It maintained the gap between the two sides, prevented a deeper understanding and exploration of the issues at hand, and essentially constituted a continuation of 500 years of colonial patronage.

In the literature documenting, explaining or analysing the legal guidelines referring to 'traditional knowledge', traditional knowledge is usually defined as 'knowledge, innovations, and practices of indigenous peoples and local communities' (from CBD Art 8j), and often described as 'inter-generational and orally transmitted' (e.g. Posey, 2002; Howell and Ripley, 2009; see also the Bellagio Declaration[35]). Its origin is hence located in the distant past, embedded in the ancestral practices of indigenous communities. The emphatic concern with origin in most contemporary dealings with traditional knowledge must have to do with the importance of origin to intellectual property law. Intellectual property protection is dependent on the origin of the intellectual work to be clearly traceable to a particular juridical person, such as *Ulysses* to James Joyce or Windows Vista to Microsoft. The assumptions underlying such originary ideology are tenuous, and an exploration of its ideological connections to creationism and doctrines of free will promises to be interesting at least. Unfortunately there is no scope for such an exploration here. Suffice here the flagging up of 'origin' as a significant discursive device in the performances of intellectual property protection and contestation.

In this context, then, what would it mean for knowledge to originate in one's relationship with a plant spirit? I realise that entertaining such an idea will be rather difficult for many readers. Nonetheless, such ways of speaking about and understanding aspects of the world encode particular attitudes and values. For example, this view of knowledge speaks of an intimate relationship between people and plants. It speaks of an understanding of plants as teachers and helpers. It speaks of the necessity to foster good relations and to learn to listen to what plant spirits may say. 'Plant spirits talk a lot', an old female healer told me while weaving a *shigra*. 'The problem is, most people don't know how to listen. They run past the little plant on their way into town. They miss the whisper of their name. "Nina, Nina" it will call you: "Nina wait and listen what I have to tell you"'. These understandings, visions and values are what 'traditional knowledge' – the knowledge of the Other – really has to contribute to the contemporary world. Another remedy for high blood pressure or obesity is merely a contribution to the wallets of the pharmaceutical industry. Indigenous activists participating in high-level fora such as the CBD can sharpen their teeth by insisting that what is at stake in the context of the protection of traditional knowledge are lifeways, values and practices to which the hegemonic constructions of knowledge and protection are blind and indeed antithetical.

Diets and charlatans

My conversation with Domingo, the *yachak*, with which I started this chapter, did of course not take into account the wider context in which the protection of traditional knowledge, as a necessity and cause, had developed. The concept of traditional knowledge and the need for its protection emerged especially in relation to developments and conflicts in the fields of nature conservation, and the wider biogenetic resource politics of the late twentieth century.[36] When I was asking Domingo about how to best protect one's traditional knowledge, I left the context within which I was posing the question as open as possible; in particular, I did not provide much indication of which threats to this knowledge I was envisioning. This of course means that the particular meaning of my question was in many ways up for grabs. Domingo interpreted it, as people usually do, according to what seemed to him the most likely way it was intended. Given that generally most of our conversations concerned shamanic practices, healing ceremonies and *ayahuasca* visions, and considering that we were sitting by a fire and a five gallon cooking pot holding the ingredients that were to turn into one of the most psychoactive substances known to humankind, it is maybe not surprising that he thought of spiritual abilities and attacks, healing and illness, the responsibilities and dangers of being a *yachak* and the intensity of the visual ('knowing') experience of an *ayahuasca* trance as the backdrop to my question, in relation to which the latter made the particular kind of sense that he took it to make.

In the company of members of ASHIN (*Asociación de Shamanes Indígenas del Napo*),[37] of which Domingo was president at the time, he could also speak very differently of the protection of traditional knowledge: 'There are fewer and fewer good shamans. The old ones, many have died. Young people don't want to learn and they break their diets. There are too many that call themselves *yachak*, and they go to the cities and they ask 500 dollars for a healing, and they don't know anything, they cannot heal and they give us a bad name. There is no control. That is why our organisation has made identification cards. See here [showing his laminated picture card]. We all carry them. They are recognised by the ministry. We are a legalised organisation, recognised by ministerial accord. We test all our members. Every member has to prove that they can heal. We go in a group, and we watch each raise [*levantar*] an ill person. If they get up at the end of the night, if they get better, then we can be sure. I have denied identity cards to some people, cousins of mine even, of whom I knew that they don't know anything, they just sing for the tourists. We have to work together, we have to unite and work collectively. We have to teach well, so that the young ones learn properly and do no harm [black magic], otherwise our medicine will die. Otherwise we will kill each other in envy and competition amongst ourselves. There is so much envy. And to make lots of money fast, we will break the diets, and forget the forest and what our grandparents told us'.

Protection of knowledge figures in this little speech as the collective adherence to a particular ethical code of practice, the respect for traditional norms and ancestral advice, as well as the use of certain techniques that the modern world affords (photographs, seals, lamination, institutionalisation) in order to create a framework within which Kichwa shamanic knowledge and practice remain unimpeded by bad reputation, failure to transmit properly to the next generation and mutual (intra-group) competition.

Through such organisations as ASHIN, common concerns and the potential for collective solutions are being explored, articulated and worked towards. Like all social movement and civil society organisations, they provide platforms for voluntary association according to shared predicaments, and for the forming of opinion about and strategies for social change. Traditional knowledge, its threats and the means of its protection are framed in this context as collective concerns, affecting each practitioner in her or his work, impinging upon the reputation and viability of the 'profession' or 'tradition' as a whole. The value of knowledge is here closely linked to the responsibility, individual and collective, to acquire it properly (observing traditional diets) and to use it properly (for healing, not for black magic or harm, nor for inflated personal gain). The greatest threat that knowledge seems to face on this account is perversion through improper conduct. Abstention from sex, alcohol and certain foods is considered an important part of a *yachak*'s so-called 'diet', especially when still in apprenticeship. Misconduct, mostly related to the breaking of one's diet, is said to lead to a loss of power/knowledge, perversion of character and the practising of black magic, usually culminating in (spiritually) injuring other people, and even madness. Protecting shamanic knowledge from perversion or distortion is thus about – collectively – ensuring right acquisition and right use. And for the leaders of ASHIN, most often represented by Domingo due to his articulateness, this was best done through a certain amount of institutionalisation and the development of a common discourse based on traditional precepts and ethics.[38]

Relevant acquaintances

The theme of loss and survival of knowledge, tradition and practices over generations ran through many conversations and encounters during my fieldwork, and not only in the context of shamanic healing. That the younger generation was not interested in the old customs, lore and bushcraft, and that too many of them did not even speak Kichwa, was a pervasive complaint. 'Young people don't realise, they don't care much about the plants. They walk elsewhere, they don't see the plants, and so they don't ask about them. Maybe ten, twenty plants they know by name, nothing more. And what they are good for, … how could they know!?' Some of the younger people I met instinctively positioned themselves in relation to this complaint: 'All my friends have moved to the city. They have employment. But they forget how to walk in the forest. I prefer to be here, I like to listen to my grandmother. She knows how to interpret dreams very well. I work with her on the *chagra* [horticultural plot] and she teaches me about the plants. The plants heal. I want to learn more, so that when she dies, I can teach my children and grandchildren' … 'I liked very much living in Quito [Ecuador's capital]. We had a lot of fun. It opened my mind. But when my son was two, his father left me, so I came back to be with my family here … I see our culture differently now, I am happy here, but I feel a lot of pain to see it disappear. All that my grandparents knew is getting lost. That's why I am learning Kichwa now and that's why I go to the dance group [group performing folkloric dance]' … 'They say all that our grandparents knew is getting lost. It is true. But I cannot change that. To improve [my family's situation], I need to study and earn some dough. That's why I live in the city … It is sad, but in my community there is not much forest left. So what use is it to know the plants!?'

Traditional knowledge ('what our grandparents told us' was the often-used idiom) in this context is understood to be slowly dying with the elders given the decreasing uptake by the next generation. The indigenous youth is seen, and sees itself, as largely uninterested in the 'life of the forest'. What was dubbed on several occasions as the 'wants and needs of the city', the (new) desires and requirements that life in or near the cities provoked, meant that for many, the everyday had to be so configured as to allow for time and energies to be directed towards the provision of money, and the creation and maintenance of those relationships which facilitated the acquisition of objects of desire, the use of the services on offer and general participation in the network of urban social relations. Since the more time one spends walking in the city, learning about its delights and treacheries, the less time one spends walking in the forest, learning about the same, it is unsurprising that one's knowledge of the forest does not only remain limited, but it also becomes increasingly irrelevant – and impossible – to acquire it in the first place.

This tension of new ways of life eclipsing older ones, and the ambivalent feelings that such changes arouse are probably ubiquitous to human history. The struggle to maintain a certain amount of permanence in the face of ever-present change, mortality and fading memory might be a contender for a universal attribute of human societies if ever there was one (cf. Weiner, 1992). However, the specific circumstances of loss and change will always be particular. They can be violent, disruptive and disorientating, or creeping, uniting and inspiring; they can be emancipatory or disempowering, sensitising or dulling down. They can be all or some of these things. Struggling to influence, and to participate in the shaping of these circumstances is a central aspect of collective self-determination.

The theme of loss and disappearance of traditional knowledge in the conversations during my fieldwork struck me as a kind of coat-hanger upon which people would hang their laments and grief about unwelcome changes to their collective lives – those perceived as too rapid, too asymmetric and too destructive. The question of how to prevent this loss – or how to create a more positive kind of change? – left many feeling mystified and powerless, and some in tears. Of course, grandchildren could listen to their grandparents, teenagers could re-learn their mother tongue after having abandoned it in the usually racism-suffused schools and parents could take their children into the forest to gather ornamental seeds for use in handicrafts, and point out a few plants and tell their stories on the way. But in the face of oil spills, toxic rivers, disappearing species and the cash barrier to participation in much of contemporary life, even in the Amazon, and in the face of all other manifestations of 'Euro-American developmentalism' (Whitten, 2003: xi), the question of the *relevance* of 'the old knowledge' looms large.

On one occasion, I asked Maria, an elderly healer and midwife who played a key role in the establishment of AMUPAKIN, what she thought about books and other ways of documenting herbal knowledge, so that her great-great-grand-children would be able to learn about what she once knew. 'My little girl' she said, 'the knowledge is in the plants themselves. Write the books! Read the books! When the plants go, the knowledge goes as well. Do what you like'… 'What about botanical gardens, then?' I wondered. After all, a medicinal plant garden was one of AMUPAKIN's long-term aims. Maybe this would be a way to carry some knowledge into the future. 'Yes, yes' Maria did not sound convinced. 'The problem is, many plants cannot be cultivated. They grow weak, and they don't heal. It's the wild ones that have the power … And the knowledge of the forest does not grow in a garden. And the lakes, and the rivers, and the hills, and the waterfalls! This knowledge cannot be known in the books'. Domingo confirmed this understanding: 'Every powerful place gives us knowledge. I have got a lot of knowledge from the lakes … There are powerful places with much energy everywhere in the forest, special places. My grandfather took me to some of them. Every healer has knowledge from these places, from rivers, from waterfalls, from big rocks, from the hills. But now … The contamination finishes these places. You go to them and there is no energy. I have analysed a lot, and it seems to me that the energies run downstream, down from the waterfalls and the hills and down the rivers, … and into the oceans. There where the contamination arrives, the energies disappear. In a short time, all that is left for us is to go to the oceans to find the energies in the sea. Otherwise, all that knowledge will be lost'.[39] (I did not have the heart to tell him that the oceans were themselves by now so contaminated and over-exploited that ninety percent of the world's biggest fish had already vanished[40]).

The knowledge valuable to these speakers is not replicable in books or other documentation. It rather speaks of an unmediated connection to certain places and plants. It is a knowledge by *acquaintance* rather than a knowledge by *description* (Russell's distinction, 1911, following Grote, 1865; von Helmholtz, 1868; and James, 1890[41]); it is through acquaintance with things that their particular powers – or energies – are imparted, a process that creates knowledge. Because such knowledge only comes into being through *experiencing* a particular place or object, through interaction and contact, a book could never transmit it. Acquaintance with a book about the Amazon, its paper, ink and glue, is on this account *knowing the book*, and not *knowing the Amazon*. (This distinction, although hidden by the equivocal character of the English word *to know*, is made in many languages, such as in Latin *noscere* and *scire*, German *kennen* and *wissen*, Spanish *conocer* and *saber* and French *connaître* and *savoir*).

The only way to 'protect' this particular knowledge, in the sense of ensuring its continued existence throughout the change of generations, is to enable people's acquaintance with these places and plants and other objects of value. The primary threat to such knowledge is the disappearance of those objects through the interaction with which it is created (think here deforestation, climate change, urbanisation, subsoil resource extraction). The deterioration of the value of these objects (through their contamination and domestication, for example) will also diminish the value of the knowledge they can impart, and thereby constitute a kind of threat to be prevented. Another threat is irrelevance. Even if valuable places and plants continue existing, if the role which they play in people's lives is eroded, acquaintance with them becomes meaningless. It is hence not just the continued existence of the places and plants, but the meaningful relationships which peoples maintain with them that is of importance in this context. As such, ways of life that integrate relationships to such objects of value ensure the relevance of this kind of knowledge, a prerequisite for any form of protection to make sense at all. This raises the question, however, whether payments through bioprospecting contracts might be a way to overcome the increasing irrelevance of 'traditional knowledge' to most people's lives. After all, if the 'life of the city' takes people away from the forest and older forms of livelihood, then maybe it is as part of the 'life of the city' that the relevance of traditional knowledge must now be revived. The ProBenefit initiative was based on a version of this view: economic benefits will provide the best, and indeed the only kind of incentive for people to value and preserve their traditional knowledge in a changing world. The problem with this perspective is that the 'value' and 'relevance' which knowledge of and acquaintance with plants and other things then carries, would be determined by its economic content. The 'power' or 'energy' which things are understood to impart is likely to get lost when the main point for getting to know them is the fact that money can be made from such acquaintance.

Patently recognised

In April 2007, ASHIN was approached by the director of the teaching module and research cluster on 'Genetic Resources and Ancestral Knowledge' of the *Pontificia Universidad Católica*. The group was enrolled in a project on the protection of ancestral knowledge that was meant to provide legal recognition to ASHIN for its members' knowledge about medicinal plants in return for a set of arrangements regarding student research opportunities and a botanical garden maintained by ASHIN as an *in situ* herbal collection for the university. Through the process of engagement with this project, and with the students and staff members of the university, Domingo's understanding and use of the idea of protecting traditional knowledge developed new facets: 'The foreign pharmaceutical companies come here and they steal our knowledge. We need to get our own patent, so that they know that it is we who are the owner, so that they cannot just take it away from us as they have always done with everything that is ours'.

So, theft as threat and patents as protection? I asked who exactly would be the patent holder, and when 'ASHIN' was the answer, raised the problem of authoritative representation of a whole people. Why ASHIN? On whose mandate could they claim ownership of traditional Kichwa plant medicines? What would those healers think who were by choice not affiliated to ASHIN? What would other associations of shamans do when they heard ASHIN had such a patent? Domingo stated in a defensive tone that they would of course hold the patent 'on behalf of the whole of the Kichwa people',[42] but that it might indeed create tensions, and that they would have to think about how best to go about this. He would call for a meeting with all the leaders of the various Kichwa federations of the lowlands. He had already thought about that, in fact. I also explained that patents were only granted for 20 years, and that the costs of filing, monitoring and enforcing a patent application could be enormous. While applying for a patent costs usually just a few hundred dollars, lawyer's fees easily extend into tens of thousands of US dollars. Moreover, in order to prevent others from copying one's invention, it is necessary to file applications in several countries: 'A rule of thumb is that it will cost approximately US$100,000 to adequately protect an invention internationally' (Carolan, 2009). And this would not be the end of one's expenses. Monitoring patent infringements is time-consuming and expensive. The biotechnology giant Monsanto is said to have an annual budget of US$10 million to police infringement (Kimbrell and Mendelson, 2004). Lastly, for patents to be useful 'protection' in cases of conflict, they would also have to be enforced in court. The American Intellectual Property Law Association has estimated that in 2000 alone, US-based companies spent US$4 billion on patent litigation (AIPLA, 2001).

Even though I did not flood him with exact numbers and references at the time, Domingo remained quiet. The point was that the director of the university programme had suggested to Domingo they make a list of the main plants known and used by members of ASHIN, and have this list attested by the public notary as a way to certify ownership until effective legislation with regard to the protection of traditional knowledge was passed nationally. Domingo had put a lot of hope in this 'patent'.[43] 'With the notarised list, we can show the proof that this is our knowledge. Nobody can come and say we don't know anything. With the help of the University we can build the clinic of natural medicine, finally. Then we can practice our medicine, and defend our knowledge. Step by step they realise what we know'.

Domingo expanded on 'the proof that this is our knowledge' by adding that 'nobody can say we don't know anything'. The racist stereotype still pervading most of Ecuador – including its recently (2013) re-elected president Rafael Correa[44] – is of course that indigenous people are generally backward and ignorant, a perspective which most of the indigenous people whom I got to know would position themselves against at one point or another. Recognition for their knowledge, for the fact *that they knew something* and for the fact that they knew things that were particular, special, and indeed characteristic of their particular existence, history and culture, was a desire expressed many times. The term 'our knowledge' was often used, it seemed to me, to express relations of identity rather than a claim to the right to dispose of such knowledge at one's will (which is the dominant interpretation of the rights that private property relations entail).

What is important to note here is the sense in which the possessive pronoun ('our') can imply a notion of property as *characteristic* as well as *ownership* (the difference is made clearer in German, in the difference between the words *Eigenschaft* – property, characteristics – and *Eigentum* – property, ownership). To find ways to 'prove' that 'this knowledge is ours' was, I believe, a way to insist on the value of (in this case) Kichwa identity, at least as much as it might have also been a way to lay claim to some of the rights that ownership confers. The struggle for *recognition* of one's value, including the value of one's ideas, one's understanding and one's creativity – one's knowledge, that is – especially in the face of discrimination, marginalisation and exclusion, easily takes on a significance that is more fundamental than the struggle for *protection*.

Highlighting the value of traditional knowledge is a major part of making the case for its protection. To call for the protection of something always entails an (implicit) claim about its value. Similarly, it cannot be ignored that the struggle for recognition might be the main driver behind calls for protection, that establishing protective strategies and putting them in place might be perceived as ways of signalling recognition, as ways of manifesting recognition in the world and that hence *recognition* is what protection is mainly about. After all, recognition is something largely intangible, and (inter-)subjective. It is hardly enough to *state* one's recognition of something ('I think you are clever' or 'I value your intellect'), unless it also reveals itself in the world, in one's behaviour (such as in my asking you for advice, or consulting you about certain subjects, promoting you if I am your boss, or maybe applauding at the end of a speech you give). Of course such manifestations (especially applause) can also feign a recognition, which really does not exist (I might ask for your advice just to make you feel valued, and maybe lend me some money somewhere down the line, and in fact, I might never act on your advice). However, for recognition to become *real* for someone, it needs to show itself in the world, it needs to leave signs and make marks that can be perceived. Passing legislation that *protects* traditional knowledge (in whichever particular sense of protection) can be, or seem like, a sign of recognition of its value. Yet this recognition could also manifest in alternative ways, and so we have to ask whether the legal protection of traditional knowledge constitutes the desired recognition, and also *what kind of value* it actually recognises.

The focus on theft

Inherent in Domingo's ideas about patents, however, was also the understanding that certain injustices ('stealing our knowledge') were being perpetrated by, for example, pharmaceutical companies, and that a patent might prevent this. This view that traditional knowledge is threatened by, and hence needs to be protected from, unapproved appropriation and subsequent commercial exploitation animates and dominates the debates in international policy-making fora concerned with traditional knowledge. As repeatedly noted, in most of the literature and activity concerned with 'the protection of traditional knowledge', protection is understood as referring to strategies and measures that prevent the unapproved appropriation and subsequent commercial exploitation of traditional knowledge. Where the threat of its erosion and loss is recognised, it is rarely treated on its own terms, but rather in conjunction with the threat of misappropriation and economic injustice, leading thus to recommendations for protection that construe and institute traditional knowledge as intellectual property of the respective indigenous community. Indigenous people are made into market actors, their knowledges into commodifiable data sets or entries into museum catalogues. As this can seem like a better deal than the predicament in which most indigenous communities find themselves – excluded from but not unaffected by an encroaching capitalist economy – it is often welcomed by indigenous peoples and their allies. However, in this way, the protection of traditional knowledge remains a mere plea for the reform of intellectual property law, leaving untouched the core beliefs of this system.

The current discourse of the protection of traditional knowledge has to be understood as a colonising discourse. It is colonising in the sense that it installs a particular meaning of its key terms, thereby invading, taking over and settling the understanding of these terms. It only articulates one particular way of understanding the protection of traditional knowledge, even though we have seen that talk of protection of traditional knowledge provokes a variety of concerns for people and in turn is used to frame and formulate these concerns.[45] Some of these alternate understandings of what the protection of traditional knowledge means and what is at stake in its realisation are not simply different to, but in fact conflict with, the colonising discourse of intellectual property and 'theft', in that they challenge some of its fundamental assumptions. When 'taken seriously' – that is, when we start to sincerely explore their implications – these challenges might force us to revise deeply ingrained ways of understanding such fundamental notions as property and knowledge, with radical consequences for contemporary social organisation in so-called knowledge-based capitalism.

Joan Martinez-Alier (2002) has argued that ecological distribution conflicts are often fought in idioms other than market valuation, making use of notions of 'the ecological value of ecosystems, the respect for sacredness, the urgency of livelihood, the dignity of human life, the demand for environmental security, the need for food security, the defence of cultural identity, of old languages and of indigenous territorial rights, the aesthetic value of landscapes, the injustice of exceeding one's own environmental space, the challenge to the caste system and the value of human rights' (ibid.). In this way, the struggle surrounding the protection of traditional knowledge is not only a struggle over access to resources, but also a struggle over meanings and values: 'in field or factory, ghetto or grazing ground, struggles over resources, even when they have tangible material origins, have always been struggles over meanings' (Guha and Martinez-Alier, 1997). However, the problem is that often the voices that are most clearly heard and whose concerns are taken most seriously are those who couch their demands in a language of valuation that resonates with the ultimate decision makers. While it can be strategically wise to encode one's message in terms of the dominant economic discourse in order to be heard, this also runs the risk of diluting one's original grievances and visions for alternatives and social change.

Domingo's sudden conviction that a 'patent' would be the solution to the wide variety of issues that he had himself previously framed and expressed through the idea of the protection of traditional knowledge leads me to the following two interrelated points in conclusion to this chapter. First, the attraction of 'private property' is not to be underestimated. Private configurations of ownership lie at the heart of the capitalist mode of production. Their appeal to individuals and defined groups is possibly the most powerful engine of capital expansion. Second, once the promise of private property appears on the horizon, alternative concerns and values seem to fade in its light. Domingo and other members of his association easily lost sight of the ways in which some of their concerns would not be addressed at all by the spurious promise of a notarised list as proof of knowledge ownership. For these reasons, this chapter is also an appeal to the indigenous movements of Ecuadorian Amazonia and beyond to not overplay the 'discourse of theft', which drowns out other ways of explaining what is at stake and other ways of demanding change. The larger and more varied the vocabularies of protest become, the more discursive possibilities there will be to illustrate the fact that values are largely irreducible to, and sometimes even incommensurable with, one another.[46] That is to say that the more ways we find to express the plurality of values which exist in the human world, the easier it will be to dispute that a singular (monetary) value can make commensurable the many goods and bads which affect people's lives as well as the more-than-human world.[47]

As this chapter has illustrated, the value of traditional knowledge – and its concomitant understanding of threat and need for protection – can take a variety of forms, all of which express people's real concerns. The discourse and practice of such initiatives as ProBenefit have the effect of silencing the diversity of values and making the protection of traditional knowledge commensurable with the global market economy. Yet without the legal, political, economic, cultural and philosophical recognition of the values of indigenous people, and especially without the value conflicts arising from such recognition, the 'protection of traditional knowledge' amounts to nothing more than a charade.

Notes

1 This chapter and the stories it contains are based on the fieldwork for my doctoral dissertation (Moeller, 2010) which took place on the fringe of the north-western Amazon region, in the Andean-Amazonian nation of Ecuador 2006–2008. My work was mainly with the Kichwa-speaking Napo Runa who inhabit the watershed areas of the upper Napo River. All names have been changed to preserve anonymity.

2 *Yachak* is Kichwa for 'one who knows', plural *yachakuna*. I use the spelling Kichwa instead of the Anglicised 'Quichua' as it is the currently most widely used spelling amongst Kichwa peoples in Ecuador.

3 *Sinzhi* means force, strength, especially spiritual/energetic in kind.

4 *Brujo* is the Spanish word for warlock/male witch, referring to *yachakuna* who practice black magic, harming others.

5 A *yachak's* diet refers to the abstention from certain foods, as well as activities, during certain periods. In particular, after *ayahuasca* ceremonies, salt, chilli, alcohol and fatty meats, such as pork should be avoided. Someone who is learning to heal is expected to abstain from sexual intercourse for several months at a time. There are times when one should not touch any object that might be either too cold or too hot.

6 This definition is to be found in the CBD's Article 8(j), available online, at www.cbd.int/traditional.

7 Especially through WIPO's Intergovernmental Committee on Intellectual Property and Genetic Resources, Traditional Knowledge and Folklore (the IGC).

8 The Doha Development Agenda's paragraph 19 concerns TRIPS, biological diversity and traditional knowledge. Available online at www.wto.org/english/thewto_e/minist_e/min01_e/mindecl_e.htm#par19.

9 The International Treaty's objectives are the conservation and sustainable use of plant genetic resources for food and agriculture and the fair and equitable sharing of benefits derived from their use, in harmony with the CBD. The centrepiece of the treaty is a 'multilateral system for access and benefit-sharing' which for certain categories of plant genetic resources guarantees facilitated access in return for benefit-sharing. In respect of traditional knowledge, the key provision of the treaty is its recognition of 'farmers' rights' through its Article 9. Available online at: www.planttreaty.org/content/texts-treaty-official-versions.

10 The text of the declaration is available online at: http://www.un.org/esa/socdev/unpfii/en/drip.html.

11 For details see http://nrsc.fr.

12 More correctly, biopiracy ought probably to be called bio-privateering. Piracy implies theft; that is the taking of someone's private property. Privateering, on the other hand, implies privatising what was hitherto not privately owned. However, to my knowledge, this more apt term has only been used by Richard Stallman (1997).

13 Bioprospecting is a relatively new term for a relatively old endeavour: it refers to the (usually corporate) development of (marketable) products based on research into and subsequent appropriation of the (commercially useful) properties of biological resources. Bioprospecting most often aims at developing pharmaceutical, nutraceutical and cosmetic products for the markets of the industrialised world, and the research phase is often aided by indigenous people and traditional farmers whose knowledge of the local biosphere is in many cases extensive and detailed. For early literature on bioprospecting, see especially Reid (1993); Svarstad (1995); Balick, Elisabetsky and Laird (1996); Shiva (1997).

14 Underlying the discussions about bioprospecting is the question of control over access to and rights to income from traditional knowledge. Who can access and use traditional knowledge, and who has the right to the economic benefits, i.e. the income which flows from such use? These are questions with regard to the property relations that characterise traditional knowledge. This is to say that in the context of bioprospecting, and in the context of ABS agreements, the question of the protection of traditional knowledge is a question of how best to configure property rights over traditional knowledge. For an extensive jurisprudential treatment of property in terms of control powers, use privileges and exchange rights (rights to income), see especially Christman (1994) and Harris (1996).

15 Some hold that depending on whether or not an access and benefit sharing agreement has been reached, bioprospecting is either legitimate research benefiting all stakeholders, or it is biopiracy (e.g. Svarstad, 1995; Balick, Elisabetsky and Laird, 1996; Schuler, 2004), others consider it to always be an instance of biopiracy simply because under current global socio-economic conditions no ABS agreement could ever be equitable (e.g. Shiva, 2007; Mooney, 2000; Takeshita, 2000, 2001).

16 In my doctoral thesis, I situate the hegemonic construction of traditional knowledge protection in the wider context of capital expansion. I identify the destruction of the conditions for people's autonomous subsistence as a vital aspect of capital expansion, and argue that the protection of traditional knowledge in its dominant form participates in this destruction. In particular, I argue that to destroy subsistence is to destroy the conditions in which traditional knowledge is created, used and reworked, and thereby the context in which it is directly meaningful and relevant to people's lives. Bioprospecting endeavours and the ABS agreements which they require constitute, no matter how fair and equitable, one of the ways in which the expansion of capital manifests today. Paradoxically, ABS agreements are also promoted and implemented as one of the key mechanisms for the protection of traditional knowledge. It is in this way that this hegemonic construction of the protection of traditional knowledge contributes to the destruction of the very foundations of traditional knowledge. For it is in the domain of autonomous subsistence that traditional knowledge is developed, made meaningful, used and changed. The domain of subsistence consists of the practices of self-provisioning through which the everyday needs of people are fulfilled, and through which their desires are shaped and addressed. It consists of the everyday lives of people and their interactions with each other and the environments they inhabit, which are not characterised by market exchange nor market rationalities and values. As the dominant form of traditional knowledge protection contributes to the expansion of capital, it also contributes to the destruction of the conditions of the very existence of traditional knowledge. See Moeller (2010).

17 I was a participant observer of ProBenefit from March 2006 until the end of its activities in the Amazon in May 2007. The misunderstandings, frictions and value clashes that characterised ProBenefit during its period of engagement with the Kichwa people of Amazonian Ecuador (represented by the indigenous federation FONAKIN) are discussed briefly below and at great length elsewhere (Moeller, 2010). As a volunteer and independent adviser to FONAKIN, I was able to work closely with ProBenefit's indigenous participants and learned about their views through extended interactions which continued until after the project's end.

18 See also http://biodiv.de/en/projekte/archiv/probenefit.html.

19 The Bonn Guidelines on Access to Genetic Resources and the Fair and Equitable Sharing of the Benefits Arising from their Utilization were adopted by the CBD sixth Conference of the Parties (COP) in 2002. These voluntary guidelines are meant to assist governments and other stakeholders in developing an overall access and benefit-sharing strategy, and in negotiating contractual arrangements for ABS. Crucially, they include the requirement to obtain prior informed consent from relevant indigenous and local communities. See https://www.cbd.int/abs/bonn.

20 I was told this in a conversation with the Schwabe representative who travelled to Ecuador with the ProBenefit team in March 2006.

21 This aim is quoted from ProBenefit's website: http://biodiv.de/en/projekte/archiv/probenefit.html.

22 The ICBG is a public grants programme sponsored by the US National Institutes of Health (NIH), the National Science Foundation (NSF), and the United States Agency for International Development (USAID) and its goals are clearly oriented to the aims of the CBD: to search for potential new drugs through bioprospecting, to promote a sustainable use of biodiversity, and to foster development through benefit sharing with developing countries – and the specific local communities involved if appropriate. Public-private sector partnerships are required by the ICBG grant protocols. One ICBG grant was implemented as an agreement between the Aguaruna of the Peruvian Amazon, Washington University, a Peruvian university and museum and Searle and Company, a pharmaceutical sub-division of Monsanto. Other grants included funding for research by the Virginia Polytechnic Institute and State University, Conservation International, Missouri Botanical Gardens, the pharmaceutical giant Bristol-Myers Squibb and a

pharmaceutical company in Suriname; research by the University of Illinois at Chicago and institutions in Vietnam and Laos; and biodiversity research in Panama.

23 The 1991 agreement between the Costa Rican quasi-governmental *Instituto de Biodiversidad* and the pharmaceutical giant Merck to exchange access to its inventories of plant samples for about US$1 million and the promise of royalties on ensuing profits from potential patents was heralded as a model at the time.

24 It is maybe worth mentioning here that it is ironic how usually public consultation is supposed to circumvent representative organisations (such as local governments, say) in order for it to be truly *public*. In the indigenous case in Ecuador, this point highlights a particular tension in the indigenous movement. It is unthinkable for an outsider to do anything 'legitimately' in indigenous territory without approaching the overarching indigenous federations of the area first. At the same time, the grassroots feel very badly represented by these federations, which are said to be corrupt, and often run over decades by members of the same families. This might simply imply that there is no one indigenous public, in the same way as there is no unified national public sphere.

25 There are two interrelated sides to this construction of the economic value of biodiversity conservation. On the one hand, biodiversity is increasingly *capitalised*. In Martin O'Connor's terms, nature 'formerly … treated as an external and exploitable domain is now redefined as itself a stock of capital' (1994: 126). In this way, it needs to be conserved and regenerated as a reservoir of capital value, rather than subjected to limitless exploitation. On the other hand, the conservation of biodiversity is itself capitalised. This is to say that conservation activities are rhetorically cast as feasible only with adequate financial return. Economic value becomes the only reason for action of any kind. This is the ideology of *homo oeconomicus* which undergirds the discourse of sustainable development and orients the CBD.

26 A badly executed consultation regarding oil exploration in the province of Napo ended in a public outcry in 2003 (Grefa, 2005). The Napo was subsequently declared a '*provincia ecológica*' by popular vote. 'Sustainable development' was to be promoted, and bioprospecting projects, such as ProBenefit, fitted this new provincial aspiration. As the disastrous consequences of an extremely irresponsible form of oil extraction had by that time increasingly been highlighted, bioprospecting projects were portrayed as a clean and just alternative to oil which would finally bring wealth to the people of the region.

27 It bears mentioning here that people drinking from polluted rivers, giving birth to malformed children, dying of cancer and losing their foodstuffs to pools of crude oil are much more radically, directly affected by the activities of the oil industry than people whose affectedness by the pharmaceutical industry is contingent on the political construction of their property relations to certain plants and of their knowledge as commodifiable. Affectedness in the case of bioprospecting needs a higher level of discursive construction than in the case of those affected by prospecting for oil and other subsoil resources (compare oil wells with plant sampling).

28 Indigenous participants repeatedly complained that the process was too fast and did not allow appropriate consultation with their community organisations. This was also the main (official) reason for the ultimate failure to negotiate an ABS agreement.

29 Only a fraction of the course participants fulfilled all the required selection criteria. From the German perspective, the capacity-building course was a good in and of itself, and the training provided would serve participants even outside of the project itself. From the indigenous delegate's point of view, participation in the capacity-building course meant absence from home without pay and without clear benefits for the future. At the end of the course, it also turned out that two signatures had been falsified, and hence two course participants were not actually the representatives of the organisations that they had claimed they were. One of the signatures had been falsified by a young man who had simply wanted his cousin to also be part of the course 'to be able to share the experience'. The other signature was interestingly falsified by one of the main leaders of FONAKIN. The 'fake' delegate purportedly represented *Salud Indígena*, the

governmental health organisation providing services in indigenous communities, staffed mainly by indigenous people themselves. The leader of FONAKIN had connections to *Salud Indígena*, but wanted a close ally to participate in the course who was not a member.

30 A parallel lack of consideration of historical context is noted by Oldham and Forero in their work with Mapuche in Chile (personal communication).

31 Association of Traditional Kichwa Midwives of the higher Napo region.

32 During my various stays, I fixed printers, set up fax machines, solved computer problems, corrected spelling mistakes, transformed handwritten notes into PowerPoint presentations, and showed my European face to visitors. I also wrote some funding proposals, and a few position papers for FONAKIN.

33 See, *inter alia*, Crosby (1972) and Denevan (1976) for reliable sources on the demographic collapse of the population of the Americas after 1492 through European violence and disease.

34 Cosmovision is the preferred term amongst indigenous peoples and rights activists, replacing the more European cosmology or myth.

35 A group of lawyers, academics and activists drafted and signed this declaration during the 1993 Rockefeller Conference 'Cultural Agency/Cultural Authority: Politics and Poetics of Intellectual Property in the Post-Colonial Era'. It can be accessed online at http://www.case.edu/affil/sce/BellagioDec.html.

36 For a genealogical exploration of traditional knowledge protection, see Chapter 2 of my doctoral dissertation (Moeller, 2010).

37 Association of Indigenous Shamans of Napo, the first legal association of shamans and traditional healers in Ecuador, founded in 1994, legalised in 1997.

38 Such processes of collective identity formation of course also produce dynamics of inclusion/exclusion and involve the normative policing of boundaries – who is a real *yachak*, who is an impostor, who knows and who does not, who is in and who is out – which have a lot to do with validation of knowledge. In this case: whose healing knowledge is valid, and who makes these decisions? ASHIN's accepted members tested new members, which runs both the risk of bias and has the advantage of grassroots agreement rather than compliance with some external standard.

39 See also Kimerling, 2006: 466–467 for an account of Huaorani beliefs in the weakening of healing powers due to environmental contamination.

40 According to a 2003 study in *Nature*. Worse things have happened to the oceans, but this was one of the numbers I had available in my memory as I was scribbling in my notebook.

41 William James explained the distinction between what he saw as two fundamentally different kinds of knowledge as follows: 'I am acquainted with many people and things, which I know very little about, except their presence in the places where I have met them. I know the color blue when I see it, and the flavor of a pear when I taste it; I know an inch when I move my finger through it; a second of time, when I feel it pass; an effort of attention when I make it; a difference between two things when I notice it; but about the inner nature of these facts or what makes them what they are, I can say nothing at all. I cannot impart acquaintance with them to any one who has not already made it himself. I cannot describe them, make a blind man guess what blue is like, define to a child a syllogism, or tell a philosopher in just what respect distance is just what it is, and differs from other forms of relation. At most, I can say to my friends, Go to certain places and act in certain ways, and these objects will probably come' (1890).

42 Not an unusual tactic, it should be noted – recall the initiative of Stuart Newman and Jeremy Rifkin to patent a chimera in order to prevent others from doing so. See, for example, Newman, 2006.

43 Such a list could arguably constitute valid documentation of *prior art,* and be used to contest a third party patent application.

44. Among other things, Correa has labelled indigenous peoples as 'infantile'. He made this statement on his weekly radio programme on 7 June 2008. See also Denvir, 2008.

45 Arturo Escobar (1995) describes the expansion of the discourse of sustainable development as the semiotic conquest of nature by capital relations. Through bioprospecting
this semiotic conquest is extended into the realm of indigenous and peasant peoples'
knowledge, practices and seeds (Brush, 1999).

46 John O'Neill (1993) calls this the 'weak comparability of values'.

47 This also addresses Bernard Williams' call: 'There is great pressure for research into
techniques to make larger ranges of social value commensurable. Some of the effort
should rather be devoted to learning – or learning again, perhaps – how to think intel-
ligently about conflicts of value which are incommensurable' (Williams, 1972).

References

AIPLA (2001) *Report of Economic Survey 2001*, American Intellectual Property Law
Association, Arlington.

Balick, M. J., Elisabetsky, E. and S. A. Laird (1996) *Medicinal Resources of the Tropical Forest:
Biodiversity and Its Importance to Human Health*, Columbia University Press, New York.

Berlin, B. and E. A. Berlin (2004) 'Community autonomy and the Maya ICBG Project in
Chiapas, Mexico: How a bioprospecting project that should have succeeded failed',
Human Organization 63 (4): 472–486.

Berlin, B., Berlin, E. A., Ugalde, J. C. F., Barrios, L. G., Puett, D., Nash, R. and M. González-
Espinoza (1999) 'The Maya ICBG: Drug discovery, medical ethnobiology, and alternative
forms of economic development in the highland Maya region of Chiapas, Mexico',
Pharmaceutical Biology 37 (4): 127–144.

Brush, S. B. (1999) 'Bioprospecting the Public Domain', *Cultural Anthropology* 14 (4): 535–556.

Carolan, M. (2009) 'A policy note on biopiracy in environment, technology and society',
Newsletter of the Section on Environment and Technology of the American Sociological Association,
Summer 2009.

Chapell, J. (2011) *Biopiracy in Peru: Tracing Biopiracies, Theft, Loss and Traditional Knowledge*, PhD
Thesis, Lancaster University.

Christman, J. (1994) *The Myth of Property: Toward an Egalitarian Theory of Ownership*, Oxford
University Press, New York.

Cooke, B. and U. Kothari (Eds) (2001) *Participation: The New Tyranny?* Zed Books, London.

Crosby, A. W. (1972) *The Columbian Exchange: Biological and Cultural Consequences of 1492*,
Greenwood, Westport.

Denevan, W. M. (ed.) (1976) *The Native Population of the Americas in 1492*, University of
Wisconsin Press, Madison.

Denvir, D. (2008) 'Wayward allies: President Rafael Correa and the Ecuadorian left', *Upside
Down World*, 25 July 2008.

Dorsey, M. K. (2004) 'Political ecology of bioprospecting in Amazonian Ecuador: History,
political economy and knowledge', In Brechin, S. R., Wilshusen, P. R., Fortwangler, C. L.
and P. C. West (Eds) *Contested Nature: Promoting International Biodiversity Conservation with
Social Justice in the Twenty-first Century*, State University of New York Press, New York.

Escobar, A. (1995) *Encountering Development: The Making and Unmaking of the Third World*,
Princeton University Press, Princeton.

Fecteau, L. M. (2001) 'The Ayahuasca patent revocation: Raising questions about current
U.S. patent policy', *Boston College Third World Law Journal*, 21 (1–4): 69–86.

Fraser, N. (1997) *Justice Interruptus: Critical Reflections on the 'Postsocialist' Condition*, Routledge,
London.

Greene, S. (2002) 'Intellectual property, resources or territory?', In Bradley, A. P. and P. Petro. (Eds)
Truth Claims: Representation and Human Rights, Rutgers University Press, New Brunswick.

Grefa, C. (2005) *La consulta previa hidrocarburífeca de los bloques 20 y 29, y su incidencia socio-organizativa en las comunidades Kichuas en la provincia de Napo*, PhD Thesis, Latin-American Faculty of Social Sciences (FLACSO), Quito, Ecuador.

Grote, J. (2008) [1865] *Exploratio Philosophica: Rough Notes on Modern Intellectual Science*, Volume 1, Kessinger Publishing, Whitefish.

Guha, R. and J. Martinez-Alier (1997) *Varieties of Environmentalism: Essays North and South*, Earthscan, London.

Harlan, J. R. (1971) 'Agricultural origins: Centres and noncentres', *Science* 174 (4008): 468–474.

Harris, J. (1996) *Property and Justice*, Oxford University Press, Oxford.

Hayden, C. (2003) *When Nature Goes Public: The Making and Unmaking of Bioprospecting in Mexico*, Princeton University Press, Princeton.

Hayden, C. (2005) 'Bioprospecting's representational dilemma', *Science as Culture* 14 (2): 185–200.

Helmholtz, H. L. F. von (1868) 'The recent progress of the theory of vision', In Helmholtz, H., *Popular Scientific Lectures*, translated by Pye-Smith, P. H., Dover Publications, New York.

Howell, R. G. and R. Ripley (2009), 'The interconnection of intellectual property and cultural property (traditional knowledge)', In Bell, C. and R. K. Paterson (Eds) *Protection of First Nations Cultural Heritage: Law, Policy, and Reform*, UBC Press, Vancouver.

James, W. (1890) *The Principles of Psychology: Volume One*, Henry Holt and Company, New York.

Kimbrell, A. and J. Mendelson (2004) *Monsanto Versus US Farmers: A Report by the Center for Food Safety*, Center for Food Safety, Washington, DC.

Kimerling, J. (2006) 'Indigenous peoples and the oil frontier in Amazonia: The case of Ecuador, ChevronTexaco, and Aguinda v. Texaco', *International Law and Politics* 38: 413.

Martinez-Alier, J. (2002) *The Environmentalism of the Poor: A Study of Ecological Conflicts and Valuation*, Edward Elgar, Cheltenham.

Moeller, N. (2010) *The Protection of Traditional Knowledge in the Ecuadorian Amazon: A Critical Ethnography of Capital Expansion*, PhD Thesis, Lancaster University.

Moghaddam, F. and S. Ginsburg (2003) 'Culture clash and patents: Positioning and intellectual property rights', In Harré, R. and F. Moghaddam (Eds) *The Self and Others: Positioning Individuals and Groups in Personal, Political and Cultural Contexts*, Praeger Publishers, Westport.

Mooney, P. (2000) 'Why we call it biopiracy', In Svarstad, H. and S. Dillion (Eds) *Responding to Bioprospecting: From Biodiversity in the South to Medicines in the North*, Spartacus, Oslo.

Newman, S. A. (2006) 'My attempt to patent a human-animal chimera', *L'Observatoire de la Génétique* 27.

O'Connor, M. (1994) *Is Capitalism Sustainable: Political Economy and the Politics of Ecology*, Guilford Press, New York.

O'Neill, J. (1993) *Ecology, Policy, and Politics: Human Well-Being and the Natural World*, Routledge, London.

Posey, D. A. (2002) '(Re)discovering the wealth of biodiversity, genetic resources, and the native peoples of Latin America', *Anales Nueva Época* 5 (12): 37–59.

PROBENEFIT (2007) *Memoria: Taller de Cierre*, Project Report, Quito, 23 October 2007.

Reid, W. V. (1993) *Biodiversity Prospecting: Using Genetic Resources for Sustainable Development*, World Resources Institute, University of Michigan Press.

Rosenthal, J. P. (2006) 'Politics, culture, and governance in the development of prior informed consent in indigenous communities', *Current Anthropology* 47 (1): 119–142.

Rosenthal, J. P. and F. N. Katz (2004) 'Natural products research partnerships with multiple objectives in global biodiversity hotspots: Nine years of the International Cooperative

Biodiversity Groups Program', In Bull, A. T. (ed.) *Microbial Diversity and Bioprospecting*, ASM Press, Washington, DC.

Russell, B. (1911) 'Knowledge by acquaintance and knowledge by description', *Proceedings of the Aristotelian Society*, XI: 108–128.

Schuler, P. (2004) 'Biopiracy and commercialization of ethnobotanical knowledge', In Finger, J. M. and P. Schuler (Eds) *Poor People's Knowledge: Promoting Intellectual Property in Developing Countries*, World Bank and Oxford University Press Co-Publication.

Scott, J. C. (1985) *Weapons of the Weak: Everyday Forms of Peasant Resistance*, Yale University Press, New Haven.

Shiva, V. (1997) *Biopiracy: The Plunder of Nature and Knowledge*, South End Press, Cambridge.

Shiva, V. (2007) 'Bioprospecting as sophisticated biopiracy', *Signs: Journal of Women in Culture and Society*, 32 (2): 307–313.

Simmonds, N. W. and K. Shepherd (1955) 'The taxonomy and origins of the cultivated banana', *Journal of the Linnean Society of London* 55 (359): 302–312.

Stallman, R. (1997) 'Biopiracy or bioprivateering?', online article available at https://stallman.org/articles/biopiracy.html.

Svarstad, H. (1995) *Biodiversity Prospecting: Biopiracy or Equitable Sharing of Benefits?* Centre for Development and the Environment, University of Oslo, Oslo.

Takeshita, C. (2000) 'Bioprospecting and indigenous peoples' resistance', *Peace Review* 12 (4): 555–562.

Takeshita, C. (2001) 'Bioprospecting and its discontents: Indigenous resistances as legitimate politics', *Alternatives* 26 (3): 259–282.

Weiner, A. B. (1992) *Inalienable Possessions: The Paradox of Keeping-While-Giving*, University of California Press, Berkeley.

Whitten, Jr., N. E. (ed.) (2003) *Millennial Ecuador: Critical Essays on Cultural Transformations and Social Dynamics*, University of Iowa Press, Iowa City.

Williams, B. (1972) *Morality*, Cambridge University Press, Cambridge.

Zeller, F. J. (2005) 'Herkunft, diversität und züchtung der banane und kultivierter zitrusarten', *Journal of Agriculture and Rural Development in the Tropics and Subtropics*, Beiheft 81: 1–104.

7

ECONOMICS

The limitations of a special case[1]

Gilbert Rist

While this work as a whole seeks to place knowledge in the service of diversity, the aim of the present contribution is to explore the reverse track: the promotion of diversity to arrive at knowledge. Let us be clear that this liberty I take with the general theme implies no criticism of the way it is formulated, for I am firmly convinced that it is necessary to practise a diversity of approaches in the social sciences (and to safeguard it in various ecosystems). Nevertheless, if such diversity is so rarely taken into account nowadays, is this not due to the hopes placed in technology – and above all to the hegemony of an economic 'science' that reduces everything to the criteria it claims to master by passing them off as laws: supply and demand, the market and growth? In other words, does not the lack of diversity at a practical level stem from the monolithism of economic theory, which often (in the last instance?) determines the actual choices according to whether it considers them 'profitable' or not, 'rational' or not? There, in a few words, is the object of this chapter.

From a base in history and anthropology, I have for a long time taken an interest in economic theory and gradually convinced myself that it bears a heavy responsibility for the dead-ends into which we have strayed (Rist, 2011). Although the age of political-economic colonization is fortunately over, or almost, the colonization of minds continues – including among the former colonizers – and the struggles needed to end it are at least as difficult as those that had to wage against earlier forms. This chapter should therefore be regarded as a modest contribution to the ongoing contest.

The provincial origin of economic 'science'

Economic 'science' was born in Europe between the late-eighteenth and mid-nineteenth centuries.[2] Of the founders of what came to be known as classical economics, only three names will be considered here: Adam Smith, David Ricardo and

Karl Marx. The first thing they share is a conception of labour as the source of value, although, of course, they differ about how it should be interpreted or measured; such differences have given rise to numerous debates, but I shall refrain from entering into them here. Something else they have in common is that each developed his theory through observation of what was before his eyes, by studying the practical activity of his contemporaries. Thus Adam Smith, right from the first chapter of his *Inquiry into the Nature and Causes of The Wealth of Nations*, marvelled at the division of labour in a pin-making factory, where 'ten persons could make upwards of forty-eight thousand pins in a day', whereas 'if they had all wrought independently and separately, [...] they could certainly not each of them have made twenty, perhaps not one pin in a day' (Smith, 1961). Ricardo, for his part, focused on exchanges of wine produced in Portugal and cloth made in England; his demonstration of the comparative advantages of trade concluded that it 'diffuses general benefit, and binds together by one common tie of interest and intercourse, the universal society of nations throughout the civilized world' (Ricardo, 1973). Lastly Karl Marx, an endless source of quotations, wrote that 'the constant revolutionizing of production, uninterrupted disturbance of all social conditions, everlasting uncertainty and agitation distinguish the bourgeois epoch from all earlier ones' (Marx, 1973b). To be sure, these brief extracts in no way summarize the thought of these authors, but they do show that, however different or even opposite their theories, each based himself on the world around him. Far from condemning their method, we should rejoice at it. Should we not start from the facts, from things we can actually establish, instead of concocting imaginary worlds?[3]

But our three authors (and those who followed them) set out to do more than just describe their world; they were aiming to establish a science with universal application. Having identified certain constants in the economic practices of their time, they imagined that these corresponded to 'laws' as true as the natural ones discovered by Newton. It is true that to introduce terms from Newtonian mechanics into the language of economics – 'monetary mass', 'financial flows', 'market forces', 'trade balance', 'elasticity of supply and demand', 'speed of circulation of money', 'market equilibrium' and so on – may give an illusion of science. But it conceals a crippling methodological error: namely, a transfer from one genre (to use an Aristotelian vocabulary) into another genre – in this case, from mechanical physics into economic and social exchanges – on the assumption that the truth value will remain the same. Alas, it is a fatal and all too common mistake, particularly in the Neoclassical school. Walras did not hesitate to claim: 'It is already perfectly clear that economics, like astronomy and mechanics, is both an empirical and a rational science' (Walras, 1954). However, it must be stressed again and again that economics belongs to the *social* sciences, and that, while these may try to account for human practices with the greatest exactitude, their conditions of 'veracity' are a long way from those of the 'hard' or 'exact' sciences.[4] Evidence of this includes the methodological debates, theoretical and above all ideological, that never cease to shake them.

As we shall see, the scientific claims dependent on Newtonian mechanics failed for another, even graver reason, but for the moment let us just underline the provincialism (that is to say, ethnocentricity) of economic theory. For the fact is that it took shape in less than a century, on the basis of local practices, and it owed its success to a particular set of circumstances that decided the future of humanity. Adam Smith published his famous work in 1776, the very year of the first (Virginia) Declaration of Rights and the American Declaration of Independence, when James Watt was putting the finishing touches to the steam engine that was about to revolutionize industry. Economics, the dignity of the individual and the energy produced by 'fire machines' were thus born simultaneously with one another. There can be no doubting the importance of the three events.

But, as regards economics, let us imagine a hypothetical situation in which it was invented elsewhere, by what were then known as 'savage' peoples. The first point to make is that they would not have produced a theory! They would have been content to tell stories about how the whole of society mobilizes to provide everyone with a place: they would have explained how its members depend on one another because the economy is completely 'embedded' in the social system (Box 7.1),[5] and how, on the same land, one person can graze goats, another enjoy fruit trees and another harvest palm dates – all in a spirit of understanding.[6] And if lessons were drawn from their accounts, it would be seen that they describe multiple practices of reciprocity and redistribution rather than individual interest; multiple precautions to avoid offending Mother Earth, rather than the exploitation of nature; attempts to contain human beings at the level of a strict minimum, rather than to imagine them with limitless needs; and warnings against the 'evil eye' threatening those who accumulate at the expense of others, rather than the promotion of individual enrichment.

BOX 7.1 EXCHANGE AS A SOCIAL PRACTICE

There is no end to the anthropological literature on the numerous forms of (non-market) exchange practised in most societies. If we find this so surprising, it is because we think of exchange as the opportunity to obtain something useful or to get ourselves 'a bargain'. All that counts is the *object* of the deal. In 'traditional' societies, however, exchange most often consists in *maintaining social ties* or acquiring prestige. Here are just a few examples.

- In Islamic desert societies, the prime necessities (water, grass, fire) are not put up for sale, because they must remain available to all.
- In Africa, as in Melanesia, there are 'compartmentalized' markets, since things are not exchanged for just anything else. Ceremonial goods intended as matrimonial compensation (loincloths, copperware, livestock) can only be exchanged for goods serving the same function, never for food or everyday goods.
- In societies organized as chiefdoms, the whole of production is destined for the chieftain, although he is obliged to distribute it generously. Elsewhere, the 'big man' and his family tirelessly amass quantities of pigs

or food, which are then consumed at great collective feasts; the prestige he gains is thus expressed in a reduction of social inequalities. The principle is always the same: power is obtained from giving, not from accumulating. The 'evil eye' always threatens the egoist.

- As Marcel Mauss showed in *The Gift*, all societies are subject to a threefold obligation (experienced as a freedom) to give, to receive and to give back (Mauss, 1966). But this circulation of the gift has nothing of an exchange between two partners: A gives to B, who gives to C, who gives to N, who 'gives back' to A (in the manner of a 'pub round'). Three points are important here: (1) to refuse to receive is tantamount to a declaration of war; (2) the 'return' of the gift is never immediate or equivalent (there is always a 'raise'); and (3) the partners go through an alternation of inequality. The giver gains power over the receiver, who becomes 'obliged to him', in accordance with the saying that 'the hand that takes is always inferior to the hand that gives'. But the receiver then regains his rank by giving in his turn.

These practices are not unknown in our market society. If some friends invite you to dinner, you take them a little gift and the mistress of the house, in accepting it, quickly interjects: 'Oh, you shouldn't have [done me this honour and put me under an obligation]!' To which you must reply: 'It's nothing, really [so you are under no obligation]!' Yet these 'little nothings' build and sustain a friendship, and the 'value' of the gift lies precisely in this 'nothing'.

Surprise is often expressed at the practice of the potlatch among the Kwakiutl, who, in order to impress the neighbouring tribe (to 'make it grovel', as Mauss would put it), smash or throw into the sea such valuable goods as boats, fabrics or copper bars. Senseless waste, in the eyes of economists. But what happens at a wedding in our own climes? Does it too not involve excessive and senseless expenditure? The bride's dress (or the groom's suit) will be used only once, the flowers will fade in a day, the savouries will be swallowed in a flash and so on. To speak like Georges Bataille, some kinds of 'useless' spending are necessary.

The presumption of generosity, characteristic of other societies, is poles apart from the self-interested egoism central to economic 'science' – a phenomenon which, in reality, is neither transcultural nor transhistorical.

These are but a few examples of what might have happened if other peoples – much more numerous than Europeans! – had set about proposing their idea of 'good practices' in the domain of exchange and the use of natural resources. Of course that is no more than a dream; it takes no account of the huge economic and military power that enabled Europe and the United States to impose their will on the rest of the world in the course of the nineteenth century. But it may also seem surprising that – in spite of this triumphant ethnocentrism – economic theory constituted itself without regard for practices that contradicted it and were still dominant across the greater part of the globe. Why should economic theory not have been more diverse and complex, so that it corresponded to the way in which most of humanity actually lived? It remains to be seen whether such different paradigms can exist side by side.

A 'science' incapable of understanding ecological questions

Let us agree that the early economists not only borrowed their vocabulary from Newtonian mechanics but genuinely wanted their scientific work to draw out the rules or 'laws' governing production, consumption, investment, distribution and exchange. That presupposed a familiarity with, or even some knowledge of, the advances being made by physics, even if the transposition of a truth from one order to another was illegitimate. Moreover, was the interest of economists in natural science not pertinent? Although exchange is a social act, the objects of production and consumption are very largely derived from nature. Whether it is a matter of wheat or coal, meat or steel, everything ultimately comes from natural resources.

Curiously, however, the only source of value for the classical economists is *labour*. Adam Smith puts it quite clearly: 'Labour alone, therefore, never varying in its own value, is alone the ultimate and real standard by which the value of all commodities can at all times and places be estimated and compared. It is their real price; money is their nominal price only' (Smith, 1776). Ricardo, less forthright, recognizes that the value of one commodity may vary in relation to others if there is a change in production techniques that makes labour easier in a particular branch. But he still speaks of 'labour as being the foundation of all value, and the relative quantity of labour as almost exclusively determining the relative value of commodities' (Ricardo, 1973). As to Marx, who concerns himself mainly with the labour time socially necessary for production, he asserts: 'Labour *alone* produces; it is the only *substance* of products as *values*. … Products can be measured with the measure of labour – labour time – only because they are, by their nature, labour. They are objectified labour' (Marx, 1973a).[7] Similar passages could be found in Proudhon or Sismondi – proof, if any were needed, of the relative unanimity of the early economists on the subject.

So, if all value derives from labour, nature plays only a marginal role in all these theories, since it is supposed to provide us free of charge with all the goods we enjoy. It is no surprise that Adam Smith, writing before the Industrial Revolution, hardly concerned himself with what we would today call non-renewable resources. But he did write at length on variations in the price of silver, which in his view were linked essentially to the cost of its extraction and to world demand.[8] The only link with the questions that preoccupy us today might be the craze for the consumption of migratory birds or rare fish. For, as he put it: 'When wealth and the luxury which accompanies it increase, the demand for these is likely to increase with them, and no effort of human industry may be able to increase the supply much beyond what it was before the increase of the demand' (Smith, 1961).

Ricardo took over the idea of the generosity of nature:

> Nature does nothing, man does all. [...] Are the powers of wind and water, which move our machinery, and assist navigation, [...] not the gifts of nature? [...] There is not a manufacture which can be mentioned, in which nature does not give her assistance to man, and give it too, generously and gratuitously.
>
> *(Ricardo, 1973)*

As to Marx, he is so optimistic about the development of the productive forces that he gives little space to the possibility that natural resources will one day start to run out. Against Malthus, he insists that any population surplus is 'purely relative', since it is 'in no way related to the means of subsistence as such, but rather to the mode of producing them' (Marx, 1973a). Nevertheless, he is more aware than his predecessors of the importance of nature. He recognizes that 'even an entire society, a nation, or all simultaneously existing societies taken together, are not the owners of the earth. They are simply its possessors, its beneficiaries, and have to bequeath it in an improved state to succeeding generations, as *boni patres familias*' (Marx, 1981). He could even pass for an ecologist *avant la lettre*: 'Capitalist production, therefore, only develops the techniques and the degree of combination of the social process of production by simultaneously undermining the original sources of all wealth – the soil and the worker' (Marx, 1976). Or again: 'As William Petty says, labour is the father of material wealth, the earth is its mother' (ibid.).[9]

Along with ethnocentrism, the early economists should therefore be reproached with anthropocentrism: the view that 'man does all'; that nature is generously offered to us on a plate; that it exists only as simple *Dasein*, to be exploited. No thought is given to the transformation of nature itself under the impact of human activity. Yet those economists (beginning with Marx) who took an interest in the physical sciences ought to have been alerted by the discovery of the second law of thermodynamics, or the law of entropy, by Rudolf Clausius; this stated that, in a closed system, the economic process transforms (and irreversibly degrades) low-entropy energy-matter (which can be used) into waste (which cannot).[10]

> Economic development through industrial abundance may be a blessing for us now and for those who will be able to enjoy it in the near future, but it is definitely against the interest of the human species as a whole, if its interest is to have a lifespan as long as is compatible with its dowry of low entropy.
>
> *(Georgescu-Roegen, 2011)*

Blinded by mechanistic certitudes and an ideology of progress that the Industrial Revolution seemed to vindicate, the Classical economists – unlike the Physiocrats – paid scant attention to nature, whose resources fed the huge material accumulation they encouraged. While claiming to found a new science in the image of physics, they haughtily ignored the discovery of the law of entropy and the radical challenge it posed to the idea of limitless economic growth.[11] With the onset of thermo-industrial civilization, a new historical era – the Anthropocene – appeared before their eyes without their noticing it![12] This is why mainstream economic 'science' – still faithful to the principles and assumptions of the nineteenth century – is incapable of understanding the ecological issues of our age (Farley and Malghan, 2016). It is hardly surprising that this 'science' continues to base itself on a paradigm that goes back two centuries, whereas all the other sciences have undergone profound revolutions during the same period.

Property at the origins of growth

Roman law is indubitably part of the heritage of 'Western civilization'. It is often invoked for the preciseness of its definition of private law (*ius privatum*), which protects individuals in civil or penal cases and sets the rules of private ownership (*res privatae*) whereby they can use the possessed object, enjoy its fruits and consume, destroy or alienate it (*ius utendi, fruendi et abuntendi*). But it is sometimes forgotten that several forms of 'property' existed in ancient Rome: there were *res sacrae et religiosae,* which, like the temples given over to them, belonged to the gods and were inalienable objects outside the sphere of commerce. There were also *res communes*, 'that is, things which belong to no one because they belong to all and by their nature elude any appropriation: the air, springs and streams and the sea. Everyone may use them, but no individual or group can be pointed to as the owner' (Durkheim, 1958). Finally, there were *res publicae* or state property, which belonged to no one in particular but to the whole political community (ports, fishing rights, rivers, highways, sea shores and so on) (Saleilles, 1889).

The details of this division are complex.[13] The point here is its diversity: we are a long way from today's dichotomy, in which goods (land, buildings and all the other things in life) belong either to private owners or to the state. Yet this multiplicity of property rights persisted for a long time. The Charter of the Forest (proclaimed by Henry III in 1217 and appended to the Magna Carta of 1215) gave to freemen the right to benefit not only from forests in the sense we understand today – which provided wood for heating – but also from pastureland and wild game. This example shows the importance of common goods, regained only through hard struggle, for the subsistence of the English population of the time.

The great turnaround came from the sixteenth century on with the enclosure movement, the privatization of the 'commons'. Rather than go back over the history, let us just focus on two of its crucial outcomes: the generalization of private property, to the detriment of age-old collective use rights; and the existential necessity for those who had previously held rights in the commons to sell their labour-power and become wage-workers in England's incipient industrialization.[14] Such were the origins of the first two factors of production recognized by economists: land and labour. Capital, which derived from them, did not take long to join them on the scene, especially as the new perspective suited the interests of the bourgeoisie and the new entrepreneurs.[15]

Today, alongside 'public property' (where the state behaves as a private owner, not hesitating to sell off things like education, health or motorways that belong to all), private property seems a matter of course: it has substituted itself for the various forms of ownership and/or possession that used to be part of the juridical repertoire. But this hegemony of private property has another consequence: the growth obligation.

What, it will be asked, is the connection between (private) property and growth? Take the owner of an asset such as land or a building: he can obtain fresh cash (which he has not earned by the sweat of his brow) through a mortgage loan from a bank, on condition that he gradually repays it, or at least – or in addition – pays interest on it to the bank. If he owns some land, he can continue harvesting its fruits, but will also have to find fresh income to meet the repayment deadlines set by the bank. If the wheat he usually grew there does not bring in enough, why not construct a building and make a higher profit? This extremely simple example shows, first, that (legally protected) property rights are a necessary condition to obtain a loan, and second that the operation enables the owner to create money: that is to say, capital. Furthermore, this practice is general in the present economic system, since everyone (farmers, industrialists, banks, governments) is forced to borrow in order to have ready cash in advance of tomorrow's profits. So arises the growth cycle, as a compelling necessity: 'It is not only that a property-based economy allows for growth, it also *imposes* growth as a result of the conditions of credit' (Steppacher and Gerber, 2012). Each player must now calculate precisely in order to honour their commitments before realizing the anticipated profit; otherwise bankruptcy will sanction their failure.[16]

We see the importance of institutional regimes in the way the economy functions; economists may ignore them (deliberately?), but they shape the whole of our system (which, in this sense too, is a really 'special case'). For economists, the property regime is a self-evident fact of life, having gradually supplanted all other legal forms that used to allow the greatest number of people (not only property-owners) to have reasonable access to resources under certain conditions. These forms may be categorized as 'possession regimes', which define 'the rights, obligations and duties regarding access to, and use and management of, resources, as well as the arrangements regulating the distribution of products that result from the exploitation of resources, including the disposal of waste and pollutants' (Griethuysen and Steppacher, 2015).[17] Extremely widespread outside the capitalist system, they ensure the sustainable exploitation of a resource by harmonizing the claims of the various rightholders, especially as collective possession is usually inalienable. We are therefore dealing with a logic of reproduction that takes account of the ecosystem and respects the social system.[18]

The property regime also has consequences of a very different kind. First of all, the credit mechanism vastly increases the money in circulation, permitting the most unbridled speculation.[19] The subprime crisis, which began in 2008 in the United States and spread to all the other capitalist economies, had the same origin: the banks took a gamble and lent money on the (wrong) assumption that the real estate market would keep moving upward, without checking on borrowers until the point when they were choked by rising interest rates and could no longer meet their obligations.[20] Then the growth compulsion – to escape the crisis! – fuelled competition among firms, which needed to release new profits and, for that purpose, did not hesitate to close or relocate unprofitable production sites, creating unemployment and increasing inequalities. Finally, all additional growth necessarily involves a new drain on energy-matter from non-renewable sources (even in the sector of the economy wrongly called 'non-material'). But 'man's continuous tapping of natural resources is not an activity that makes no history. On the contrary, it is the most important long-run element of mankind's fate' (Georgescu-Roegen, 2011).

According to the classification of Joan Martinez-Alier, the economy has three levels: the 'financial level', which speculates with money created by credit; the 'real economy', which produces everything from wheat to cars by using resources of capital, labour and land; and the 'real-real sector', which corresponds to the flows of energy-matter and the polluting emissions that accompany it (Martinez-Alier, 2012). Mainstream theory focuses on the first two levels but ignores the third, the most important (because of its intergenerational dimension), even if it is expressed in the categories of physics, chemistry, geology or biology. This is perhaps what Engels meant when he wrote:

> According to the materialist conception of history, the *ultimately* determining factor in history is the production and reproduction of real life. Neither Marx nor I have ever asserted more than this. Hence if somebody twists this into saying that the economic factor is the *only* determining one, he transforms that proposition into a meaningless, abstract, absurd phrase.
>
> *(Marx and Engels, 1975)*

It is a pity that his disciples gradually moved away from this 'materialist conception of history', which addressed the reproduction of real life that is today in danger.

What is to be done?

At the end of this brief survey of origins and assumptions, it is clear that mainstream economics is unlikely to contribute anything to the extrication of humanity from its present impasse; in fact, it bears the principal responsibility for the false turns and deviations that have come about. True, certain economists say they are 'appalled' by the subservience of neoliberal policies to the financial markets, and advocate palliative measures against the calamitous effects of the current crisis.[21] But although some of their proposals are interesting, they do no more than criticize current policies in Europe without challenging their underlying assumptions. On the other hand, some members of the Association française d'économie politique forthrightly declare:

> Economic theory has gradually become necrotic, in so far as it has moved away from its original posture of a 'social science' and modelled itself on the

practices and criteria of the natural or 'normal' sciences. This profound deviation has gone too far for there to be any hope that the mere good will of some will be enough to reverse the course.[22]

This statement is a cause for rejoicing. It remains to be seen how the authors think the course can be reversed and how far back they aim to go. But in any case a more radical approach is necessary.

Expanding the economic field

This requirement arises from a critique of the ethnocentrism that attended the birth of economics. There is no point in blaming its founders for their ignorance of social anthropology, which emerged considerably later,[23] but one may be legitimately surprised that their successors took no account of it. In any event, if contemporary economists were to read Marcel Mauss's *The Gift* (1966) or Polanyi's *The Great Transformation* (1944) or the most recent Bolivian development plan,[24] they might draw useful lessons concerning the limits of their 'science'. What is at issue here is not only the provincialism of its origins but the reductionism of the marginalist theory that prevails today: 'The marginalists shrank the field of *political economy* to construct an economic "science" that excluded everything that had been a problem and had required answers at the fundamental level of the question posed' (Jorion, 2012). This means that the essential themes that ought to be the object of economics (division of the commons, redistribution of wealth, exploitation of natural resources) have been replaced with a model of *homo oeconomicus* as a 'natural' (hence eternal and universal) being completely geared to the satisfaction of his needs (or 'preferences'),[25] whose 'rationality', synonymous with cost-benefit calculation, is supposed to provide him with maximum return for a minimum effort, all within a complete social and ecological vacuum. Although mainstream economists will think this an outrageous simplification,[26] it really is a fair summary of the theoretical foundation of the model.

In 'real life', things happen very differently. No one ever decides alone, but always in a definite social and ecological context that imposes norms and constraints, whether written or unwritten. The free autonomous individual is a fiction. It is society — and the position of an individual within it — which decides his or her tastes and preferences. 'Poor people' are in principle never isolated:[27] they are part of a network in which they can count on their 'brothers and sisters' (by blood, classification or self-declaration), who in a way are in their debt. Similarly, 'the rich' can keep their position only through a degree of redistribution — otherwise they are gradually excluded from their group and are in danger of becoming 'poor'. A canny peasant knows that he risks exhausting his land if he works it beyond a certain limit. In other circumstances, the state — the main enemy for mainstream theory! — imposes redistribution by means of a progressive income tax and social insurance. People everywhere form ties with one another through gratuitous acts: gifts, invitations, community projects or mutual services, and all these displays of generosity must be reciprocated, sometimes to excess.

It would be easy to list many other social practices that testify to generosity and a concern for others; they are as frequent as acts of supposed egoism, their main function being to build up society, to make 'living together' possible and to preserve the ecosystem. They all radically contradict the methodological individualism that characterizes mainstream economic theory. Or, to put it differently, if mainstream theory was actually applied, it would spell the end of society[28] and of the existing ecosystems.

For all these reasons, it is time to rethink the assumptions of economics so that it becomes a genuine social science, one based on the full range of actual practices involving the production, consumption, exchange, redistribution and management of natural resources. As things stand, those who invoke the 'laws' of economics to run the world's affairs are like a traveller who, eyes glued to a bad map, thinks he is walking in a plain when he is actually a few steps from the edge of a precipice. Enlarging the field of the economic therefore means enriching the input from social practices and refusing to accept that abstract principles are enough to define them.[29] The point is not to invent a host of economic theories corresponding to all the different situations or societies, but above all to understand the reasons (or assumptions) that ground these practices. This requires asking (before answering) the questions that marginalist theory has deliberately ignored in the name of methodological individualism: how to 'make society' and maintain the 'social fabric'? how to preserve wealth, including natural wealth? how to distribute such wealth? how to maintain across generations the 'services' rendered by the ecosystem? how to reduce economic inequalities? how to limit property rights? It is a vast programme, no doubt, and it reverses our customary way of posing the problems. Instead of thinking we are governed by natural laws, we might thus ground new laws on the principles we constantly invoke: the rights of the human person, to be supplemented with those of nature so as to preserve their indispensable diversity. Many societies tried to do this for centuries, with varying success, and the task today is to make the programme topical again, by going beyond the imperialism of mainstream economics. To be specific, the study of economics needs to be totally recast, so that it includes not only the (largely neglected) history of the discipline but also – perhaps mainly, thereby questioning our own assumptions[30] – an anthropological approach to exchange systems such as reciprocity and redistribution (or a Buddhist-style extinction of desires, in an 'economy of enough'), as well as an ecological perspective that draws out the limits of thermo-industrial society. In other words, the task is to take seriously Polanyi's idea that the economy is always *embedded* in society, although perhaps this should be extended to include the embedding of society in the biosphere on which it depends.[31] Hence economic 'science' is on the wrong track when it dreams up 'laws' independent of society and nature – laws which exist only in the set of equations that constitute them, and whose variables scrupulously avoid what economists dismiss with contempt as 'non-economic factors'.

Today's North American orthodoxy, reinforced by the awarding of Nobel prizes to its leading lights, seeks to follow up battles in the field of economics by heaping discredit on the above way of looking at things. It is not a new perspective, of course: its long history goes back to Durkheim and Mauss or – to mention a few names closer to our own time – Nicholas Georgescu-Roegen, André Gorz, Cornélius Castoriadis and Ivan Illich, or members of the (French) Anti-Utilitarian Movement in the Social Sciences (D'Alisa *et al.*, 2014). In addition to the initiatives mentioned above,[32] we should also mention the movement 'for a post-autistic economics', founded in Paris in 2000 and since then echoed by US-based pluralist economics (Fullbrook, 2008).[33] So, there are a few reasons for hope: knowledge through diversity.

Restoring common goods

To limit the damage caused by the growth obsession of mainstream economics, it is necessary to widen the scope of property regimes and to overcome the pseudo-dichotomy between private and public (or state) property.[34] As we have seen, the latter difference is so minimal that it can be ignored, since the state, instead of wisely managing property created by and for the community as a whole, behaves more and more often as an owner entitled to alienate the property in question – witness the privatization of public spaces, railways, water, motorways or universities, to take just a few examples. In a way, such practices expropriate the rights that the community used to have over what belonged to it, without paying it equitable compensation (as the law provides in the case of private individuals) (Mattei, 2011). Hence the importance of re-establishing other property (or possession) regimes and inscribing them in the law in the same way as private property.

Not all authors are agreed on definitions here, since these usually flow from assumptions bound up with a particular economic theory. Strictly speaking, a *public good* is generally available and no one is excluded from the benefit it provides; a street lamp or a lighthouse is a good case in point.[35] According to the neoclassical vulgate, such goods have to be produced by a public authority, which uses tax revenue to make up for a 'market failure' (since no individual would be prepared to finance the object of public utility). Some consider 'public goods' as synonymous with collective goods, while others reserve that term for natural resources which, though freely accessible, may be appropriated privately (for example, the fish in the sea).[36] As to 'common goods' ('the commons'), they are a community possession: that is, a resource that 'belongs' to a particular group, whose members make use of it within the limits of self-established (collectively formulated and accepted) rules; it is handed down from generation to generation and cannot be alienated.[37] Examples of this kind are communal pastureland and forest, irrigation systems and the household water supply. The category is sometimes defined so as to include any 'general interest' good: for example, the quality of the air, which the public authorities cannot delegate to a private enterprise (as they can with street lighting), but which is part of the collective heritage that everyone participates in producing and managing (Gadrey, 2012; Cordonnier, undated).

The appearance of new concepts such as 'global public goods', 'global commons' or 'common heritage of mankind'[38] has recently made the picture more complex.

Like 'classical' public goods, they are available to all and cannot be denied to anyone – for example, the atmosphere, the airwaves or biodiversity – but, unlike a street lamp or a lighthouse, they are not funded by anyone and therefore do not constitute a market failure (Coussy, 2002). In an international context, public goods – that is, goods over which no individual or state holds sovereign power – must be protected by international conventions, and the theory of negative externalities (concerning damage such as pollution caused by the use of a resource) is most often used to justify the costs for those who benefit from the resource (Kyoto Protocol). In the course of time – and with the degradation of the 'common heritage of mankind' – the list of global public goods has grown considerably longer. It includes not only 'material' goods (oceans, climate, ozone layer, biodiversity), but also 'immaterial' ones (cultural heritage, security, monetary stability). A short quotation from Joseph Stiglitz might give some idea of thinking on these subjects: 'To remedy the problem of the global commons […] the only sensible and workable remedy is some form of global public management of global natural resources, some set of global regulations on usage and on actions giving rise to global externalities' (Stiglitz, 2006). As the subtitle of the French edition of this work indicates,[39] the aim is to combat 'market fanaticism' and thus to create a new 'property' regime for the global commons. All that remains is to make it happen!

These quarrels are not only semantic: they show how some seek to justify common goods by wrapping them in the respectability of mainstream economics. Beyond these quarrels, however, the commons may be defined as a possession regime that allows the needs and basic rights of a collective to be satisfied according to the available resources, which have to be used sparingly. In other words, we must not dissociate the goods from the social groups that use them.[40] The etymology of the word 'commons', going back to the Latin *munus* (charge, function, gift, service rendered), actually contains the idea of reciprocity, inscribing it in a social relationship (Lipietz, 2010)[41] as well as a specific ecological context. Thus, common goods escape the prescriptions of mainstream economics, which focus only on the maximization of satisfaction (that is, on constant enrichment regardless of resource depletion), especially considering that 'Western capitalist development is based on the plunder of the commons' (Mattei, 2011).

What, then, should enter into the category of the commons? As we have seen, the list differs from author to author. In addition to air and water, which are essential for life, one might include (in no particular order) forests, fishing grounds, the countryside, seeds, land, aquifers, underground resources, parks, public space, culture, health,[42] even money[43] – everything necessary for a particular collective to live well, while providing itself with the means to preserve it for future use.[44] It is all the more difficult to identify and seriously consider these goods because they most often 'go unnoticed', and because their importance and the wellbeing they induce tend to be appreciated only after they have been lost.[45] To restore the commons – which usually involves political struggles over access to basic goods[46] – means reversing the course of history. This, especially in Europe and more recently elsewhere, may be summed up as the step-by-step commodification of nature and social relations (Rist, 2008): not in order to 'turn the clock back', which would make no sense at all, but to reclaim less perilous, or even suicidal, modes of existence than those demanded by the prevailing orthodoxy. One way of doing this is to (re)invent legally established regimes of possession, in addition to (or sometimes in place of) those which today guarantee private or state property. Suffice it to mention the land reforms overdue in a large number of countries. These could protect the social and ecological 'life spaces' that everyone could use without misusing them, but which are too often threatened by the market or regalian rights. Once again, the right approach is to bank on diversity.

Going for degrowth

No doubt a kind of temerity (or insouciance) is required to call for 'degrowth' when all the media are publicizing government efforts to 'revive growth'.[47] It should be said at once, however, that the word 'degrowth' is a trap, because it suggests something like the recession that has driven large numbers of the unemployed and homeless to the margins of society. The aim, then, should not be to do the exact opposite of what politicians advocate – the opposite of what we criticize is not necessarily right! – but rather to sift through all the 'revive growth' appeals with which we are bombarded and to ask ourselves what they actually contribute to our own and others' wellbeing. 'Growth objectors' cannot be reduced to individuals who have freely chosen a simpler way of life, although no doubt that would be one of the consequences. The real point is to bring about radical political and economic change of the whole society, including its imaginative dimension; this has become necessary not only because of acute ecological threats but also because of a drive for social justice, both nationally and internationally. In other words, no degrowth is possible in a growth-oriented society, especially in one that has been globalized. The approach of growth objectors is therefore not unrelated to the argument in earlier paragraphs about a refounding of economics and the importance of the commons. Let us dare to admit it: there will be less than before; fewer cars, fewer fruit and vegetables out of season, less publicity or air travel, smaller amounts of pointless and polluting trade.

Of course – and this is an important objection – such choices would further restrict the number of jobs, whose limited supply is rightly a worry in the present system. It is also possible that they would reduce the length of paid work, following a trend that has been palpable for some time. The norm in the nineteenth century was a twelve-hour day, or a seventy-two hour week. Today it is between thirty-five and forty hours a week. Why not keep moving down this road? No one would object to it – except that 'we've got to live' (to pay for the rent, food and other necessities)! What would be the pay for a much shorter working week? The question is certainly pertinent. Perhaps the answer would be a universal benefit that allowed everyone to live in dignity without depending only on the labour market,[48] but also left open the possibility of employment. But what will people do with all that free time? When 'comfort goods' have disappeared from the TV ads and the supermarkets, there will be not only less but also more: people will rediscover the commons, interaction with neighbours, collective gardens providing vegetables in season, walks in the forest, political meetings to save a place that property developers have their eyes on or the government wants to convert into a parking lot, discussions on new laws to ring-fence the commons and on forms of possession other than private or state property, as well as a thousand other activities of which we feel constantly deprived today (Latouche, 2011; Jackson, 2017).

A utopia? No doubt – for the moment, anyway. But why not try to achieve it, before radical lifestyle changes are suddenly forced upon us, amid a disorder that penalizes the weak? No one can be unaware that climate change will transform our environment, affecting both agricultural production and the landscapes to which we are attached; that fossil reserves are inexorably running out;[49] or that biodiversity is being eroded. And that is just to mention a few well-established instances. Jean-Pierre Dupuy put it well: 'We think catastrophe is impossible, but meanwhile the data at our disposal tell us that it is likely and even certain, or semi-certain' (Dupuy, 2002). In other words, we know but we do not believe what we know; we are content to pay lip service with terms such as 'sustainable development' or 'green growth', but without changing anything in the dominant paradigm (Rist, 2008). That is another reason why growth objectors have such a hard time convincing people, why they are thought passé or even antediluvian ('Do you want to go back to candlelight?'), whereas in reality they are decidedly 'ahead of their time'.

'Growth objection' therefore entails social changes that are incompatible with mainstream economics, an exclusive regime of private property or government tutelage. It involves much more: not only individual resistance to the seductions of consumer society (a first step, necessary though difficult!), but a remoulding of society as a whole, drawing citizens closer to their local habitat, evaluating the drain on (renewable and non-renewable) resources and the effects of the resulting pollution, establishing redistributive mechanisms to prevent the growth of inequalities, recognizing and protecting the commons from which everyone benefits (D'Alisa et al., 2014). Is this a strategy for withdrawal into autarkic 'sanctuaries'? Certainly not. If generalized, it would protect the autonomy of all and directly counter the present form of globalization, which encourages pillage of the world's resources and the exploitation of workers in the name of market 'laws'.[50] But since all countries obviously do not enjoy the same advantages (especially in terms of raw materials), trade will always be a necessity. The effect of the policy in question would therefore not be to discourage, but only to reduce, international trade: we would no longer see lorries carry Polish peat to Andalusia in exchange for courgettes and tomatoes, and Swedish prawns would no longer be transported for shelling in Morocco before they are sold in Denmark. On the other hand, oil tankers from the Middle East would continue for some time to unload their cargoes in European ports.

Curbing finance?

The financial markets are a sad example of what happens when the private property regime does not go according to plan. It can be accepted that, in order to found or develop a company, the entrepreneur will seek to assemble capital in the form of shares held by various parties. These, in their turn, will hope that the company makes a profit and distributes some of it to them as annual dividends, but also that its total value will appreciate and allow them to sell their shares at a price higher than they paid for them. Such an outcome is possible thanks to the stock exchange, since the shares in question are considered 'liquid': that is, immediately negotiable in so far as they find a purchaser. So far, so good – even though the investor is not faced with a good (a 'utility') that he can acquire immediately, as in the case of a market for goods and services, but rather with a relationship to time, since both the next dividend and the value of the share a week or two years hence are unknown quantities. Neoclassical theory (in accordance with the hypothesis of rational expectations) postulates that a consensus will emerge among rational, well-informed players and dominate the market, equilibrium supposedly being established around the 'fundamental' value of the security that in a way exists prior to the market (Orléan, 2011). The problem is, however, that the stock exchange does not serve only to provide funds for entrepreneurs; it may also, or chiefly, be a locus of speculation, where the players' main interest is not in company profits or dividends but in the evolution of the share price or market value (which conjures up all kinds of new runaway products, especially derivatives). The question then becomes how to anticipate the market, even in the very short term, by selling if it is expected to move down, and later perhaps reversing the operation if it is expected to move back up.[51] Now it is the mimetic, self-referential logic of liquidity that prevails: 'If we keep to this analysis, the delinking of finance and production seen in self-referential speculation or speculative bubbles ceases to appear an irrational event' (Orléan, 2011). To win at this game, it is not enough to be clever at working things out on your own: you have to imagine what others are imagining, as Keynes pointed out in his famous metaphor of the beauty contest.[52] So arise collective beliefs which, though resting upon no justification, are fulfilled because the 'right' price is the one that everyone else considers 'right', and because the only possible 'rationality' for a market trader is to follow the opinion of the majority. So it is these positive feedback loops that fuel runaway markets. 'In the most rational way in the world [accepting that the market depends on collective beliefs], the market is thus disconnected from objective reality; a gulf opens up that may reach the degree of "madness" before it subsides into panic' (Dupuy, 2003). The system is all the more perverse in that, just when the speculative bubble is ready to burst, speculators collectively gamble on its continuation in the hope of making their biggest profits. The resulting evaporation of billions of dollars severely penalizes not only investors but citizens as a whole. However, as Jean-Pierre Dupuy puts it, 'a social system tends all the more to disintegrate into panic because the evil is already contained within it' (ibid.).

This brief review of the role of speculation and crises shows at least three things: (1) classical economic theory is incapable of explaining speculative bubbles; (2) financial markets function in a way that is disconnected from the real economy; and (3) this 'exuberance' of finance is a direct result of the private property regime, which involves the 'privatization of money' by the banks, whereas its creation used to be a state monopoly. In other words, the crisis we have been experiencing since 2008 is due neither to shady dealers nor to a sudden computer overload nor to a lack of morality among bankers: it is the outcome of a system which, as long as it is not transformed, cannot fail to reproduce the same kind of disasters.

This aim of this chapter is not to narrate the history of the present crisis, which has forced European and North American governments to inject more than a trillion dollars into the system to keep it afloat, and which is continuing with unprecedented austerity measures in most EU countries in support of the Stability and Growth Pact (1997) limiting budget deficits to 3% of GDP and total national deficits to 60%.[53] In so far as the responsibility for this collapse lies largely with the theories and practices that encouraged useless, indeed damaging, speculative activities (Jorion, 2012), 'it follows that the capacity of the financial system for self-correction is close to zero' (Orléan, 2011). As for the national governments, they are particularly badly placed to impose controls on banking operations because they were behind their deregulation in the first place. As far as the EU is concerned, the Single European Act (1986) states that the Council 'shall endeavour to attain the highest possible degree of liberalization. Unanimity shall be required for measures which constitute a step back as regards the liberalization of capital movements' (Art. 16/4). The Maastricht Treaty (1992) states that 'all restrictions on the movement of capital between Member States and between Member States and third countries shall be prohibited' (Art. 56, repeated in subsequent treaties). The United States, for its part, rescinded in 1999 the Glass-Steagall Act of 1933, which provided for a partition separating commercial (deposit) banks and investment banks. Under these conditions, it is hard to see that the way back envisaged by some would be feasible, since the unanimity rule in the EU leaves little scope for such action, and the system has allowed such a degree of interconnection among banks that it has become difficult to disentangle their activities.[54] As regards the Basel Accords (Basel III/2013), their proposal to raise the 'tier one' capital level of banks (to a timid 7%) has met with resistance on the part of bankers, who find it excessive. In addition to this depressing list, let us note that, in order to save the banking system, heavily indebted countries face the dual constraint of austerity (which penalizes their people) and growth (which increases environmental problems).[55] Although it is easy (too easy!) to create money to rescue the banks, it will be incomparably more difficult (or impossible) to reconstitute the natural resources destroyed on a world scale.

What is to be done, then? Apart from the ideas mentioned above, the Tobin tax might discourage 'instant speculation' by increasing transaction costs, but to be effective it would have to be the object of an international agreement (which is very difficult to obtain), so that non-participating countries or tax havens did not act as a deadweight.[56] Support may also be given to Mary Mellor's proposal for the 'statization' or 'communalization' of money: 'The principle must be that if the public, via the state, stands guarantee for the capitalist financial system, then that system must be in public hands' (Mellor, 2010). It is an original idea and links up with the institution of complementary currencies that are owned in common and favour the local economy – although it is not clear how the capitalist system could pass into public hands, as she seems to wish. Others advocate equally radical solutions, such as a partial or total default on sovereign debt, or an injection of European Central Bank liquidity that would lead to inflation. The point here is not to decide whether confrontation or compromise is the better strategy; the former is hard to envisage, while the latter seems more likely. Whichever is chosen, it must preserve the interests of the greatest number rather than of those who champion a (purely illusory) 'moralization' of the system (Lordon, 2003).

Conclusion

This chapter started from the idea of promoting diversity to arrive at knowledge. Perhaps we should have added: 'and to arrive at wisdom'. But that can happen only when market monotheism has come to an end. A lot of arrogance is required to convince oneself that mainstream economics is the only way of understanding the world. Not only do other ways of conceiving of production, consumption and distribution exist elsewhere in the world, but the assumptions of the standard model prevent it from grasping actual practices.

> One is tempted to argue, in a reversal of Marx, that economists have shown too much interest in changing the world, and that it would be good if they now took more trouble interpreting it. This is one of the goals of our proposed refoundation: to put forward an economic approach that pays more attention to the facts and prioritizes an understanding of what exists in reality.
> *(Orléan, 2011)*[57]

Such an initiative would naturally be welcome, but the necessary 'refoundation' is so difficult because mainstream economics has become a faith, or even a dogma (Lebaron, 2000). The broad consensus surrounding it has turned it into a religion, in Durkheim's sense of the term, that is, 'a system of beliefs and practices relative to sacred things [...] – beliefs and practices which unite into one single moral community [...] all those who adhere to them' (Durkheim, 1965). As with any religion, economics may be expressed in scholarly form (in the 'major' US journals charged with guarding orthodoxy), in semi-scholarly form (among International Monetary Fund experts or economy ministers), or in a more popular form (among teachers at universities and other institutions, journalists in charge of the economics pages in the press or corporate economists). As with any religion, there are various chapels: neoliberals and neo-Keynesians savage each other as Jesuits, Dominicans or Franciscans once used to do. But the dogma is never on the line: all agree that the economic 'facts'[58] are 'beyond us', and that we must yield to their 'constraints' or laws', as if they emanated from a divine Providence representing 'the religion of the calculated optimum' (Polanyi, cited in Latour, 2012). True, economics *resembles* a science, since it is expressed in mathematical language, but this resemblance is deceptive: it simply *makes us believe* in the scientific character of economics, while masking assumptions that rest upon an ethnocentric social imaginary with little or no understanding of the constant practices of other societies. Unfortunately, dogmas and religious beliefs change much more slowly than the paradigms of the genuine sciences.

This theoretical (ideological and religious) fundamentalism might be simply treated alongside all the others, as one more refuge from the realities of the world, if it did not involve so many dangers for the future of humanity and its living environment. It is therefore essential to deconstruct the fundamentalism centred on growth and productivism, which, as we have tried to show, is incapable of being reformed. So long as it prevails, all well-meaning efforts to 'save the planet'[59] will be doomed to failure.

For the moment, discordant voices are stifled by the particular interests of a minority clinging to its privileges. The battle is therefore not only theoretical but also political. The hegemony of private property (in natural resources or biodiversity as well as the land) will not give way as if by magic to other regimes of possession or the establishment and maintenance of the commons. It is through the diversity of perspectives, theories and practices that we may collectively acquire a certain wisdom, and that will necessarily lead to the safeguarding of the diversity that underpins it.

Notes

1   This title is a kind of allusion to Dudley Seers, 'The Limitations of the Special Case' (1963). In that article, the author shows that the economic policies which opened the way for the development of Europe and the United States, beginning in the nineteenth century, cannot be transposed as such to the countries of today's Third World. I am especially grateful to Marie-Dominique Perrot, Edouard Dommen and Jean-Noël DuPasquier for their critical remarks and various suggestions.

2   Some would trace its origins much further back, treating Aristotle as one of its first architects. But those who read Aristotle today are few and far between, whereas Adam Smith is most often thought of as the real founding father of the discipline. On the other hand, the neo-classical economists (William S. Jevons, Carl Menger and Léon Walras), whose theories are dominant today, made utility (rather than labour) the source of value – a radical change of perspective, even if the two approaches are both forms of 'value fundamentalism'. Cf. André Orléan, *L'Empire de la valeur. Refonder l'économie*, Paris: Le Seuil, 2011.

3   Of, course the facts are always constructed within a theory, even if this remains implicit. In the case of these authors, the 'observation of facts' generally took place within an optimistic vision of progress.

4   It should not be forgotten, though, that the 'hard' or 'exact' sciences also undergo major paradigm shifts in the course of history. Contemporary physics, for example, has put a lot of space between itself and Newton, or even Einstein.

5   See the account of Tamati Ranaipiri in Marcel Mauss, *The Gift: Forms and Functions of Exchange in Archaic Societies* (1966, p14); see also Karl Polanyi, *The Great Transformation* (1957).

6   In these frequent cases, one might speak of 'shared' or 'layered' ownership.

7   Emphases in the original.

8   'The great market for silver is the commercial and civilized part of the world. If by the general progress of improvement, the demand of this market should increase, while at the same time the supply did not increase in the same proportion, the value of silver would gradually

rise in proportion to that of corn. [...] Labour, it must always be remembered, and not any particular commodity or set of commodities, is the real measure of the value both of silver and of all other commodities.' Smith, *The Wealth of Nations*, vol. 1, pp 196, 207.

9   Or again: 'Labour is *not the source* of all wealth. Nature is just as much the source of use-values (and surely these are what make up material wealth!) as labour' (Marx, 1974, p341; emphasis in the original).

10   The two laws, formulated between 1850 and 1865, state: 'The energy of the universe remains constant. The entropy of the universe moves at all times toward a maximum' Georgescu-Roegen (1999, p129).

11   Of course, these criticisms cannot be made of the authors of the first half of the nineteenth century, but they do apply to all their successors down to the present day, with the sole exceptions of Thorstein Veblen (1898) and William Stanley Jevons, whose essay of 1865 raised questions about the possible depletion of coal by showing that the growing efficiency of machines, far from reducing its consumption, actually expanded it by a knock-on effect. Mainstream economic theory does not completely ignore these problems (known as 'negative externalities'), but it claims to solve them through the laws of the market: 'Faced with externalities, the issue for most economists is how to invent the missing markets or, failing actual markets, how to imitate the market mechanism' (Marglin, 2008, p51.) 'Unfortunately, ecological economics has focused attention on the calculation of the prices of natural capital. In doing so, non-market values must be collapsed into prices so that these values are commensurable with purely economic values. [...] Our problem is not a problem of finding the right price; our problem is one of fundamentally reconstructing economic arrangements' (Krall and Gowdy, 2012, p144).

12   'This concept of the Anthropocene denotes the latest geological era, marked by the impact of global industrial and demographic growth, which clearly differs from the relative stability of the Holocene (that is, the last 10,000 years roughly corresponding to the "immobile history" of the agrarian societies of the Neolithic)' (Grinevald, 2007, p31).

13   To complete the picture, we would also need to mention *res nullius*, things belonging neither to the state nor to any individual but simply available to the first occupier or conqueror (perhaps a deer killed by a hunter in the public forest), and *res derelictae* (waste capable of being reused).

14   Today it is hard to appreciate the scale of this transformation. When I tried to explain it in a course I gave at the University of Bamako, the Malian professor who had invited me reminded his students – who were there in the hope of landing a well-paid job – that the colonial power had had difficulty recruiting their grandparents to work for a wage in the enterprises that 'exploited the colonies'. The conversion of free men (and women) into wage-workers never proceeded without coercion and suffering. In our part of the world too, until the Industrial Revolution, soldiers in receipt of 'pay' (a *solde*, in French) and mercenaries ('hired' for money that often cost them their lives) were the only people paid for their 'work'.

15   To these three 'classical' factors of production, we might add today the strategic resources comprising knowledge, petroleum, pollution rights, water and biogenetic resources. See Steppacher, 2003, p182.

16   'The capitalist requirements of solvency, profitability and time pressure are essential features of property-based capitalist expansion. Agents who fail to meet these requirements are eliminated (through the foreclosure or acquisition of their property), a process that further concentrates economic power' (Griethuysen, 2012, pp262–9).

17   I am not happy with the use of the term 'possession' (which connotes belonging or appropriation) in opposition to 'property', but I will retain it here so as not to complicate the debate. Possession regimes most often concern the use (or enjoyment) of 'biotic' resources, whose pace of renewal is determined by nature, and which do not lend themselves to exponential growth.

18   Of course, this is not always the case. In a famine emergency, some populations are virtually forced to sacrifice their environment in order to survive.

19 Today 3% of the money in circulation has been issued by the central banks; the rest comes from bank loans (Mellor, 2010, p33).

20 See Rist (2011, p179).

21 *Manifeste d'économistes atterrés. Crise et dettes en Europe. Dix fausses évidences, 22 mesures en débat pour sortir de l'impasse,* Paris: Les liens qui libèrent, 2010. For an English translation of the manifesto, see http://www.paecon.net/PAEReview/issue54/Manifesto54.pdf, in the online *Real-World Economics Review* No. 54.

22 'Pour un retour au pluralisme de la pensée', *Le Monde,* 31 July 2012 (signed by, among others, M. Aglietta, L. Boltanski, R. Boyer, J.-P. Dupuy, A. Orléan and J. Galbraith).

23 The beginnings of social anthropology are usually traced to Lewis H. Morgan and his *Ancient Society* (1877), whose evolutionism is today unanimously criticized.

24 Decreto supremo 29279: *Bolivia digna, soberana, productiva y democrática para vivir bien.*

25 This supposition of naturalism or universalism is, of course, the economists' first line of defence against the accusation of provincialism.

26 A history of economic 'science' could be written in terms of its growing complexity (limited rationality, decision-making in a state of uncertainty, asymmetry of information, etc.), but this always aims to safeguard the standard model.

27 In Africa, the 'real poor' are thought of as social orphans abandoned to themselves, and therefore an extreme rarity. See Rahnema, 2005, pp121. and Seyni Ndione, 1987, p154.

28 One thinks of Margaret Thatcher's boastful assertion: 'There is no such thing as society', delivered in a speech to the Conservative Party Conference (9 October 1987) and repeated in an interview with *Woman's Own,* 31 October 1987.

29 'Economic principles are daily refuted by the facts: practice. Things accomplished pursuant to principles are disastrous: theory. By constantly excusing theory by practice, or practice by theory, one eventually removes common sense from the picture, and arbitrariness is certain to be always right' (Proudhon, 1982, p49).

30 See Rist (2011).

31 This probably corresponds to Polanyi's profound idea that land (together with labour and money) is a 'fictitious commodity', which was not originally supposed to be the object of market exchange. Let us just recall that for a long time, even in Europe, landed property escaped the market: it could be acquired only by inheritance, marriage – or war.

32 See notes 40 and 41.

33 See also the site www.paecon.net.

34 In both public and international law, the state is defined as a juridical person – a fiction which hides the fact that it is not a person or a 'player' but the result of political struggles. Depending on the coalition in power at a given time, the state is thus given over to the interplay of particular interests that makes it incapable of acting as an impartial umpire. Neither a neutral nor an abstract entity, the state is a battlefield where various players confront one another and strive to gain control of it.

35 Technically, these are non-rival goods (useable by a number of persons without losing any of their effectiveness), from which no one can be excluded. By contrast, market goods (subject to property law) are exclusive rivals; the apple I bought and am now eating is not available to my neighbour.

36 These are rival goods (the fish I catch are no longer available to another fisherman), but they are not exclusive (free access is guaranteed, in the absence of special rules or quotas).

37 This is exclusive (inaccessible to non-members of the group), but it is not a rival good (all members can enjoy it equally).

38 This term is part of international public law, based on conventions governing space and the Law of the Sea. It was reaffirmed in the UN Charter of Economic Rights and Duties of States of 12 December 1974 (Res. 3281, Art. 29): 'The sea-bed and ocean floor and the subsoil thereof, beyond the limits of national jurisdiction, as well as the resources of the area, are the common heritage of mankind.'

39 *Un autre monde: contre le fanatisme du marché.*

40 The founding texts on the question of the commons are: Hardin's 'The Tragedy of the Commons', 1968), which implicitly justifies the enclosures and private property by confusing the commons with open-access places devoid of rules; and (a critique of Hardin's thesis) Ostrom's *Governing the Commons: The Evolution of Institutions for Collective Action* (1990), for which she received the Nobel Prize for Economics in 2009, with the implication that her neo-institutionalist perspective remains dependent upon mainstream economic theory.

41 See also 'Managing the Global Commons', an interview with Alain Lipietz by Barbara Lipietz, at www.bartlett.ucl.ac.uk/dpu/documents/Managing_the_Global_Commons.pdf.

42 Digital networks (Wikipedia, Linux, Creative Commons, various open sources) are frequently mentioned as examples of common goods by Elinor Ostrom and her supporters (cf. Helfrich *et al.*, 2010), but this is sometimes contested, since the internet may be placed at the service of the market rather than the commons (Mattei, 2011, pp89ff.)

43 'The money system needs to be reclaimed from the profit-driven market economy and socially administered for the benefit of society as a whole as a public resource' (Mellor, 2010, pp3, 159).

44 Mattei (2011, pp52ff.) stresses the necessary link between the 'good' and its social context. Thus, an Italian trade union federation organized a demonstration under the slogan 'work: a common good', to show that work is not something contractual but an activity that has meaning only when it is common, and that it should be protected collectively, not only by provisions of the law.

45 One need only think of the ozone layer depletion and the skin cancers to which it has given rise, or of the pollution of water tables in Brittany resulting from intensive pig-breeding. Doubtless we will soon be lamenting the exhaustion of fish resources.

46 Emblematic here is the huge popular mobilization in 2000 in Cochabamba, Bolivia, which eventually saw off the US multinational Bechtel and its commission to privatize water distribution. Just a few years ago in Paris, in 2009, water distribution was similarly taken out of the hands of the Veolia and Suez groups and entrusted to a public authority. More generally, the Zapatista experience in Mexico may be considered an example of how a particular community may regain control of its commons.

47 So as not to overburden the text, I will refer the reader to the works of Serge Latouche, especially *Farewell to Growth* (2008), which clearly addresses all the aspects of degrowth. See also Chapter 9 in my *The Delusions of Economics* (Rist, 2011).

48 Exhaustive calculations for Switzerland, using a monthly basic income of 2,500 francs (2,000 euros) per person, prove that there is nothing utopian about this idea: see Müller and Straub (2012). It was presented to the Swiss electorate in the form of a citizens' initiative.

49 Some still deny it, though. Christophe de Margerie, the CEO of Total, stated: 'Peak oil is no longer just around the corner. Recent discoveries and the development of new technologies have made it possible to increase the long-term reserves available to the world, thanks especially to shale oil and gas. We reckon that, at current consumption levels, we have more than a hundred years of oil reserves'. *Le Monde*, 11 January 2013.

50 The term 'globalization' (unlike the French *mondialisation*, which refers mainly to the free circulation of goods and services among a few key economic centres) encompasses ecological phenomena that affect the whole 'globe', such as pollution and climate warming – which is not to say that these are usually uppermost in the minds of English speakers who use it.

51 Without getting too complicated, let us add that it is possible to buy forward by depositing a small part of the sum required at the due date (leverage effect), and that many transactions are carried out by computerized systems in a fraction of a second.

52 'Professional investment may be likened to those [...] competitions in which the competitors have to pick out the six prettiest faces from a hundred photographs, the prize being awarded to the competitor whose choice most nearly corresponds to the average

preferences of the competitors as a whole; so that each competitor has to pick, not those faces which he himself finds prettiest, but those which he thinks likeliest to catch the fancy of the other competitors, all of whom are looking at the problem from the same point of view. [...] We have reached the third degree where we devote our intelligences to anticipating what average opinion expects the average opinion to be. And there are some, I believe, who practise the fourth, fifth and higher degrees' (Keynes, 1961, p156).

53 There is such a narrative in Paul Jorion, *Misère de la pensée économique* (2012). The author, quasi anthropologist and computer specialist, worked in various American banks in the first decade of this century. For another version (which does not contradict the first), see Mary Mellor, *The Future of Money, From Financial Crisis to Public Resource* (2010). For the sake of impartiality, let us note that some of the government money used to bail out the banks has since been repaid.

54 See the dossier 'Le découpage des grandes banques en question', *Le Monde éco & entreprises*, 15 January 2013.

55 'When growth goes together with multiple tendencies that are destroying the planet [...] a GDP in excellent shape may be the sign of an acceleration of the destructive process. [...] From this point of view, the fact that growth has been made a determining factor in the managing of public debt [...] may be considered a monumental blunder with incalculable consequences. [...] Moreover, the ill-advised austerity policies of European leaders, in 2011 and 2012, have a negative impact on growth (the much-feared scissors effect), since the growth rate declines while the rate demanded for new debt by the capital markets increases' (Jorion, 2012, pp318, 322–3).

56 Such a tax might still be adopted by the EU, however: not to assist countries in the South but to boost the budget of the Union.

57 Also: 'Simiand is especially opposed to this constant tendency of economists to interpret the gap between model and reality as a defect in reality rather than an error in the model!' (ibid., p317).

58 '*Les faits sont faits*', the facts are already on the table, as Bruno Latour constantly repeats in *On the Modern Cult of the Factish Gods* (2011).

59 Strictly speaking, the planet has no need to worry; it will continue spinning, as imperturbable as ever. The only question is which species will survive an alteration of the biosphere.

References

Cordonnier, L. (undated) 'Eclairage sur les biens communs', Blog, *Alternatives Economiques*, http://alternatives-economiques.fr/blogs/gadrey/files/laurent-bc-v2.pdf.

Coussy, J. (2002) 'Biens publics mondiaux: Théorie scientifique, réalité émergente et instrument rhétorique', In Constantin, F. (ed.) *Les Biens Publics Mondiaux. Un Mythe Légitimateur pour l'Action Collective?*, L'Harmattan, Paris.

D'Alisa, G., F. Demeria and G. Kallis (2014) *Degrowth. A Vocabulary for a New Era*. Routledge, London.

Dupuy, J-P. (2002) *Pour un Catastrophisme Eclairé. Quand l'Impossible est Certain*, Le Seuil, Paris.

Dupuy, J-P. (2003) *La Panique*, Les empêcheurs de penser en rond, Le Seuil, Paris.

Durkheim, É. (1958) *Professional Ethics and Civil Morals*, Free Press, Glencoe, Ill.

Durkheim, É. (1965) *The Elementary Forms of the Religious Life*, The Free Press, New York.

Farley, J. and D. Malghan (2016) *Beyond Uneconomic Growth: Economics, Equity and the Ecological Predicament,* Edward Elgar Publishing, Cheltenham.

Fullbrook, E. (ed.) (2008) *Pluralist Economics*, Zed Books, London.

Gadrey, J. (2012) 'Des biens publics aux biens communs', Blog, 24 April 2012, *Alternatives Economiques*, http://alternatives-economiques.fr/blogs/gadrey/2012/04/24/des-biens-publics-aux-biens-communs.

Georgescu-Roegen, N. (1999) *The Entropy Law and the Economic Process*, Harvard University Press, Cambridge, MA.

Georgescu-Roegen, N. (2011) 'The entropy law and the economic problem', In Bonaiuti, M. (ed.) *From Bioeconomics to Degrowth: Georgescu-Roegen's 'New Economics' in Eight Essays*, Routledge, London.

Griethuysen, P. van (2012) '*Bona diagnosis, bona curatio*: How property economics clarifies the degrowth debate', *Ecological Economics*, 84(C): 262–9.

Griethuysen, P. van and R. Steppacher (2015) 'Propriété et possession: Une approche économique', In Travési, C. and M. Ponsonnet (Eds) *Les Conceptions de la propriété foncière à l'épreuve des revendications autochtones: Possession, propriété et leurs avatars*, Pacific-credo Publications, Aix-Marseille University, Marseille.

Grinevald, J. (2007) *La Biosphère de l'Anthropocène, Climat et Pétrole, la Double Menace, Repères Transdisciplinaires (1824–2007)*, Georg, Geneva.

Hardin, G. (1968) 'The tragedy of the commons', *Science*, 162(3859): 1243–8.

Helfrich, S., Kuhlen, R., Sachs, W. and C. Siefkes (2010) *The Commons – Prosperity by Sharing*, Heinrich Böll Foundation, Berlin.

Jackson, T. (2017) *Prosperity without Growth: Foundations for the Economy of Tomorrow*. Routledge, London.

Jorion, P. (2012) *Misère de la Pensée Economique*, Fayard, Paris.

Keynes, J. M. (1961) *The General Theory of Employment, Interest and Money*, Macmillan, London.

Krall, L. and J. M. Gowdy (2012) 'An institutional critique of natural capital', In Steppacher, R. and J-F. Gerber (Eds) *Towards an Integrated Paradigm in Heterodox Economics, Alternative Approaches to the Current Eco-Social Crises*, Palgrave Macmillan, Basingstoke.

Latouche, S. (2008) *Farewell to Growth*, Polity, Cambridge.

Latouche, S. (2011) *Vers une société d'abondance frugale: Contresens et controverses de la décroissance*, Editions Fayard/Mille et Une Nuit, Paris.

Latour, B. (2011) *On the Modern Cult of the Factish Gods*, Duke University Press, Durham, NC.

Latour, B. (2012) *Enquête sur les Modes d'Existence, Une Anthropologie des Modernes*, La Découverte, Paris.

Lebaron, F. (2000) *La Croyance Economique. Les Economistes entre Science et Politique*, Le Seuil, Paris.

Lipietz, A. (2010) 'Questions sur les biens communs', *Esprit* 2010/1.

Lordon, F. (2003) *Et la Vertu Sauvera le Monde. Après la Débâcle Financière, le Salut par l'Ethique'?*, Raisons d'agir, Paris.

Marglin, S. A. (2008) *The Dismal Science. How Thinking Like an Economist Undermines Community*, Harvard University Press, Cambridge, MA.

Martinez-Alier, J. (2012) 'Social metabolism, environmental cost-shifting and valuation languages', In Steppacher, R. and J-F. Gerber (Eds) *Towards an Integrated Paradigm in Heterodox Economics, Alternative Approaches to the Current Eco-Social Crises*, Palgrave Macmillan, Basingstoke.

Marx, K. (1973a) *Grundrisse: Foundations of the Critique of Political Economy*, Penguin/NLR, London.

Marx, K. (1973b) 'Manifesto of the Communist Party', In Marx, K. (1973) *The Revolutions of 1848*, Penguin/NLR, London.

Marx, K. (1974) 'Critique of the Gotha Programme', In Marx, K. (1974) *Political Writings*, vol. 3, *The First International and After*, Penguin/NLR, London.

Marx, K. (1976) *Capital Volume 1*, Penguin/NLR, London.

Marx, K. (1981) *Capital Volume 3*, Penguin/NLR, London.

Marx, K. and Engels, F. (1975) 'Engels to Joseph Bloch, 21 September 1890', In Marx, K. and F. Engels (1975) *Selected Correspondence*, Progress Publishers, Moscow.

Mattei, U. (2011) *Beni Communi. Un Manifesto*, Laterza, Rome.

Mauss, M. (1966) *The Gift: Forms and Functions of Exchange in Archaic Societies*, Cohen & West, London.

Mellor, M. (2010) *The Future of Money. From Financial Crisis to Public Resources*, Pluto Press, London.

Müller, C. and Straub, D. (2012) *Die Befreiung der Schweiz. Über das bedingungslose Grundeinkommen*, Limmat Verlag, Zurich.

Orléan, A. (2011) *L'Empire de la Valeur. Refonder l'Economie*, Le Seuil, Paris.

Ostrom, E. (1990) *Governing the Commons: The Evolution of Institutions for Collective Action*, Cambridge University Press, Cambridge.

Polanyi, K. (1957) *The Great Transformation*, Beacon Press, Boston, MA.

Proudhon, P.-J. (1982) 'Système des contradictions économiques, ou, philosophie de la misère', In Proudhon, P.-J. (1982) *Œuvres Complètes*, Slatkine, Geneva.

Rahnema, M. (2005) *Quand la Misère Chasse la Pauvreté*, Fayard/Actes Sud, Paris.

Ricardo, D. (1973) *The Principles of Political Economy and Taxation*, Dent, London.

Rist, G. (2008) *The History of Development. From Western Origins to Global Faith*, 3rd ed., Zed Books, London.

Rist, G. (2011) *The Delusions of Economics. The Misguided Certainties of a Hazardous Science*, Zed Books, London.

Saleilles, R. (1889) *Le Domaine Public à Rome et son Application en Matière Artistique*, Larose et Forcel, Paris.

Seers, D. (1963) 'The limitations of the special case', *Bulletin of the Oxford Institute of Economics and Statistics*, 25 (2): 77–98.

Seyni Ndione, E. (1987) *Dynamique Urbaine d'une Société en Grappe: Un Cas, Dakar*, ENDA, Dakar.

Smith, A. (1961) *The Wealth of Nations*, vol. 1, Methuen and Co. Ltd, London.

Steppacher, R. (2003) 'La petite différence et ses grandes conséquences: Possession et propriété', *Brouillons pour l'avenir*, Nouveaux Cahiers de l'IUED, n 14, Presses universitaires de France, Paris.

Steppacher, R. and J.-F. Gerber (2012) 'Meanings and significance of property with reference to today's three major eco-institutional crises', In Steppacher, R. and Gerber, J.-F. (2012) *Towards an Integrated Paradigm in Heterodox Economics: Alternative Approaches to the Current Eco-Social Crises*, Palgrave Macmillan, Basingstoke.

Stiglitz, J. (2006) *Making Globalization Work: The Next Steps to Global Justice*, Penguin, London.

Veblen, T. (1898) 'Why is economics not an evolutionary science?', *The Quarterly Journal of Economics*, 12 (4): 373–397.

Walras, L. (1954) 'Preface to the Fourth Edition', In Walras, L. (1954) *Elements of Pure Economics*, George Allen & Unwin, London.

8

DEMOCRATIZING KNOWLEDGE AND WAYS OF KNOWING FOR FOOD SOVEREIGNTY, AGROECOLOGY AND BIOCULTURAL DIVERSITY

Michel P. Pimbert

Introduction

Achieving food sovereignty, agroecology and biocultural diversity is a transformative process that seeks to expand the realms of democracy and freedom by regenerating a diversity of autonomous food systems based on social justice and ecological sustainability (Pimbert, 2008). As part of this transformation,[1] social movements are increasingly challenged to develop more *inclusive and participatory ways of knowing* to produce knowledge that is not only ecologically literate and socially just, but which also embodies the values of a new modernity as well as plural visions of the 'good life'. This is a huge challenge. To paraphrase the philosopher of science, Thomas Kuhn (1962), nothing less than a paradigm revolution is necessary to generate knowledge for food sovereignty, agroecology and biocultural diversity.

Previous chapters in this book have highlighted the need for a fundamental transformation of knowledge in the following areas:

- Reductionist science, which unlike more holistic knowledge and ways of knowing, has largely failed to promote the sustainable management of agro-ecosystems, natural resources or landscapes. For example, peasant farmers who want to grow their crops and rear their animals using agroecological approaches clearly need very different technical knowledge than that currently provided by the dominant agricultural research system, which focuses on the delivery of pesticides, growth hormones, food additives and other external inputs marketed by agri-chemical companies (Chapters 2, 3, 4 and 6).
- Crisis narratives that blame rural communities for environmental degradation and justify standard environmental management packages which neglect people's knowledge, priorities, locally adapted management systems and local institutions. Yet recent research has shown that several orthodox views on

people-environment interactions are myths that are often based on a-historical views and erroneous Malthusian assumptions (Chapters 3, 4 and 5).

- Current economic disciplines that underpin policies for growth and competitiveness in the food system. This is leading to the economic genocide of unprecedented numbers of small-scale producers (family farmers, pastoralists, fishers, farm workers, etc.) and rural livelihoods throughout the world. It is now imperative to re-think mainstream economics on the basis of radically different principles such as reciprocity, solidarity, respect, freedom, equity and sustainability (Chapters 5, 6 and 7).

More generally, scientific research offers an increasingly dismal picture with its daily reports of retractions, fraudulent peer reviews, 'fake science' and misinformed science-based policies (Benessia *et al.*, 2016). A growing number of citizens no longer trust scientists and the institutions they work for:

> Worldwide, we are facing a joint crisis in science and expertise … Today, the scientific enterprise produces somewhere in the order of 2 million papers a year, published in roughly 30,000 different journals. A blunt assessment has been made that perhaps half or more of all this production 'will not stand the test of time'…. Meanwhile, science has been challenged as an authoritative source of knowledge for both policy and everyday life … Perhaps nutrition is the field most in the spotlight. It took several decades for cholesterol to be absolved and for sugar to be re-indicted as the more serious health threat, thanks to the fact that the sugar industry sponsored a research program in the 1960s and 1970s, which successfully cast doubt on the hazards of sucrose – while promoting fat as the dietary culprit.
>
> *(Saltelli, 2016)*

During the second half of the twentieth century, universities in the USA have redefined their mission to serve private business and become much more profit-oriented in their operations and strategic objectives (Heller, 2016a). This business model has since spread worldwide. And most universities today increasingly embrace 'what we might call a cognitive capitalism, which pursues new forms of knowledge that can be more or less immediately commodified as intellectual property: patents, inventions, copyrights and even trademarks' (Heller, 2016b). Corporations and large financial investors increasingly control the directions and outcomes of research in the social and natural sciences, as well as in the humanities. Private-public sector partnerships, funding priorities, patents and other intellectual property rights – in addition to widespread corporate control and corruption of science – all ensure that mainstream research selectively favours the production of knowledge that reflects and reinforces the interests of company shareholders and financial institutions – from patented seeds and new natural products derived from indigenous and peasant knowledge to neo-liberal food policies and trade agreements (Chapters 1, 5 and 6; Heller, 2016b). Moreover, a set of legal rules is in place to protect investors' rights in the frame of the World Trade Organization and in bilateral investment treaties and clauses in free-trade agreements. These 'trade and investment agreements have equipped private corporations with extraordinary and powerful tools for asserting and defending their commercial interests. Thus, foreign investors have been accorded the unilateral right to invoke binding investor-state dispute settlements (ISDS) to claim damages for violations of the broadly framed rights they now enjoy under these treaties' (Monsanto Tribunal, 2017). These provisions undermine the capacity of states to maintain policies, laws and practices protecting human and environmental rights, *including the freedom indispensable for academic research.*

In this context of free trade and binding ISDSs, research and innovations favoured by transnational corporations and financial investors all help fuel today's historically unprecedented concentration of wealth and power by a tiny minority of super-rich individuals (Chomsky, 2017). At the time of writing, the super-rich comprise less than 100 people who own and control more wealth than 50% of the world's population (Beaverstock and Hay, 2016). Contemporary dynamics of knowledge creation and use play a key role in the processes that underpin the generation of super-wealth and its unequal distribution within and between societies (Chomsky, 2017; Harvey, 2014; Hay and Beaverstock, 2016; Noble, 1995).

In sharp contrast, counter-hegemonic practices by peasant networks, indigenous peoples and social movements seek to reframe food, agriculture, biocultural landscapes and the 'good life' in terms of a larger vision based on radical pluralism and democracy, personal dignity and conviviality, autonomy and reciprocity and other principles that affirm the right to self-determination and justice (for example, see Esteva and Prakash, 2014). Making these other worlds possible requires the construction of radically different knowledge from that offered today by mainstream universities, policy think tanks and research institutes.

More than ever before, new ways of knowing are needed to construct knowledge for social inclusion, economic justice, environmental sustainability and cultural diversity. This is a formidable challenge because we need to simultaneously 'confront the question of what kinds of knowledge we want to produce, and recognize that that is at the same time a question about what kinds of power relations we want to support – and what kind of world we want to live in' (Kamminga, 1995).

Constructing knowledge for food sovereignty, agroecology and biocultural diversity entails reversing top-down research and the hegemony of scientism, as well as the current privatization of research and commodification of knowledge. It also means reversing the current democratic deficit in the governance of research by enabling more direct citizen control over the priorities and conduct of scientific, social and technological research. Transformation thus partly depends on making a radical shift from the existing top-down and increasingly corporate-controlled research system, to an approach which gives more agency and decision-making powers to peasant farmers, indigenous peoples, food workers, pastoralists and citizen-consumers in the production and validation of environmental, economic, social and technical knowledge.

This closing chapter first briefly identifies key moments at which previously excluded farmers and citizens can intervene in the politics of knowledge to decide what knowledge is produced, and for whose benefit. Next, the chapter critically explores two complementary approaches for democratizing the construction of knowledge(s) for food sovereignty, agroecology and biocultural diversity. The first emphasizes the potential of grassroots innovation and self-managed research to de-institutionalize research for autonomous learning and action. It focuses in particular on critical education and strengthening horizontal networks of farmers[2] and other citizens who actively produce knowledge in the many 'living campuses' where they derive their livelihoods. The second approach focuses on democratizing and transforming public research to better serve the common good rather than narrow economic interests. Particular attention is given here to institutional, pedagogical and methodological innovations that can enable citizen participation and agency throughout the entire research cycle – from deciding upstream strategic research and funding priorities to the co-production of knowledge and the framing of risk assessments.

Democratizing the politics of knowledge

Issues of power and whose knowledge counts are at the heart of the governance of research and its impacts on society. Nuanced and scholarly analysis of the politics of knowledge show how research is influenced by powerful combinations of political interests, dominant policy discourses and effective actor networks that span local, national and international levels (e.g. Apffel-Marglin *et al.*, 1990; Apffel-Marglin and Marglin, 1996; Dominguez Rubio and Baert, 2012; Meusburger *et al.*, 2015).

A few simple questions can help shed light on these politics and the processes that construct knowledge and innovations: Which actors are involved? Whose knowledge is included and whose is excluded? Where is 'knowledge-making' actually taking place? How is knowledge circulated and applied? Who has the final control and say? Whose interests are served? Is someone held accountable? If so, to whom and how? Asking these questions helps to shift attention from an analysis of knowledge per se (Is the knowledge produced by research addressing the relevant issues? Is scientific and technical knowledge good or misguided?) to the analysis of the *processes* of knowledge generation and its politics (Whose perspectives, priorities, values, knowledge, interests and aspirations are embedded in research and its products, and whose are excluded? Where, why and how is knowledge constructed and applied, for whose benefit and with what effects on environment and society?).

In practice, expanding knowledge democracy calls for institutional and methodological innovations that enable the *direct* participation of farmers and other citizens in research and development (R&D) and, more generally, in the construction of knowledge. A focus on the entire R&D cycle allows for a shift from narrow concepts of co-inquiry and participatory research that confine non-researchers (peasant farmers, food workers, consumer-citizens) to 'end of the pipe' technology development (e.g. participatory technology development) to a more inclusive approach in which farmers and other citizens can directly define the *upstream strategic priorities* of research and the overarching national policies for research and development. Instead of being seen as passive beneficiaries of trickle-down scientific development or technology transfer, farmers and other citizens are viewed as knowledgeable and active actors who can be centrally involved in both the 'upstream' choice and design of scientific innovations, and their 'downstream' implementation, spread and regulation. In this context, science and the construction of knowledge are seen as part of a bottom-up, participatory process in which citizens take centre stage in decisions on *what* knowledge is produced, *why, how* and *for whom.*

Democratizing knowledge through grassroots innovation and self-managed research

> In our interactions with the world, we are all involved in the production of knowledge about the world—in that sense, there is no single group of experts.
> *(Kamminga, 1995)*

Historically, a great deal of knowledge has been produced by people who have not received any professional university training. Well before scientific institutions and agricultural research stations existed, farmers and livestock keepers generated a huge diversity of locally adapted crop varieties and livestock breeds by working with nature. This agricultural biodiversity is an embodiment of peoples' knowledge and their labour. This is an important peasant heritage on which modern plant and animal breeding depends to develop pest resistant crops and livestock, as well as adaptations to climate change (FAO, 2010; FAO, 2015).

Even today in advanced industrial societies, farmers as well as ordinary citizens are engaged in the production of knowledge on a significant scale *outside* universities and research institutes. People without any specialized professional training are increasingly involved in creating new knowledge and innovations in many different areas, including (Callon *et al.*, 2001; Charvolin *et al.*, 2007; Irwin, 1995):

- victims of pollution developing a people's epidemiology in rural and urban areas (Brown, 1992; Irwin 1995).
- citizens affected by HIV/AIDS or other illnesses engaging in scientific activism to contest medical expertise and discrimination (Lorway, 2017).
- amateur naturalists and gardeners involved in national surveys or biodiversity conservation plans (www.naturescalendar.org.uk/science).
- computer game players contributing to the enrichment and design of new games.
- the world's open source community developing non-proprietary software and internet programmes (O'Mahony and Ferraro, 2007).
- community-based digital fabrication workshops known as hackerspaces, fablabs and makerspaces. These spaces allow people to come together to learn about and use modern technologies of digital design and manufacturing. Skills

and tools are freely available to people who directly participate in design as well as peer-to-peer production. Hackerspaces, fablabs and makerspaces share knowledge through both social media and face-to-face meetings (Maxigas, 2012; Troxler, 2014; www.hackerspaces.org).

Self-organizing grassroots research and innovation networks play an increasingly important role in the practice of the larger social movements working for food sovereignty, agroecology and biocultural diversity. Some of the more emblematic examples of self-organized peasant-led research and grassroots innovation networks are briefly described in Box 8.1. By focusing on processes of knowledge creation and innovation, these networks of small-scale producers are generating an alternative material culture. Indeed, such grassroots research and innovation movements are 'distinctive because the principal means of social change is the development of new or alternative forms of material culture, a means of change that is often associated with calls for significant institutional and policy changes as well' (Hess, 2005).

BOX 8.1 EXAMPLES OF SELF-MANAGED RESEARCH AND GRASSROOTS INNOVATION MOVEMENTS CONSTRUCTING KNOWLEDGE FOR FOOD SOVEREIGNTY, AGROECOLOGY AND BIOCULTURAL DIVERSITY

Peasant farmers, indigenous peoples, pastoralists and other citizens engaged in grassroots research and innovation rarely work alone. They are usually members of a collective of peers, an affinity group or an association. Self-organized peasant-led research and innovation processes are typically part of horizontal socio-cultural networks that usually span large geographical areas.

The *Campesino a Campesino* (CAC) movements in Central America and Cuba. *Campesino a Campesino,* or Farmer to Farmer, is a grassroots movement which originated in the early 1970s in Guatemala and spread through Mexico, Nicaragua and Cuba. Mayan *campesinos* in Guatemala pioneered methods of soil and water conservation as well as an innovative farmer-to-farmer pedagogy which they initially shared with each other, and then with small peasant farmers in Mexico. This 'peasant pedagogy' has been well described by Holt-Giménez (2006). It is notable that the CAC process generated effective site-specific agroecological solutions as well as empowering forms of non-hierarchical communication and local social change which peasants themselves eventually spread throughout Central America and the Caribbean. Using their own farms as classrooms, the peasant farmers rely on principles of popular education and peer-to-peer learning to build local capacity, autonomy and empowerment. As a social process methodology, CAC has achieved a significant impact in Cuba where the National Association of Small Farmers (ANAP) has adopted it along with the explicit goal of building a grassroots movement for agroecology (see Machín Sosa *et al.*, 2010, 2013). In less than a decade (from the mid-1990s), the transformation of conventional farms into diverse agroecological systems spread to more than one third of all peasant families in Cuba (Rosset *et al.*, 2011).

The Peasant Seeds Network in France. In 2003, the *Réseau Semences Paysannes* was created in France by the *Confederation Paysanne*, the National Coordination of Defenders of Farm Seeds and several organic farmers' associations. The *Réseau Semences Paysannes* comprises over 70 member organizations. This peasant network focuses not only on vegetable seeds but also on cereals, oilseeds, fruit and grapevines. The main objective of the network is for peasant farmers 'to regain total autonomy over seeds, which means being able to do our own plant breeding, and select plants in our own fields' (personal communication Guy Kastler, 24 November 2016). In this context, selecting and producing one's own seeds not only represents a fundamental rejection of the 'commercial and industrial productivist system', but is also a quest for autonomy, peasant identity and meaning. Members of the network engage in participatory and evolutionary plant breeding and facilitate grassroots research and innovations in agroecology. They also co-produce knowledge needed for their political advocacy work in defence of farmers' rights to save and exchange seeds, as well as for the recognition of their collective rights as innovators in national plant breeding programmes (see http://www.semenc-espaysannes.org).

Autonomous research and learning networks in Bangladesh. The *Nayakrishi Andolan*, or New Agriculture Movement, comprises over 300,000 farming families in 19 districts of Bangladesh. As an autonomous network for learning and action, the *Nayakrishi Andolan* builds on rural peoples' systemic art and science of combining and integrating all aspects of life. Its holistic orientation aims to re-unite those 'dimensions that civilisation has systematically broken into institutional and social silos, including livelihood (labour), wealth (capital), reciprocity (market), governance (government), spirituality (religious institutions), knowledge (science), aesthetics (arts), love (family) and pleasure (sex and entertainment)' (Mazhar *et al.*, 2006). This grassroots innovation movement has developed biodiversity-rich agriculture(s) based on ten simple rules derived from the day-to-day experiences and knowledge of male and female family farmers. These rules for the design and adaptive management of agroecosystems are reviewed every year to incorporate new peasant-generated knowledge and agroecological practices (see http://ubinig.org/index.php/nayakrishidetails/showAerticle/5/6).

The Zero Budget Natural Farming movement in India. Zero Budget Natural Farming (ZBNF) is both a set of farming methods and a grassroots peasant movement that has spread to various states in India. It been very successful in southern India, and particularly in the State of Karnataka where it first evolved. The ZBNF in Karnataka has been actively promoted by the *Karnataka Rajya Raitha Sangha* (KRRS), a member of *La Via Campesina*. 'Zero budget' natural farming seeks to significantly reduce production costs by ending dependence on all outside inputs and loans for farming. The word 'budget' refers to credit and expenses. Thus, the phrase 'zero budget' means not using any credit, and not spending any money on purchased inputs. 'Natural farming' means farming with nature and without synthetic pesticides and other chemicals. Peasant farmers are the main protagonists of the movement and they have relied on self-organized processes with a strong pedagogical content. The farmers mostly come from the middle peasantry – i.e. they own land. Most practising ZBNF farmers are informally linked to each other and engage in both organized and spontaneous farmer-to-farmer exchange activities. The main centrally organized activities are five day-long training workshops where farmers learn about philosophy, ecology, ZBNF practices and successful farmer experiences. As a grassroots innovation movement, ZBNF has been very effective in bringing agroecology to scale. For example, it is estimated that in the State of Karnataka some 100,000 small and family farmers practise ZBNF (FAO, 2016b).

Brazil's Landless Workers' Movement (o Movimento dos Trabalhadores Rurais Sem Terra or MST)

The MST is the largest and most powerful agrarian reform movement in Latin America. Since the 1980s, this grassroots movement has been struggling for equitable land redistribution by occupying unproductive land estates in Brazil. Transformative education is central to the MST's struggle for social justice and ecologically sustainable peasant farming. The MST has actively engaged in a deep discussion on the political role of education in collectively constructing a different model of development for the countryside. The emergent concept of *Educaçao do Campo* – 'education for the countryside' and 'education by the countryside' – is important is this regard. *Educaçao do Campo* is an 'education by and for the countryside, by and for its historic actors, and by and for the peasant people; also an education for the collective transformation of reality in the countryside' (Barbosa, 2016). The MST's more specific education programmes on agroecology and food sovereignty grew out of a recognition that agricultural knowledge – like land itself – is highly concentrated in the hands of élites, and inaccessible to marginalized communities. Most notably, MST has developed critical place-based agroecological education which promotes counter-hegemonic agricultural practices as alternatives to the dominant, capital-intensive model of large-scale farming in Brazil (Meek, 2015).

Grassroots networks for food sovereignty and biocultural diversity in India, Indonesia, Iran and Peru.
Indigenous communities in the Peruvian Andes, women peasant collectives in the drylands of south India, nomadic tribes of pastoralists in Iran, and peasant farmers from Java's rice

producing areas in Indonesia have all been involved in multi-scale networks working for food sovereignty and self-determination. These grassroots horizontal networks have participated in power-equalizing research that has involved both researchers and non-researchers in close co-operative engagement, jointly producing new knowledge on agroecology, biocultural diversity and food sovereignty. Peer-to-peer dialogues, farmer field schools and farmer exchanges for mutual learning within and between countries are some of the empowering pedagogies that have enabled the bottom-up construction of knowledge (Fakih *et al.*, 2003; Pimbert *et al.*, 2017).

URGENCI and community supported agriculture. URGENCI, the international network for Community Supported Agriculture (CSA), emphasizes the need to consider citizen-consumers as key subjects in peer-to-peer learning on agroecology and food sovereignty. Popular education about the realities of farming and the entire food system is at the heart of the CSA movement. The URGENCI network emphasizes mutual assistance and solidarity as well as direct connections and shared risk between farmers and the people who eat their food; agroecological farming methods (sometimes requiring organic certification); the importance of biocultural diversity; and high-quality and safe food that is accessible to as many people as possible and based on negotiated prices that are fair to producer and consumer. A recent survey of CSAs in 22 European countries estimates that there are currently 6,300 CSA initiatives producing food (vegetables, fruit, meat, dairy products, etc.) for over one million consumers (URGENCI, 2016).

L'Atelier Paysan **in France and Farm Hack in the USA.** These communities of farmers and mechanics use internet platforms to share knowledge about farm tools and machinery they design and build on their farms or in community workshops and co-operatives. These grassroots communities of innovators are interested in developing and sharing open-source tools for a resilient agriculture. They also sometimes assemble offline in the form of face-to-face meetings, workshops and hands-on building events. For example, *L'Atelier Paysan* not only distributes free plans on its website, it also organizes winter self-help training sessions, during which farmers train in metalworking and build tools which they can then use on their own farms. Lastly, these networks are inclusive of different types of knowledge holders and comprise not only farmers but also people with common interests: engineers, designers, architects, tinkerers, programmers and hackers. See http://farmhack.org/tools and http://www.latelierpaysan.org

These decentralized and distributed forms of peasant and peoples-led research and innovation sharply contrast with the organization and practice of mainstream science and technological R&D (Table 8.1). This is mainly because they seek to expand

TABLE 8.1 A comparison between grassroots innovation movements and institutions for science, technology and innovation

	Grassroots research and innovation movements	Science, technology and innovation institutions
Predominant actors	Local communities in rural and urban areas, indigenous peoples and peasant networks, civil society organizations, worker co-operatives, social entrepreneurs, NGOs and social movements	Universities, research centres, corporations, venture capital, science ministries and business entrepreneurs
Priority values	Social justice, autonomy, convivial communities, environmental justice, not necessarily focused on for-profit innovation, sustainable livelihoods and human well-being	Scientific advance, economic growth, for-profit innovation/ not necessarily focused on competitiveness
Type and quality of participation	Interactive participation in joint analysis, which leads to action plans and the formation of new local groups or the strengthening of existing ones. Self-mobilizing participation involving people taking initiatives independent of external institutions to change systems	Participation as a means for increasing the effectiveness of research and reaching pre-determined objectives decided by scientists. Mostly consultative, instrumental and passive participation in which scientists and professionals are under no obligation to take on board peoples' views
Incentives and drivers	Social needs, co-operation, community empowerment, mutual aid and solidarity, claiming citizens' right to participate in decision making	Market demand, corporate agendas, expert authority, career progression and reputation
Sources of investment	Community finance, donations, crowd source funding, state finance, development aid and grassroots ingenuity	Public funds, corporate investments, venture capital
Resources	Peoples' knowledge and assets, local organizations and the networks they form	Scientific and professional expertise, including technical infrastructure
Location of activity and sites of innovation	Villages, fields, forests, factories, backyards, co-operatives, neighbourhoods, online, community projects and social movements	Laboratories and R&D institutes, boardrooms and ministries, market-based firms and large corporations
Predominant forms of knowledge	Local, situated knowledge/ indigenous and peoples' knowledge	Scientific and technical knowledge

(*continued*)

TABLE 8.1 Continued

	Grassroots research and innovation movements	*Science, technology and innovation institutions*
Education	Emphasis on critical education that focuses on political and practical dimensions of change. Often counter-hegemonic educational activities based on plural knowledge systems	Banking model of education. Reflects dominant categories and epistemology of knowledge
Appropriation	Freely shared practices. Not appropriated by individuals – seen as common goods (e.g. knowledge commons)	Intellectual property framework strongly biased towards patent-based innovation and proprietary technologies
Emblematic fields of activity	Agroecology, alternative food networks and economies, small-scale renewable energy, community health and sanitation, housing and low-impact human settlements	Biotechnology, medicine, nanotechnology, synthetic biology and geo-engineering, surveillance technologies, weapons

Source: Adapted and modified from Fressoli *et al.*, 2014.

'knowledge democracy' which firstly 'acknowledges the importance of the existence of multiple epistemologies or ways of knowing such as organic, spiritual and land-based systems, frameworks arising from our social movements and the knowledge of the marginalized or excluded everywhere, or what is sometimes referred to as subaltern knowledge. Secondly it affirms that knowledge is both created and represented in multiple forms including text, image, numbers, story, music, drama, poetry, ceremony, meditation and more. Third, and fundamental to our thinking about knowledge democracy is understanding that knowledge is a powerful tool for taking action to deepen democracy and to struggle for a fairer and healthier world'

(Hall and Tandon, 2015)

Similarly, the grassroots research and innovation movements described here are fundamentally different from 'citizen science' initiatives in which people act as amateur scientists and/or helpers to the scientific community (Haklay, 2015). These initiatives are usually large-scale scientific projects in which goals are reached more effectively and cheaply thanks to a mass of citizen contributors who 'participate' in surveys or experiments designed by a small number of scientists and the institutions that employ them.

In sharp contrast, networks of peasant farmers, pastoralists, indigenous peoples, fishers, food workers, forest dwellers and other citizens create knowledge and innovations through self-organizing processes under *their* control (see Table 8.1 and Box 8.1). For example, the grassroots innovation network *l'Atelier Paysan* in France is based on a vision of food sovereignty and democracy in which peasants *directly* control technological research and the development of farm machinery for agroecological and organic farming (InPACT, 2016).

Depending on history and context, grassroots ways of knowing and knowledge-creation processes unfold in different ways. Some horizontal networks for autonomous learning and action clearly distance themselves from the state and rely on self-mobilization and self-financing. But most peoples' networks promoting food sovereignty, agroecology and biocultural diversity often consciously adopt a dual power approach to transform existing knowledge, policies and practices. For example, farmers, pastoralists and indigenous peoples engage with formal scientists in participatory research on the basis of clearly negotiated roles, rights and responsibilities, while also maintaining a decentralized network of safe spaces for more autonomous and plural ways of knowing (experiential, local, tacit, feminine, empathizing, phenomenological etc.). This dual approach reflects an awareness of the partial and incomplete nature of all knowledge systems. Indeed, rather than uncritically valuing traditional knowledge and romanticizing the past, grassroots networks of peasant farmers generally embrace scientific knowledge and new technologies when they are appropriate to local needs and context. In the words of Alberto Gomez Flores:

> We *campesinos* have what is called the school of life, knowledge of life. We have the imagination and the ability to know what to do, but not the capacity to translate all of this in writing, or to technically support all of this. So there should be integration between the capacity of technical professionals from the universities and the everyday, practical knowledge that we have accumulated over generations as small farmers. We should try to integrate these different capacities.
>
> *(Alberto Gomez Flores, in Cohn et al., 2006)*

Worldwide, new entrants to farming and younger farmers are increasingly conscious that the development of miniaturization, multipurpose machines, computer-assisted technology for community design and manufacture, multimedia communication and open source software, new knowledge on the dynamic complexity and resilience of socio-ecological systems, reimagined economics and efficient renewable energy systems all have the potential to enhance local autonomy and self-determination (Bookchin, 1986; Cooley, 1982). When under the control of citizens, these innovations can help regenerate local ecologies and economies, minimize pollution and carbon footprints and expand the realms of freedom and culture by eliminating needless toil.

Many grassroots networks of small-scale food producers thus selectively incorporate external concepts and modern technologies to produce social and technical knowledge for autonomy, cultural affirmation and self-determination. For example, in the drylands of Andhra Pradesh, India, grassroots collectives of women *dalit*[3] farmers use modern digital video technology to co-create, document and share knowledge on agroecological farming, biocultural diversity, and food sovereignty. Their community and participatory video films have been an integral part of research processes in which university-trained professionals and non-literate, marginalized peasants and rural people have worked as co-inquirers – producing new knowledge that challenges the dominance of western science. This video making has transformed the lives of the people involved, empowered marginalized people – especially the *dalit* women – and facilitated counter-hegemonic social and ecological change. An important impact of this decentralized knowledge creation process and reliance on modern global communication technologies is that the videos travel across borders and cultural boundaries to inspire a younger generation of peasant farmers, scholars and practitioners to find better ways of doing research *with, by* and *for* people, not just on people (The Community Media Trust *et al.*, 2008).

These grassroots networks operate at different scales – local, national, regional and, increasingly, at the global level. They work from the bottom up and tend to be organized based on a more horizontal and egalitarian logic. The knowledge and innovations they develop can either be conceptual, methodological, technical and/ or institutional. They often rely on forms of adult and critical education to build the capabilities and confidence of participants in grassroots networks. Farmers and other citizens are part of non-hierarchical 'peer-to-peer' collectives which typically seek to go beyond the concepts, categories, criteria and epistemology of dominant knowledge in the natural and social sciences, as well as in the humanities. Most notably, grassroots networks aim to strengthen farmer/citizen-led research and innovation as a key strategy for spreading food sovereignty, agroecology and biocultural diversity to more people and places.

Equally important, self-managed research and grassroots innovation networks help *democratize* the politics and production of knowledge. For example, autonomy, democratic control, endogenous solutions and solidarity remain central objectives for peasants and other small-scale producers involved in the Farmer-Scientist Partnership for Development in The Philippines (MASIPAG). Reversals from normal practices ensure that peasants – rather than scientists alone – determine research priorities and oversee a power-equalizing process of knowledge creation in farmers' fields and villages (Vicente, 1993; Bachmann *et al.*, 2009; www.masipag.org).

Deepening democracy in the construction of knowledge for food sovereignty, agroecology and biocultural diversity depends on further strengthening grassroots research and innovation networks. This can be done through actions that amplify and reinforce several transformative processes, described in the sections which follow.

Education for critical consciousness and place-based learning

Critical education is at the heart of what makes self-managed research and grassroots innovation networks successful and capable of 'going to scale'. For many actors involved in these horizontal networks, education is about raising critical consciousness for transformation and peoples' empowerment. This educational philosophy is strongly grounded in traditions of popular education – particularly Paulo Freire's work on critical pedagogy, dialogic education and agricultural extension (1970, 1973). Critical pedagogy helps students learn 'to perceive social, political, and economic contradictions, and to take action against the oppressive elements of reality' (Freire, 1973). Grassroots networks such as those described in Box 8.1 use education as a tool for developing critical consciousness and encouraging peasant farmers to learn from their own reality, to recognize the power structures that shape their lives and to transform social and economic injustice in their communities and wider society. Values such as shared knowledge and shared learning, spirit, struggle, solidarity and love often motivate progressive social change and transformation (hooks, 2003). Lastly, education for critical consciousness usually reflects a deep commitment to radical democracy and human rights. It emphasizes hope and a politics of possibilities (Amsler, 2015; hooks, 1994, 2003).

Within the larger movements for food sovereignty, agroecology and biocultural diversity, pedagogies of hope believe in peoples' ability to make history and transform society. In the first instance, emancipatory and deeply enabling pedagogies seek to dissolve the mental blockages or prejudice that often translate into the disabling practices of agricultural extension and natural resource bureaucracies. Indeed, the history of grassroots self-managed research and innovation networks has been – and still is – partly about breaking down deeply embedded mental stereotypes that cast farmers and rural people – and especially women – in subservient and helpless roles. A large body of field-based evidence strongly supports the view that hope and trust in people's abilities is not just a naïve act of faith: peasant family farmers, farm workers, fishers, indigenous peoples and pastoralists – men and women – can transcend their limitations when given a chance (see Box 8.2; and also Holt-Giménez, 2006; Machín Sosa et al., 2013; Meek, 2015; Rosset et al., 2011; Pimbert et al., 2017).

BOX 8.2 CRITICAL EDUCATION HELPS DISPEL MYTHS ABOUT FARMER IGNORANCE IN INDONESIA

Following a devastating pest outbreak induced by the use and abuse of pesticide applications in rice farming in Indonesia, the government introduced a national integrated pest management (IPM) programme in 1989. A co-operative programme between the Food and Agriculture Organization of the United Nations (FAO) and the Indonesian Government specifically pioneered methods for training rice farmers to understand the agroecology of plant-pest relations. The FAO programme centred on farmer field schools (FFS); schools without walls, based on Paulo Freire's pedagogy for critical education (Freire, 1973). The aim was to make farmers experts in their own fields, enabling them to replace their reliance on pesticides and other external inputs with their skills, knowledge and labour. Over time the emphasis of the programme shifted towards community organization, community planning and management of IPM, and became known as Community IPM (see also Box 8.4 below). According to the former Director of the FAO Community IPM Programme in Asia, the outcomes of the FFS pedagogy and its horizontal spread at the community level helped disprove at least four enduring myths about peasant farmers:

1. **Farmers are ignorant and scientists are the experts.** At the outset, few believed that farmers could even identify insects, let alone deal with something as abstract as field ecology. But soon, most of the disbelievers had seen with their own eyes that farmers could indeed master 'complex' agroecology.

2. **Farmers cannot train other farmers.** The Community IPM programme postulated that if farmers could master the process of 'discovery learning' in their own fields, they could also facilitate other farmers in their learning. The first 'Farmer to Farmer' IPM field schools emerged spontaneously. They were then built in as an integral part of the programme. By 1999, nearly 50% of all IPM Farmer Field Schools were organized and run by IPM Farmer Trainers. Over 20,000 field school graduates have gone on to be trained as Farmer Trainers and conduct field schools for other farmers.

3. **Farmers cannot do research.** Most believed that farmers would be limited to simple experiments and 'demplots'. However, in hundreds of locations farmers have engaged in field-based scientific investigations of complex local problems. Farmers have undertaken activities previously thought impossible, such as the rearing, breeding, spreading and maintaining of biocontrol agent complexes (parasitoids, virus, bacteria) while training other farmers in their use. Now, 'farmer researchers' are often invited to national research meetings to present their findings and their programmes on participatory plant breeding, ecological approaches to soil fertility management, IPM and agroecology in rice, vegetables and other crops. Researchers unfamiliar with the independence, intelligence and diligence of peasant farmers are initially shocked. These same researchers found that a significant number of farmers were out-producing research stations. This flew in the face of the opinions of many experts who viewed farmers as the main problem in agriculture production instead of recognizing them as potential problem solvers.

4. **Farmers are incapable of strategic planning and organizing complex programmes.** Farmer-led planning and organizing activities now extend from the neighbourhood to the national arena. There are many examples of farmers holding dialogues with government ministers and creating farmer forums to develop advocacy on peasant rights. From the late 1990s, organized grassroots farmer networks in Indonesia slowly gained increasing access and much greater leverage over local, regional and even national policies. Here too, patronizing views that cast farmers as passive actors in need of professional help and guidance were proven wrong.

(Modified from Dilts, 1999; Fakih *et al.*, 2003)

Moreover, education for critical consciousness values people's experiential knowledge and place-based learning. The detailed and intimate knowledge of the places where one lives and works matters, as does the tacit knowledge that comes from learning by doing. In the *Nayakrishi Andolan* movement for example, this living knowledge is co-generated and distributed in multiple spaces: in fields and farming landscapes, in the workshops of mechanics and carpenters and in the many village campuses inhabited by men and women peasant farmers as well as by potters, artisans and healers, fishers and hunters, leaders and priests, story tellers and musicians (Box 8.1 and Mazhar *et al.*, 2006). Peasant farmers and other citizens involved in this way of knowing rely on their senses (smell, sight, taste, touch, hearing...) to perceive and interpret phenomena. Most notably, observations and sense-making activities are carried out in real-life situations – in the field and *in vivo*. Careful observations and inclusive conversations help map, analyze, understand and respond to complex and ever-changing natural and social phenomena in place-specific situations. In contrast with most science, technology and innovation institutions (Table 8.1), peasant farmers, indigenous peoples, pastoralists, forest dwellers and other citizens tend to be involved as full and whole human beings, with all their senses engaged in a relation of empathy with living beings, minerals and the wider environment. Intimate conversations as well as emotional and spiritual bonds with plants, animals, ecosystems, landscapes and human communities are viewed as legitimate sources of knowledge and ways of knowing for many indigenous and peasant communities (Chapter 6 in this book; Posey, 1999; Toledo and Barrera-Bassols, 2009). For indigenous peoples in Peru for example,

> the Andes is a world of affectionate conversationalists because it is love for the world which allows life to flow ... A pre-requisite in this nurturance is that we all be disposed to listen perpetually and in each circumstance to the 'speaking', to the sign of each one ... In the conversation each member of the Pacha[4] is recognized as a sensible organism in constant speech. Here language is not only a human attribute but one belonging to all members of the Pacha and communication takes place through the senses – which are like the 'windows' of life. It is through them that one converses with everyone ... Conversation is thus an attitude, a mode of being in unison with life, a knowing how to listen and knowing how to say things at the appropriate moment.
> *(Rengifo Vasquez, 1998)*

When and where grassroots research and innovation networks create such safe spaces for communication and action, theory and interpretive frameworks are often built from knowledge that echoes and reflects the *sensuous* and *sensitive* qualities of human beings and their intimate relationship with place.

Horizontalism and dialogic knowledge production

Grassroots peasant networks often rely on an explicitly horizontal form of participatory knowledge production and sharing which breaks down the dichotomy between learners and teachers. This 'horizontalism' mediates 'democratic communication on a level plane and involves – or at least intentionally strives towards – non-hierarchical and anti-authoritarian creation rather than reaction' (Sitrin, 2006). New knowledge is produced among equals through face-to-face communication in a process of 'dialogic education' (Freire, 1970) in which everyone has something to share, and each perspective is seen as valuable. In this democratic process of knowledge co-production and sharing, participants are the subjects of their own process of discovery, innovation, learning and agency. Such horizontal knowledge production is exemplified by the *Campesino a Campesino* (CaC) approach, and to varying degrees by the other peasant networks described in Box 8.1.

For many peasant and indigenous networks working for food sovereignty, agroecology and biocultural diversity, horizontal knowledge production takes the form of a *diálogo de saberes*: a 'dialogue among different knowledges and ways of knowing' (Martínez-Torres and Rosset 2014). Different knowledges within peasant networks have been able to dialogue with each other, and also with the knowledge of external scientists and technicians invited to participate in dialogues with farmers, fishers, forest dependent people and pastoralists. For example, *diálogos de saberes* have allowed indigenous communities in the Peruvian Andes to gain the confidence to engage in intercultural dialogues with scientists and extension agents on practices for food sovereignty. Four different topics were critically explored in the intercultural dialogues between indigenous and scientific knowledges in Peru: (1) adaptation to climate change in the Andes; (2) fisheries management in Lake Titicaca; (3) animal breeding programmes for alpaca and llamas; and (4) the conservation and use of genetically diverse Andean crops in indigenous farming (Salas, 2013).

This process is usually empowering for marginalized groups because it:

> embraces subaltern knowledges, especially those that sustained traditional cultures and today re-signify their identities and position themselves in a dialogue of resistance to the dominant culture that imposes its supreme knowledge. *Diálogo de saberes* is a dialogue with interlocutors who have been stripped of their own words and memory, traditional knowledges that have been buried by the imposition of modernity, and the dialogue becomes an investigation, an exegesis, a hermeneutics of erased texts; it is a therapeutic politics to return the words and the meaning of languages whose flow has been blocked.
>
> *(Leff, 2004, my translation)*

More generally, horizontal networks value and work with the diversity of peoples' knowledge. As such, grassroots research and innovation seeks to reverse 'cognitive injustice' and 'epistemicide' (Boaventura de Souza Santos, 2014). The idea of cognitive justice emphasizes the right for different forms of knowledge – and their associated practices, livelihoods, ways of being and ecologies – to coexist (Visvanathan, 1997). As Visvanathan argues, cognitive justice is 'the constitutional right of different systems of knowledge to exist as part of a dialogue and debate'. This implies the continued existence of 'the ecologies that would let these forms of knowledge survive and thrive not in a preservationist sense but as active practices' (Visvanathan,

2005). For example, the successful protection of biocultural heritage in the Potato Park in Peru has grown out of local communities' affirmation of their sovereign right to sustain their *entire* knowledge system, including the landscape and territories that renew biodiversity, culture and livelihoods (Box 8.3).

BOX 8.3 INDIGENOUS COMMUNITIES CLAIMING COGNITIVE JUSTICE IN PERU

The concept of Indigenous Biocultural Heritage Territories (IBCHT) has guided a successful community-led initiative in the Potato Park in Cuzco, Peru. Located in a biodiversity hotspot for potatoes, the park is an IBCHT centred on the protection of potato biodiversity and related knowledge. The area is home to more than 4,000 varieties of potato as well as other traditional Andean crops, including quinoa and oca. The Potato Park provides an alternative approach for protecting traditional knowledge. It protects not only the intellectual, but also the landscape, biological, economic and cultural components of knowledge systems, thereby halting loss of traditional knowledge as well as misappropriation. Communities' collective control over their knowledge has been strengthened by systematically affirming the holistic and indivisible nature of their rights to land, territories and self- determination. Cognitive justice is being claimed as the concept of IBCHT is increasingly recognized in national and international negotiations on the protection of biodiversity and knowledge.

(Argumedo and Pimbert, 2008; http://biocultural.iied.org)

The collective construction of technical, practical and political knowledge

Closely related empowering pedagogies allow people to participate in the joint production of *collective* knowledge throughout their horizontal networks. These pedagogies usually encourage radical visions for food sovereignty, agroecology and biocultural diversity. They include the social process methodology used in constructing sustainable peasant agriculture and food sovereignty in Cuba (Rosset *et al.*, 2011); the *Campesino a Campesino* approach in Central America (Holt-Giménez, 2006); peasant-run Farmer Field Schools in Indonesia (Fakih *et al.*, 2003; Pontius *et al.*, 2002); 'phenomenon-based learning', which engages students in an innovative pedagogical model for agroecological teaching and learning in real-world situations (Francis *et al.*, 2011; Francis *et al.*, 2013); decolonizing pedagogies and methodologies for research by indigenous peoples (Chilisa, 2012; Smith, 2012; Zavala, 2013); and the social learning methods of the thousands of villagers who are *gono gobeshoks* (people's researchers) for whom participatory research has 'sharpened their minds' and helped them develop self-reliance in Bangladesh (Wadsworth,

2005). In many ways, these empowering pedagogies seek to develop an attitude of inquiry which enhances people's awareness and understanding that they are part of a social and ecological order, and are 'radically interconnected with all other beings, not bounded individuals experiencing the world in isolation. Thus, an attitude of inquiry seeks active and increasing participation with the human and more-than-human world' (Marshall and Reason, 2007).

It is also noteworthy that these pedagogical processes generate not only technical knowledge needed to solve agronomic problems like pest outbreaks and soil erosion. They also facilitate the construction of practical and political knowledge. This is largely because these pedagogies are integrated and integrative forms of critical education. They usually unify different domains of learning (technical, practical and political), rather than contain them in separate categories that break down education into the usual cognitive, affective and psychomotor areas. Most of these empowering pedagogies are based on the taxonomy of learning put forward by Jurgen Habermas (1971). In this Critical Theory Perspective, people approach knowledge with an 'orientation toward technical control, toward mutual understanding in the conduct of life, and toward emancipation from seemingly "natural" constraint' (Habermas, 1971). Habermas' learning framework thus simultaneously addresses three fundamental human interests: the technical domain of work, the domain of interaction and communicative action and the domain of emancipatory action for empowerment (Ingram, 1987).

By integrating these three domains of learning, grassroots research and innovation networks construct new knowledge on environmental, institutional, social and technical issues. Practical and political knowledge as well as holistic and phenomenological understandings of complex dynamic realities emerge from specific places throughout these peer-to-peer networks and dialogues among different knowledges. Peoples' inclusion and participation in the creation of new knowledge thus provide the concepts, practices and institutional innovations needed for the horizontal spread of agroecology, biocultural diversity and food sovereignty. Some noteworthy examples are given in Box 8.4.

BOX 8.4 SELECTED EXAMPLES OF PRACTICAL AND POLITICAL KNOWLEDGE GENERATED BY HORIZONTAL NETWORKS OF PEASANT FARMERS

Participatory research by the *Movimiento Campesino a Campesino* on the impacts of Hurricane Mitch. In October 1998, Hurricane Mitch dumped 20–50% of the average annual rainfall on parts of Central America in only five days. Mitch's torrential rains destroyed natural vegetation and standing crops ready to be harvested. Millions of tons of topsoil were washed down from hillsides into rivers. While first reports indicated massive agricultural damage, closer observation showed that small farms usually described as 'sustainable' appeared to have suffered less damage than their 'conventional' neighbours (Bunch, 1998; Ernst, 1998). These farms generally belonged to peasant farmers working within the *Movimiento Campesino a Campesino* (Farmer to Farmer Movement or MCAC) on the Central American hillsides. The farming practices often used in this movement included a wide range of soil conservation and agroecological methods, tested and promoted by these smallholders for over 20 years. A more systematic survey carried out by the MCAC confirmed that farms using agroecological diversification practices such as agroforestry, cover crops and intercropping were less damaged by the hurricane. With the help of 40 non-governmental organizations and 100 farmer-technician teams, 1,743 farmers measured key agroecological indicators on 1,804 plots paired under the same topographical conditions. The study included 360 communities and 24 departments in Nicaragua, Honduras and Guatemala. This coverage, and the massive mobilization of farmer-technician field research teams, was made possible by the existence of the MCAC and its capacity to mobilize farmers. The MCAC study found that agroecological farms had 20–40% more topsoil, greater soil moisture as well as less erosion and gully formation. They also experienced significantly lower crop and economic losses than their conventional monoculture neighbours (Holt-Giménez, 2002). The findings of this grassroots-led study emphasized the importance of increasing plant diversity and complexity in agroecosystems to reduce vulnerability to extreme climatic events. The knowledge generated by the MCAC in the late 1990s thus supports more recent scientific evidence which shows that agroecosystems are more resilient to shocks and stresses when they are part of a complex landscape matrix (Perfecto *et al.*, 2009). In particular, the resilience of a complex landscape matrix depends on it being made up of genetically heterogeneous and diversified cropping systems that use appropriate techniques to increase soil organic matter and conserve water.

Innovations in plant breeding by the Réseau Semences Paysannes (RSP). In sharp contrast with mainstream science, members of the French Peasant Seeds Network (RSP; see Box 8.1) clearly reject the reductionist, utilitarian and mechanistic view of the living world. For example, in the early phases of participatory plant breeding work, farmers criticized researchers for using such terms as 'genetic material', 'weeds' and 'quantifiable selection criteria'. In contrast, the RSP farmers described their relationship with their crops as *living* plants and companions. They never view the plant as an *object*. The peasant farmers have a strong emotional attachment to plants and see them as a source of knowledge and inspiration, provided one has a friendly and empathizing relationship with them. This emotional bond with plants clearly positions the farmers outside the positivist scientific paradigm which values a cool 'objective detachment' in the pursuit of knowledge. The farmers' ways of knowing are thus radically different from the epistemological norms of mainstream plant genetics and breeding. The RSP's more holistic understanding of dynamic complexity and participants' engagement with the living world is leading to new forms of evolutionary plant breeding, enabling this peasant network to generate crop varieties that are resilient to climate change and suited to a diversity of unique situations. It is becoming apparent that the peasant farmers' experiential knowledge and phenomenological understanding of the living world resonate with new insights of modern genetics and biology. This is true, for example, in the areas of fluid genomes and indeterminate relations between genes and the environment; non-linear dynamics, plasticity and the emergence of new forms; epigenetic effects in which the environment modulates genetic expression and leads to heritable phenotypic changes; emergent properties and the self-organization of the living world (Commoner, 2003; Mae Wan Ho, 2013; Pouteau, 2007a/b). Yet this knowledge creation is happening largely outside universities and the national agricultural research system where there is no, or very little, work on evolutionary plant breeding (Réseau Semences Paysannes, 2004; Pimbert, 2011; www.semencespaysannes.org).

Indonesian farmer networks develop knowledge for a Peasant Rights Charter in Indonesia. As more farmers trained other farmers, the Farmer Field School (FFS) programme in Indonesia (described in Box 8.2) was able to be relatively flexible and responsive in developing new curricula to meet the evolving needs of farmers. A wide range of FFS curricula were developed by and for this grassroots network – ranging from *How to Strengthen Farmer Trainers* to a curriculum on *Participatory Ecology Training and Soil Management*. The principles of FFS were extended from rice to other crops such as vegetables and cotton, and from IPM to integrated nutrient management, plant breeding and participatory health monitoring. FFS pedagogies also broadened from the technical to the empowerment domain as farmers' social learning and action focused on building the knowledge they needed to engage in advocacy, policy processes and governance (Pontius *et al.*, 2002). Critical education in the form of FFS for advocacy and political literacy soon led to the formation of a large farmer movement asking for agrarian reform and fundamental changes in agricultural policy. By May 2000, an alliance for Peasant Rights had emerged from below. This grassroots network of FFS and local organizations mobilized farmers' collective knowledge to develop a Peasants' Rights Charter. The charter was used as early as April 2001 to argue for the protection and fulfilment of farmers' basic rights in national fora and policy dialogues with the Indonesian Human Rights Commission. The charter emphasized eight key areas in particular (Fakih *et al.*, 2003):

1. Livelihood rights (rights to sufficient and healthy food and a reasonably good job)
2. Resource control rights (rights to fertile land, rights to biological diversity)
3. Production rights (including technology choices)
4. Consumption rights (including the right to choose what to produce)

5.   Marketing rights (including market access rights), quality protection and property rights
6.   Political and social rights (including the right to organize themselves/ build their own organizations)
7.   Reproductive rights (as they relate to both human reproduction and maintaining biological diversity)
8.   Rights to free expression (including the rights of language, culture, religion and arts).

In many ways, the Indonesian Peasants' Rights Charter thus anticipated and prefigured thinking that has since spread throughout the international food sovereignty movement – see for example *La Via Campesina*'s 'Declaration of Rights of Peasants' and its call for an International Convention on the Rights of Peasants (La Via Campesina, 2009). It is noteworthy that the collective knowledge developed by Indonesian grassroots peasant networks 15 years ago has helped frame today's discussions on the recent United Nations Draft Declaration on the Rights of Peasants and Other People Working in Rural Areas (Claeys, 2015; http://www.ohchr.org/EN/HRBodies/HRC/RuralAreas/ Pages/3rdSession.aspx).

Building extended peer communities to validate and protect collective knowledge

All members of grassroots horizontal networks are viewed as knowledge producers and users who act as an 'extended peer community'. This peer group not only creates a space for conviviality and meaningful exchanges of opinion, it also plays a key role in validating new knowledge and innovations. As active participants in the construction of knowledge, peasant farmers, pastoralists, food workers, indigenous peoples, fishers, forest dwellers, urban farmers and other citizens introduce 'facts' and sources of knowledge which scientists working in more standardized and ideal research conditions simply cannot 'factor in' and/or assess. The subsequent cross-checking of facts and opinions, analysis of collected information, questions about the quality and validity of knowledge, farmer and citizen deliberations and peer-to-peer reviews are all involved in the *in situ* validation of useful knowledge. This is essentially an 'extended peer review' process and the practice of a post-normal science[5] in horizontal networks of grassroots research and innovation (Ravetz, 1971, 2006; Funtowicz and Ravetz, 1990 and 1993). Unlike peer reviews that only involve 'scientific experts', extended peer communities also include farmers, pastoralists and other citizens. The diversity of perspectives and interests represented in extended peer groups will vary depending on the complexity and scale of the issues dealt with. Their decentralized and distributed nature enhances community and socio-ecological resilience because 'extended peer communities' enable the *in situ* validation of knowledge-based solutions needed for local adaptive responses to social and environmental challenges. As such, horizontal networks of locally rooted extended peer communities are particularly well suited for the validation of knowledge in a fast-changing, unpredictable and uncertain world (e.g. climate change, spread of new diseases, unstable markets, political change etc.).

The kind of knowledge that emerges from this process of social learning has been well described by James Scott in his book 'Seeing like a State' (1998). He speaks of 'forms of knowledge embedded in local experience' (*mêtis*) and sharply contrasts them with 'the more general, abstract knowledge displayed by the state and technical agencies'. *Mêtis*, says Scott, is 'plastic, local and divergent ... It is, in fact, the idiosyncrasies of *mêtis*, its contextualities and its fragmentation that make it so permeable, so open to new ideas'. As he suggests,

> *mêtis*, with the premium it places on practical knowledge, experience and stochastic reasoning, is of course not merely the now superseded precursor of scientific knowledge. It is a mode of reasoning most appropriate to complex material and social tasks where the uncertainties are so daunting that we must trust our (experienced) intuition and feel our way.
>
> *(Scott, 1998)*

This production of *collective* knowledge is intimately linked with the nurturing of human relations and reciprocity within grassroots networks of self-managed research and innovation. People are involved in a deeply sense-making activity through the co-construction of collective knowledge for food sovereignty, agro-ecology and biocultural diversity. Moreover, there is usually a strong commitment to ensuring that knowledge, genetic resources and other innovations remain *accessible to all*. The enclosure and privatization of knowledge in particular is seen as incompatible with the ethos of sharing that characterizes many horizontally organized networks of self-managed research and grassroots innovation. For example, patents on seeds make it illegal for farmers to save and exchange seeds, and thus deeply undermine the collective nature of peer-to-peer knowledge production within peasant networks (Box 8.5). More generally, these emergent community economies (Gibson-Graham 2006) and knowledge commons (Bollier and Helfrich, 2012, 2015) fundamentally reframe the 'economy' and 'knowledge' to emphasize certain ethics and values (solidarity, fairness, co-operation, ecological etc) over others (individualism, competition, compulsive acquisition, enclosure and monopoly control).

BOX 8.5 PEER-TO-PEER KNOWLEDGE CREATION IN THE *RÉSEAU SEMENCES PAYSANNES*: RECIPROCITY AND THE NEED TO PROTECT COLLECTIVE KNOWLEDGE

Members of the *Réseau Semences Paysannes* (RSP; see Boxes 8.1 and 8.4) are organized into horizontal networks that link many people and places throughout France. Within the RSP, seeds are exchanged among peasant farmers who have the capacity to observe and experiment, who have a caring relationship with plants and who are sincere in their motivations. The farmers exchange seeds as gifts in the sense defined by Marcel Mauss in his classic work 'The Gift' (1990). This gift exchange leads to a mutual interdependence between giver and receiver. The giver does not merely give an object but also part of himself, because the object is indissolubly linked to the giver: 'the objects are never completely separated from the men who exchange them' (Mauss, 1990). Because of this bond between giver and gift, the act of giving creates a social bond with an obligation to reciprocate on the part of the recipient. It is the fact that the identity of the giver is invariably bound up with the object given that causes the gift to have a power which compels the recipient to reciprocate. According to Mauss (1990), solidarity is achieved through the social bonds created by gift exchange. This is a deeply sense-making activity for people involved.

By affirming the importance of reciprocal peer-to-peer exchanges of seed and knowledge among members of their network, the RSP is essentially developing a solidarity-based economy that is clearly distinct from today's more anonymous commodity exchanges. This solidarity-based moral economy thus creates an autonomous space in which the de-commoditization of seeds and farmer knowledge becomes possible though a *diálogo de saberes* (see above). Indeed, RSP not only rejects the modern forms of enclosure that increasingly privatize and commodify seeds and farmers' knowledge—for example, the new European Union seed regulations and World Trade Organization (WTO) compatible intellectual property rights legislation (patents and plant breeders' rights). It also actively works to develop new knowledge on seed legislation and policies that recognize the secular rights of farmers to freely save and exchange seeds, as well as their rights to collective knowledge. The peasant network's understanding of 'good' economics is thus radically different from the neo-liberal model of commodity exchange and privatization which is, implicitly or explicitly, an integral part of the normative framework adopted by most professional plant breeders, research institutes and policy makers.

(www.semencespaysannes.org)

Strengthening local organizations to scale out grassroots research and innovation

Expanding knowledge for food sovereignty, agroecology and biocultural diversity partly depends on ever more people and communities engaging in grassroots self-managed research and innovation over ever larger territories. This process of geographical spread and numerical increase ('scaling out') is distinct from the process of 'scaling up', which means institutionalizing enabling policies and practices for research, education, extension and credit (see IIRR, 2000; Pachico and Fujisaka,

2004; Pimbert, 2004). The changes needed in public research and education are discussed later in the chapter.

The horizontal scaling out of grassroots innovation and research is driven in large part by the processes identified thus far in this chapter: critical education and empowering pedagogies, nurturing a sense of place through experiential knowledge, dialogues between different knowledges (*diálogos de saberes*), horizontal networks for peer-to-peer learning, emphasis on practical and political knowledge, extended peer communities to validate knowledge and collective knowledge sharing. All these processes depend on co-ordinated action made possible by local organizations that bring people together for joint activities – from resource management, labour-sharing, marketing and other activities that would be too costly, or impossible, if done alone.

Local organizations[6] play a key role in facilitating collective action and co-ordinated knowledge creation within grassroots research and innovation networks. In the first instance, knowledge is generated as part of the day-to-day activities mediated by local organizations that have been set up for different purposes within communities (Pimbert, 2009a), such as:

1. sustaining the ecological basis of food systems – including producing knowledge and joint actions for the local adaptive management of land and the development of reliable bio-physical indicators to track and respond to change;
2. co-ordinating human skills, knowledge and labour to generate both use values and exchange values in the economy of the food system;
3. governing food systems – including decisions about people's access to food and natural resources as well as collectively generating the political knowledge needed to shape policies and institutions.

In the second instance, new organizations can be especially created to co-ordinate local processes of social learning based on critical education, empowering pedagogies, *diálogos de saberes,* as well as the peer-to-peer production of knowledge. Notable examples include Farmer Field Schools based on Freirian principles (Fakih *et al*, 2003; Pontius *et al.*, 2002); food sovereignty and agroecology schools run by peasant themselves (Meek, 2015; Rosset and Martinez Torres, 2012); permaculture schools for radical transformation (Beckie and Berezan, 2017); women's *Sangham*[7] networks for autonomy and food sovereignty (Women Sanghams *et al.*, 1999); *comunidades de base* as decolonizing organic structures among agricultural indigenous communities in Mexico, Nicaragua and Colombia (Fals Borda, 1987); social movements based on horizontal links between local peasant organizations (Holt-Giménez, 2006, 2002; Rosset *et al.*, 2011); and educational organizations that work with the Brazilian Landless Workers' Movement (MST) to promote critical place-based education for agroecology, food sovereignty and *Educação do Campo* (Meek, 2015; Meek and Tarlau, 2015).

Several organizations with different functions, powers and responsibilities are thus usually involved in facilitating the construction of knowledge for food sovereignty, agroecology and biocultural diversity. Such 'nested organizations' and their polycentric networks operate at different scales and act in complementary ways (Ostrom, 2005, 2010). These interlinked organizations not only provide the institutional landscape that is needed to manage the social and ecological realms in which food systems are embedded, they also provide the organizational fabric that enables the co-ordinated and timely production of collective knowledge by grassroots networks of peasants, indigenous peoples, fishers, pastoralists and urban farmers (Pimbert, 2009a, 2009b). At the FAO's Regional Agroecology Symposium in Budapest in 2016, the role of local organizations in renewing the commons of collective knowledge was strongly emphasized by Guy Kastler (a French peasant farmer and member of *La Via Campesina,* see Box 8.1): 'there is no agroecology knowledge without strong farmer organisations'.

Nested local organizations also often work to holistically integrate knowledge on the ecology, economy and culture of places. This makes it possible to express the unity of all knowledge beyond disciplines: a key aim and claim of transdisciplinary research today (Nicolescu, 2008, 1994; Lang *et al.,* 2012). For Andean indigenous communities living in the Potato Park (Peru), agricultural production, landscape management, economic exchanges and spiritual life are mediated by interacting networks of local organizations which include producer associations, women organizations responsible for running barter markets for food and medicinal plants, farmer groups with specialist knowledge on crop breeding, the women's restaurant collectives and groups of shamans (Argumedo and Pimbert, 2005; www. andes.org). This polycentric network of local organizations supports indigenous ways of knowing, seeing and thinking that holistically link different areas of life while also generating new knowledge (Argumedo and Pimbert, 2010; Marti and Pimbert, 2007).

Nested local organizations and the horizontal networks they form thus play a crucially important role in the construction of knowledge for food sovereignty, agroecology and biocultural diversity. And without exception, all major success stories in these areas depend on nested local organizations to facilitate and coordinate collective action at different scales (see Box 8.1 for examples). Webs of interacting local organizations provide the basis for autonomous learning and action, self-managed research and grassroots innovations (conceptual, technical, social and political innovations). They also provide the institutional landscape and social organization that allows for the potentially more decentralized, horizontally distributed and democratically controlled production of knowledge. Horizontal networks of local organizations of farmers and other citizens should therefore be strengthened to enhance their capacity to 'scale out' the processes described above.

Yet on all continents, local organizations and their capacity for self-administration have been undermined by a toxic cocktail of large-scale authoritarian state development plans (Scott, 1998) and a capitalist modernity that thrives on 'accumulation by dispossession' (Harvey, 2004). As farmers and farms have dramatically declined in numbers, land and capital have become concentrated into larger and larger farm holdings.[8] The net result of these trends is that there are simply fewer and fewer farmers around to build local organizations and engage in participatory processes. Without people there is, by definition, no participation possible and no capacity to develop strong local organizations. As a result, many rural communities are no longer in charge of managing their local food systems, economies and environments. Most significantly, they are not 'trusted' by state bureaucracies to be able to do so (see Scott, 1998, 2009). In many places, communities continue to be actively disempowered, and their local organizations are becoming incapable of co-ordinating collective action in the social, economic, ecological and political realms. Re-creating and strengthening local organizations and their polycentric webs is thus a key priority for citizens who seek to democratize research and exercise their right to construct knowledge for the society they want. Wider policy and economic changes for food sovereignty are also required to provide farmers and other citizens with the *free time* and *material security* needed to build local organizations, engage in participatory decision making and sustain their knowledge systems (Chapters 1 and 7 in this book). As Pedro Magana Guerrero – a former peasant leader of UNORCA[9] in Mexico – says, the 'consolidation of alternatives rests completely on what is happening at the local level, it depends on the development of organisations in their regions, in their countries. This gives viability to a global process' (Pedro Magana Guerrero, cited in Desmarais, 2007).

However, local organizations and the networks they form should not be roman-tically idealized and viewed as unproblematic. They are not always welcoming spaces for women, nor inclusive of the marginalized, nor free from manipulation by more powerful actors. Community-based local organizations can sometimes be overwhelmed by internal inequities and social injustices, with decisions taken by the powerful (the men, the landowners, the 'upper' castes and privileged classes). This is often at the expense of the relatively powerless (women, landless farm work-ers, pastoralists, forest peoples, urban slum dwellers etc.). These shortcomings in relation to equity, gender, social inclusion, race and entitlements of the very poor and marginalized clearly need to be acknowledged and tackled by social move-ments (hooks, 1994; Masson et al., 2017). Nurturing a conscious social commit-ment to a politics of freedom, equity and gender inclusion is key in ensuring that grassroots research and innovation networks do not reproduce overt or subtle forms of exclusion, including enduring forms of homophobia, misogyny and patriarchy. Fortunately, local organizations that facilitate knowledge creation for food sover-eignty, agroecology and biocultural diversity are well positioned to harness criti-cal place-based education and horizontal social methodologies for transformation 'within'. Through their counter-hegemonic practice, they can collectively decide and organize to promote critical education and decolonizing pedagogies that deepen freedom from patriarchy and injustice. By becoming safer spaces for com-munication and action, local organizations and collective structures can also culti-vate a non-hierarchical sensibility, empathy and mutual respect, diversity and social inclusion, as well as citizenship and the art of participatory democracy.

Last but not least, local organizations and collective structures that facilitate the 'scaling out' of grassroots research and innovation also have a potentially key role in 'scaling up' policies and practices designed to democratize public research on food, agriculture, environment and society. When part of larger federations and social movements, local organizations of peasant farmers and other citizens can greatly influence decisions on national research priorities and funding. More specifically, federated networks of local organizations help build the countervailing power needed for citizens to claim and realize their rights to democratically participate in the governance of national research systems and universities. This is further discussed in the next section, which focuses on the transformations needed in public research for the widespread construction of knowledge for food sovereignty, agroecology and biocultural diversity.

Democratizing and transforming public research

University-based scholars and researchers can produce critical and counter-hegemonic knowledge which grassroots networks of farmers value and do use. Contributors to this volume exemplify this trend. Similarly, the *Centre for Agroecology, Water and Resilience* at Coventry University has recently published over 35 examples of critical research projects in Africa, Asia, the Americas and Europe that offer alternatives to dominant knowledge on food, agriculture and human well-being (People's Knowledge Editorial Collective, 2017). These diverse case studies show how monopolies of knowledge by élites can be contested through collaborations between academics and social movements as well as community-university partnerships. Co-inquiry, participatory action research (PAR), feminist PAR inquiry and other forms of liberatory inquiry are shown to be effective in openly challenging racism, sexism, colonialism and hierarchies of knowledge in science and its practice.

A simplistic rejection of all research and science as a whole will therefore not do. Instead, the issue here is how to transform existing research systems (universities, research institutes, policy think tanks, research extension services...) so that they can contribute more appropriate knowledge for food sovereignty, agroecology and biocultural diversity? Under what conditions can alliances and complementarities be built between self-managing grassroots research networks and public research institutions? Which institutional and methodological innovations are required to develop new forms of collective intelligence that combine the partial and incomplete knowledges of scientists, peasant farmers and indigenous peoples? What transformations will help decolonize research so that indigenous ways of knowing are reclaimed and can flourish? How can critical, deviant and disobedient knowledge thrive in research organizations rather than be disciplined and punished? How can public research and education be reinvented to generate many more 'organic intellectuals' (*sensu* Gramsci, 1978) rather than traditional academic intellectuals and liberal scholars who prop up the leading scientific organizations that 'do little except chase money and reinforce the ruling nexus of politics and finance' (Macilwain, 2016)?

Grassroots movements for food sovereignty, agroecology and biocultural diversity also recognize the liberating potential of modern science and technology. But as argued in the previous section, organized networks of farmers, indigenous peoples and other citizens should *directly* decide what new knowledge and innovations are needed, for whom, when, where and under what conditions. This means re-embedding farmers and other citizens in the production of transdisciplinary knowledge in ways that fundamentally democratize research organizations and decolonize research methods in the social and natural sciences as well as the humanities. Some of the radical transformations required in the governance, culture, organization and professional practice of public research are highlighted below.

Putting citizens at the heart of decision making in research

Throughout the world, the governance and funding of higher education, science and technological research and development (R&D) are largely controlled by upper-middle class men who are increasingly distant from diverse local realities as they align themselves more and more with corporate interests (Beder, 2006a, 2006b; Chomsky, 2017; People's Knowledge Editorial Collective, 2017; Benessia *et al.*, 2016). Several institutional and methodological innovations can help reverse the current democratic deficit in the governance of research and development (R&D). By putting farmers and other citizens at the centre of decision making for R&D, the social innovations highlighted below can also help (re)construct knowledge for food sovereignty, agroecology and biocultural diversity.

At one level, existing governance and funding bodies for R&D can be reformed and opened up to more citizen participation by including more gender-balanced representation of peasant farmers, indigenous peoples, pastoralists, fisherfolk, farm workers, artisanal food processors and citizen-consumers. However, this more equitable representation of citizens in structures that govern research (boards, funding bodies, expert committees …) will also need to be complemented by more transformative and direct forms of democracy that create space for the voice and agency of hitherto excluded people. This is consistent with the food sovereignty paradigm and its central emphasis on the fundamental right of citizens to decide their own food and agricultural policies.

There are four key moments or stages at which direct citizen participation can occur in the research and development cycle:

1. the framing of national policies for science and development;
2. the choice of upstream strategic priorities for R&D, including decisions on budget allocations by funding bodies;
3. during scientific and technological research – the production and validation of knowledge in the natural and social sciences, as well as the arts and humanities;
4. in evaluating research results and impacts, including risk and sustainability assessments.

Participatory methods and deliberative processes that genuinely include different actors are important in 'opening up' the entire research cycle to greater citizens' oversight and democratic control over *what* knowledge is produced, *for whom, how, where* and *with* what likely effects. In practice, a range of methodological approaches and processes can be used to facilitate direct engagement and participation of farmers and citizen-consumers in different stages of the R&D cycle. For both scientific and technological research, as well as risk and impact evaluations (Stages 3 and 4 above), a suite of methods for participatory inquiry can be combined in different sequences. Such participatory methods and systems of inquiry (see Box 8.7 further below) enable farmers and citizens to use their own knowledge to analyze their conditions and participate in co-inquiries with outsiders (e.g. scientists, planners and other professionals) (Chambers, 1993, 1996, 2008; Chilisa, 2012; Pretty and Chambers, 1993; Salas, 2013; Salas *et al.*, 2007; Zavala, 2013).

A range of institutional and methodological innovations can also be used to enhance citizen deliberation and inclusion in the governance of national research systems (Pimbert, 2010; Testart, 2015). These innovations are particularly appropriate for involving farmers and citizens in agenda setting and the upstream definition of research priorities, the framing of national policies for scientific research and development, decisions on research funding and budget allocations, as well as in risk and sustainability assessments (Stages 1, 2 and 4 above). Examples of these methods for deliberative and inclusive processes (DIPs) include citizens' juries, scenario workshops, public hearings, multi-criteria mapping and visioning exercises (see Coote and Lenaghan, 1997; Lowndes and Stoker, 1998; Pimbert and Wakeford, 2003; Stirling and Maher, 1999; Wakeford and Pimbert, 2004).

When these participatory methods are used well and are not designed to close down debates,[10] they are part of a process in which new practical and political knowledge can be constructed for food sovereignty, agroecology and biocultural diversity (Pimbert, 2010). An example of the latter is the global initiative known as Democratising Agricultural Research (www.excludedvoices.org). Here participatory methods enabled deliberative and inclusive processes in which small-scale producers and other citizens were invited to decide on the kind of agricultural research they want in the Andean region (Bolivia, Ecuador and Peru), South Asia (India, Nepal and Sri Lanka) and West Africa (Burkina Faso, Benin, Mali and Senegal):

- The *Raita Teerpu* (the 'farmer's verdict') took place in the State of Karnataka in 2009. It brought peasants, especially women, *Dalits* and indigenous peoples in a citizens' jury set up to analyze the relevance of agricultural research for small farmers. After carefully listening to evidence given by specialist witnesses from government, the private sector, research institutes, activists and peasants, the jury of marginalized small farmers and landless farm workers presented their policy recommendations to decision makers and the media in Bangalore, the capital of the State of Karnataka. The extensive use of media (radio, TV, newspapers, recordings in local language ...) before, during and after the *Raita Teerpu* ensured that over 10 million households followed these farmer deliberations and heard the jury's recommendations on what kind of agricultural research is needed for marginalized peasants, who represent the majority of the population in Karnataka and rural India (http://www.raitateerpu.com).

- An international workshop in 2013 shared lessons from Africa, Asia and Latin America with a wider community of European peasants, policy makers and representatives of the donor community. Known as the St. Ulrich Workshop on Democratising Agricultural Research for Food Sovereignty and Peasant Agrarian Cultures, this international workshop brought together 95 participants from a total of 17 countries (www.excludedvoices.org). Inspired by their peers from the global South, several European participants followed up by organizing networks of farmer-to-farmer exchanges, *diálogos de saberes,* and actions aimed at democratizing agricultural research in Europe (see DARE at www.agroecologynow.org). More generally, the multimedia resources[11] and other outcomes generated by these emblematic farmer-citizen deliberations and *diálogos de saberes* continue to inspire the food sovereignty movement and its struggles to democratize agricultural research.

- A high-level policy dialogue on agricultural research and the future of farming brought together West African family farmers and representatives from the Alliance for a Green Revolution in Africa (AGRA) and its main funders: The Gates Foundation and the UK's Department for International Development (DFID). Chaired by the UN Special Rapporteur on the Right to Food this three-day event was organized in Ghana in 2012 (Pimbert, 2012). The West African farmers had previously participated in a series of citizens' juries on the governance and priorities of agricultural research (Pimbert *et al.*, 2011). They had been mandated by their peers to discuss the citizens' juries' recommendations for policy and practice with AGRA and its donors. While the views of AGRA scientists and farmers converged on some points, this dialogue of different knowledges highlighted strongly divergent visions for the future of farming in West Africa and on the kind of agricultural research needed by the small-scale producers who produce most of the food in the region (IIED *et al.*, 2012). Guided by their vision of food sovereignty and family farming, the West African farmers continue to organize and argue for a fundamental rethinking and reorientation of the research done in their name.[12] This long-term participatory process in West Africa is thus enabling hitherto excluded and subaltern farmers – men and women – to mobilize their knowledge and build the countervailing power needed to democratize and re-invent agricultural research for food sovereignty, agroecology and biocultural diversity.

Putting citizens at the heart of decision making in research depends on successfully scaling up and institutionalizing people's participation in the policies and practices of national systems. However, institutionalizing participation (see Box 8.6) can have substantially different outcomes depending on whether the process and methods are primarily used to enhance control by powerful actors and justify their decisions or whether, instead, they aim to devolve power away from dominant institutions to strengthen peoples' sovereignty and autonomous decision making.

BOX 8.6 INSTITUTIONALIZING PARTICIPATORY APPROACHES AND PEOPLE-CENTRED PROCESSES

The term 'institutionalization' describes the process whereby social practices such as participation become regular and continuous enough to be called institutions. The dynamics of institutionalizing participation and people-centred approaches imply long-term and sustained changes which recognize conflicts between different agendas, interests, values and coalitions of power. In practice, the process of institutionalizing participatory approaches emphasizes several interrelated levels of change:

* Spreading and scaling up change from the micro (e.g. project/local) to the macro (e.g. policy/national) level.
* Scaling out from a single line department, sector or initiative to catalyze wider changes in organizations (e.g. government and donor agencies, non-governmental organizations, civil society groups and federations, private corporations) and in policy processes.
* Changing attitudes, behaviour, norms, skills, procedures, management systems, organizational culture and structure, as well as policy change.
* Including more people and places through lateral spread, from village to village, municipality to municipality, district to district and so on.

At one end of the spectrum, the notion of 'institutionalizing participation' is used only as a discourse or rhetorical label to make projects and proposals attractive to donors and policy makers, while actions continue to be disempowering and extractive (Arnstein, 1969; Pretty, 1994). At the other end of the 'institutionalizing participation' continuum, participatory approaches, methods and processes are used as part of a strategy of policy and organizational transformation as well as local institutional development that decentralizes and redistributes power in the hands of peasant farmers, indigenous peoples, pastoralists, landless farm workers and other citizens (Pimbert, 2011). 'Institutionalizing participation' in this context depends on the capacity of social movements and federated citizens to exert the counter-hegemonic and countervailing power needed to put *direct* democracy at the heart of the governance of research (Pimbert, 2009).

Embracing transdisciplinarity and methodological pluralism in research

It is now increasingly recognized that complex environmental and social phenomena, behaviours and dynamics often cannot be understood, nor solutions to societal challenges found, without fundamental changes in how research is carried out. As long as four decades ago, De Rosnay (1975) was arguing that science and society need to take a macroscopic as opposed to a microscopic view of phenomena, embracing transdisciplinary approaches. Jean Piaget (1972) and the International Center for Transdisciplinary Research had previously introduced the concept of transdisciplinarity to describe ways of knowing that stress the fundamental unity of all knowledge beyond disciplines (Nicolescu, 1994, 2008). Transdisciplinary research thus works simultaneously *between* the disciplines, *across* the different disciplines and *beyond* each individual discipline. Transdisciplinarity embraces the hybrid nature of knowledge production (Bernstein, 2015; Latour, 1987, 1993), and responds to the need to integrate 'both the science of parts and the science of the integration of parts' (Holling, 1998).

Fundamentally pragmatic and relational, transdisciplinary inquiry is directed towards finding integrated solutions to complex and critical environmental and social challenges (MacGregor, 2014). In addressing messy (Gharajedaghi, 2011) and wicked societal problems (Rittel and Webber, 1973; Xiang, 2013), it stresses the evolutionary potential of the present as well as adaptive innovations based on new forms of collective intelligence that bring different knowledges together. Transdisciplinary research simply cannot be done from within the narrow boundaries of single disciplines and in isolation from society as this will only generate partial solutions based on specialized and therefore incomplete knowledge. It requires instead the engagement of all relevant academic disciplines and 'ordinary' citizens in the identification of issues and research priorities, the framing of research questions and the execution of the research, including the interpretation, dissemination and uptake of findings. Deeply participatory, this transdisciplinary approach to knowledge creation calls for the meaningful involvement of all relevant actors, academic and non-academic, in the co-design, co-production, co-validation and co-dissemination of research, in a joint effort to address common and complex problems.

Transdisciplinarity is not a new science. It is an emerging new methodology for doing science *with* society. As such, transdisciplinary ways of knowing emphasize the importance of methodological pluralism to integrate different traditions of knowledge and multiple sources of evidence. Novel mixes of methodologies are needed to dismantle boundaries between disciplines, disrupt knowledge hierarchies, foster intercultural dialogue between different knowledge systems, remove siloes around disciplinary turfs and co-produce knowledge with different social actors. The methodological landscape encompasses quantitative, qualitative and transformative research methods which can be appropriately combined to construct knowledge, policies, organizational cultures and practices (Bergmann *et al.*, 2013; Gibbs, 2015; Haire-Joshu and McBride, 2013; Pohl and Hirsch Hadorn, 2007; Reason and Bradbury, 2008; Scholz and Tietje, 2002). For example, the contributors to this volume *collectively* point to a range of methodological approaches that can help contest harmful myths about people-environment interactions and economics as well as construct knowledge for food sovereignty, agroecology and biocultural diversity:

- A simultaneous analysis of social *and* environmental history, combining structural and agency-focused analysis of change across space and time. In this context, methodological designs focus on micro-scale understandings, endogenous conceptions and local experiences of social and environmental change, emphasizing community rights, participation, people's agency and everyday forms of struggle and resistance to ecological destruction and social exclusion (Chapters 4 and 5; Peet and Watts, 2004; Forsyth, 2004).

- A focus on how international, national and local sets of practices interact and interlock with each other – to reveal their interconnections and how they might be mutually constitutive at different scales as determinants of innovation. Multiple layers of politics that extend from the local to international levels are examined here along with how public and corporate policies are dialectically linked with ecological and social dynamics in the construction of gendered knowledge on people and the land (Chapters 1, 2, 4 and 6; Adger *et al.*, 2001; Walker, 2006; Harcourt and Nelson, 2015).

- Mapping these processes and the role(s) of different actors relies on a multi-sited ethnography that (1) contextualizes a locality in its wider national and global contexts; and (2) helps do research at different sites to explore their connections and relations, by following stories, following people, following finance and following networks across them (Chapters 2, 5, 6 and 7; Soyini Madison, 2012).

- Methodological pluralism and the complementary use of participatory, quantitative and qualitative methods e.g. combining gender sensitive methods for historical and social analysis, ethnographic methods, quantitative tools from the natural sciences along with decolonizing research methods that build on local knowledge, analysis, diverse perspectives and cosmovisions[13] (Chapters 3, 4 and 6; Salas *et al.*, 2007; Smith, 2012).

The construction of holistic knowledge for food sovereignty, agroecology and bio-cultural diversity depends on such methodological diversity and complementarity. However, the co-creation of knowledge by scientists and peasant farmers should increasingly be part of a participatory process driven by a transformative logic of changing society – rather than just interpreting it. More specifically, a transformative methodology is required to frame and firmly locate transdisciplinary practice in an *overarching,* flexible, open-ended, participatory and iterative process of action and reflection. Transformative methodologies typically include methods from Action Research, Participatory Action Research (PAR), Artful Inquiry, Participatory Video (PV) and Participatory Learning and Action (PLA) (Box 8.7) and decolonizing indigenous research methodologies (Chilisa, 2012; Smith, 2012).

BOX 8.7 TRANSDISCIPLINARY AND PARTICIPATORY METHODOLOGIES: SOME EXAMPLES

Action research is a participatory process concerned with developing practical knowing in the pursuit of worthwhile human purposes. It seeks to bring together action and reflection, theory and practice, in participation with others, to generate practical solutions to issues of significance concerning the flourishing of human persons, their communities and the wider ecology in which we participate (adapted from Reason and Bradbury, 2008). Moreover, like participatory action research (PAR), action research involves a whole range of powerless groups of people – the exploited, the poor, the oppressed, the marginal – as well as the full and active participation of the community in the entire research process. The subject of participatory action research originates in the community itself and the problem is defined, analyzed and solved by the community. The ultimate goal is the radical transformation of social reality and the improvement of the lives of the people themselves. The beneficiaries of the research are the members of the community. The researcher is a committed participant and learner in the process of research, i.e. a militant rather than a detached observer (modified from Hall, 1992, 1997).

Artful inquiry enhances the transformative practices of action research by cultivating the imagination, non-verbal, holistic and embodied experiences (Seeley, 2011). Artful knowing is 'learning by doing, cultivating the body and the senses as explicit seats of knowing, being concerned with the evolution of society, ecosystems and consciousness knowing' (Seeley and Thornhill, 2014).

Other participatory systems of inquiry include decolonizing research methodologies (Smith, 2012), Participatory Learning and Action (PLA Notes, 2002) and Social Analysis Systems for collaborative inquiry and social engagement (Chevalier and Buckles, 2008).

This overarching transformative methodology provides the context in which both quantitative research methods (empirical experiments, mathematical modeling, Geographic Information Systems – GIS, statistical methods etc.) and qualitative methods (ethnography, interviews, surveys, discourse analysis etc.) are combined in specific sequences with participatory methods (PLA, VIPP, citizens' juries etc.) to construct knowledge(s) for food sovereignty, agroecology and biocultural diversity. When respectfully inclusive of different cosmovisions and knowledge systems, this transformative process is 'a participatory way of knowing that transcends the dichotomies of man-nature, subject-object or mind-matter, which are so ingrained in the Western mind and form the bedrock of object thinking' (Holdrege, 2013). Knowledge integration is key in this participatory process: finding appropriate ways and means of integrating theoretical, practical and political knowledge as they emerge. As discussed above, this entire participatory process of transdisciplinary inquiry should also be based on principles of cognitive justice (Santos, 2014; Visvanathan, 2005); be bottom-up and subject to deliberation to question underlying assumptions; and rely on extended peer communities of scientists and peasant farmers to validate knowledge *in situ* (Ravetz, 2006), and expand knowledge democracy (Hall and Tandon, 2015).

It is encouraging that more universities and donors are declaring their interest and commitment to transdisciplinary approaches to education and research (Hirsch Hadorn *et al.*, 2008; Mauser *et al.*, 2013; Van Breda and Swilling, in press). Transdisciplinary research is also increasingly mentioned today in new funding calls put out by the European Union (EU), national research councils, NGO donors and foundations. However, there are major structural constraints to the widespread adoption of transdisciplinarity and participatory knowledge creation in higher education and research. These are described below.

Lack of expertise within academia

Transdisciplinary co-inquiry is about transgressing boundaries (Nowotny, 2006). As such, it creates challenges for university departments that have historically been engaged in relatively specialized education and research. Building internal capacity to 'walk the talk' of transdisciplinarity first requires recruiting more staff familiar with its theory and practice. Second, the uptake and spread of transdisciplinarity in universities and research centres also requires a large-scale effort to re-orient, re-skill and train currently employed researchers and teaching staff – both in the natural and social sciences as well as the arts and humanities (Gibbs, 2015). Much of this internally directed educational effort in universities and research institutes would need to focus on reversing enduring systemic biases against the knowledge of women, indigenous peoples, under-represented ethnic groups and other disadvantaged groups such as the lesbian, gay, bisexual and transgender (LGBT) community. For example, the training and re-orientation of researchers in transdisciplinary approaches and methodologies would necessarily have to focus on how gendered knowledge interacts with class, caste, race, culture and ethnicity to shape processes of ecological change, access to and control over resources and the multi-scalar dynamics of food systems and land use in which women play central roles (Mollett and Faria, 2013; Rocheleau *et al.*, 1996; Harcourt and Nelson, 2015). Gendered relations of ecologies, economies and politics would have to be systematically explored through at least three complementary lenses: (1) gendered science, including local knowledge on food, agriculture and environment (Keller, 1985; Harding, 1991, 2006; Lederman and Bartsch, 2001); (2) gendered rights and responsibilities, including the right to food and nutrition (Agarwal, 1995; Rocheleau *et al.*, 1996; Bellows *et al.*, 2016); and (3) gendered environmental and food politics (Merchant, 1992 and 1996; Saunders, 2002; Harcourt and Nelson, 2015).

Education for professional re-orientation is a pre-requisite for a decisive shift from well-established research traditions (mono-disciplinarity, inter-disciplinarity and multi-disciplinarity) to a new paradigm that embraces transdisciplinarity, methodological pluralism and peoples' knowledge. Internal capacity building is also required to nurture the more respectful attitudes and behaviours needed to work with subaltern groups and 'ordinary' citizens (Bainbridge *et al.*, 2000); to decolonize knowledge and research methodologies (Smith, 2012); as well as to reject racism and sexism (hooks, 2000) in universities and the production of knowledge. This is essential because most universities and research institutes continue to be dominated today by a culture that is primarily white, upper-middle class and male.

> When people from communities that have previously been excluded are asked to take part in research – even participative research – they are seldom able to do so on equal terms … A person's race, class, gender, sexuality, health status or disability, a lack of formal training, or a different mode of expression, can all prevent their insights from being accepted as potentially valid. The expertise people gain from life experience is routinely ignored by professionals, even those whose job it is to engage with such people.
>
> *(People's Knowledge Editorial Collective, 2017)*

Indeed, perhaps the biggest challenge today is to go beyond an emerging shallow practice of 'transdisciplinary research' that *only* includes well-known tribes of 'trusted' disciplinary scientists, towards natural and social scientists engaging and working with peoples' knowledge in all its diversity. For both ethical and practical reasons

> all citizens on earth deserve to be as significantly involved in judgments about future developments in agriculture as possible, in ways that historically they have never been. Under these circumstances of participation and deliberation, the need is for the academy to engage *with* the citizenry and not just work for it or on it or extend out to it.
>
> *(Bawden, 2007)*

Inappropriate definitions of 'research excellence'

A less obvious but equally important pre-requisite for change is the need for new definitions of 'research excellence' that can allow transdisciplinarity to thrive. 'Excellence is the holy grail of academic life' (Lamont, 2009), as evidenced by the proliferation of 'excellence frameworks'.[14] Excellence, as most research frameworks define it, focuses on the ability of scholars to publish in prestigious international journals, their ability to gain external grants and other metrics of scholarly output including research impact. Measures of 'excellence' achieved are then used to rank and reward universities for the quality of their research – high scoring universities receive more government funding.

However, transdisciplinary research that works with farmers and other citizens is usually unrecognized and/or under-valued by research excellence frameworks and their metrics. This is partly because the fetishization of 'excellence' in research encourages conformity rather than a transformative shift to transdisciplinary co-inquiry that is inclusive of diverse forms of people's knowledge. By restricting the *types* and *styles* of scholarship, the rhetoric of 'excellence' effectively marginalizes transdisciplinary and participatory ways of knowing in academia:

a focus on 'excellence' impedes rather than promotes scientific and scholarly activity: it at the same time discourages both the intellectual risk-taking required to make the most significant advances in paradigm-shifting research … It encourages researchers to engage in counterproductive conscious and unconscious gamesmanship. And it impoverishes science and scholarship by encouraging concentration rather than distribution of effort.

(Moore et al., 2017)

According to these authors, 'administrators captured by neo-liberal ideologies, funders over-focused on delivering measurable returns rather than positive change, governments obsessed with economic growth at the cost of social or community value …' (Moore *et al.*, 2017) are partly responsible for this obsession with metrics-driven excellence. But this is not the only reason:

the roots of the problem in fact lie in the internal narratives of the academy and the nature of 'excellence' and 'quality' as supposedly shared concepts that researchers have developed into shields of their autonomy. The solution to such problems lies not in arguing for more resources for distribution via existing channels as this will simply lead to further concentration and hyper-competition. Instead, we have argued, these internal narratives of the academy must be reformulated.

(Moore et al., 2017)

New definitions of 'excellence' are needed to allow transdisciplinary knowledge and ways of knowing to thrive. The following transformative actions seem particularly relevant here:

1. Within the academic sphere, instilling a prefigurative politics. This could help to (a) position all academic knowledge as situated (Haraway, 1988); (b) demonstrate the liberatory potential and impact of participatory and transdisciplinary approaches for the co-production of knowledge – and their direct relevance for effectively addressing major societal and environmental crises; (c) actively engage in research that seeks to 'understand better, change, and re-enchant our plural world' (Fals Borda, 2001); and (4) emphasize the transformative effects on theory of 'having our ideas critiqued by social movements live and direct' (Mason, 2013).
2. Broadening out and opening up the entire research and development cycle to farmer and citizens' direct participation – from setting national research priorities and deciding on budget allocations to risk and impact assessments. Reversing the current democratic deficit in R&D goes hand in hand with what is required for the practice of a post-normal science in a fast-changing world (Funtowicz and Ravetz, 1993). Uncertainty has indeed become the norm in political and environmental affairs and 'normal' puzzle-solving science is utterly inadequate as a method for solving the great social and environmental

crises of our times. Post-normal science recognizes that the facts are uncertain, values are often in dispute, stakes are high and decisions are urgent. Its core ideas include an 'extended peer community' and the recognition of a diversity of legitimate perspectives on every issue. In particular, extended peer communities of farmers and other citizens can no longer be relegated to second class status, and people's knowledge can no longer be dismissed as 'unscientific', inferior or bogus (see Ravetz, 2006).

3. Adopting much broader criteria of excellence, validity and quality to assess the process of knowledge creation and its outcomes. Final objective answers and so-called 'research excellence' matter far less than processes of emerging democratic engagement. And 'scholarly detachment, creating knowledge that denies or suppresses our embodied, connected being in the world, seems ill suited to the issues of our times' (Marshall and Reason, 2007). One important criterion of quality could focus on the extent to which researchers' self-reflective practice and reflexivity are alive and disciplined. Marshall and Reason (2007) describe this process as 'taking an attitude of inquiry' and suggest that it is enabled by the following qualities: curiosity; willingness to articulate and explore purposes and values; humility; developing a sense of self-irony, playfulness and lack of ego attachment; participation; living research as an emergent process; and a radical empiricism that relies on multiple sources of evidence. Another criterion for quality and validity is whether or not a process of knowledge creation has opened up new communicative spaces for democratic inquiry to take place. The process of constructing knowledge for food sovereignty, agroecology and biocultural diversity thus aims to 'shift the dialogue about validity from a concern with idealist questions in search of truth to concern for engagement, dialogue, pragmatic outcomes and an emergent, reflexive sense of what is important' (Bradbury and Reason, 2000).

Disabling donor practices

Another challenge relates to the culture and practice of donors (EU commissions, governments, NGOs, foundations etc.). For example, participants at the FAO Regional Symposium on Agroecology for Europe and Central Asia highlighted the mismatch between EU Horizon 2020 calls for process-oriented, multi-actor research on the one hand and, on the other, the EU's inflexible and standardized internal project management procedures based on a focus on quantitative outputs, logical frameworks and other simplistic assumptions about complex and fast-changing realities. Participants at the FAO Symposium on Agroecology spoke about the urgent need to reverse the deep mismatch between donors' twenty-first-century aspirations for transdisciplinary research and their outdated twentieth-century project administration and financial management practices.[15]

However, current donor attempts to 'reform the system for the twenty-first century' remain deeply problematic and wedded to a top-down culture of command and control. For example, the UK's Department for International Development

(DFID) argues that it makes sense for it 'to take a tougher, more business-like approach by *requiring results up front before payment is made*. Better sharing of risk in this way will drive value for money as partners become more incentivised to deliver' (DFID, 2014, my emphasis). According to the Secretary of State, 'DFID is … becoming a world leader in pioneering innovative Payments by Results (PbR) programmes for tackling complex development problems' (DFID, 2014). Chambers has analyzed the perversities of PbR – including its misfit with complexity, unpredictability, flexibility and adaptability (Chambers, 2014). He laments that it provides incentives to do shoddy work by focusing on what the numbers demand instead of local ownership, empowerment and long-term sustainability. These donor practices fundamentally undermine participation and transdisciplinarity in research and development.

Professional reversals and organizational transformation

Transdisciplinary co-inquiry calls for power reversals and new roles for research, donors and development professionals. In essence, people – their knowledge and the diverse environments that sustain them – become central, instead of university research centres, government departments, scientific peer groups and the narrow 'research excellence metrics' used to evaluate academic papers and their impacts.

> Professionally, this means putting people before things … Bureaucratically, it means decentralizing power, de-standardizing and removing restrictions. In learning, it means gaining insight less from 'our' often out-of-date knowledge in books and lectures, and more from 'their' knowledge of their livelihoods and conditions which is always up-to-date … In behaviour, it means the most important reversal of all, not standing, lecturing and motivating, but sitting, listening and learning.
>
> *(Chambers, 1993)*

These reversals in roles and locations all require a new professionalism with new behaviours, concepts, methods and values (Pretty and Chambers, 1993). The challenge is to make the shift from the old professionalism to the new (Table 8.2).

A significant shift to a new professionalism and participatory praxis for transdisciplinarity requires profound transformations in the governance, culture, operational procedures and reward structures of research organizations and their donors. This is the major conclusion of a substantial body of studies on how to institutionalize people-centred processes, mainstream gender justice, enable transdisciplinary approaches and embed participation and self-organizing processes in bureaucracies and research institutions (Arnold and Cole, 1987; Bainbridge *et al.*, 2000; Calas and Smircich, 1997; Clegg *et al.*, 2006; Crozier and Friedberg, 1977; Goetz, 1997a, 1997b; Guijt and Shah, 1998; Laloux, 2016; Macdonald *et al.*, 1997; Mauser *et al.*, 2013; Pimbert, 2004; Rao and Stuart, 1997; Wheatley, 2006).

TABLE 8.2 Changing professionalism from the old to the new

	From the old professionalism	To the new professionalism
Who sets priorities?	Social and natural scientists, as well as other professionals, set priorities for research and development. They are in charge of decisions on research priorities – but only within the framework and boundaries defined by donor agencies, research council funds, government bureaucracies and corporations that fund research.	Peasant farmers and other citizens set priorities, including upstream strategic priorities for public research. Scientists sometimes work together with citizens to do this. But on other occasions indigenous peoples, peasant farmers and other citizens are the ones who decide priorities for public research *after* listening to the specialist knowledge of different scientists and other knowledge holders (food consumers, farm workers, government officials etc.). They then carefully deliberate on the pros and cons of possible research priorities. In this latter scenario, scientists are invited as resource people, and they provide information which farmers and citizens use to decide research priorities and resource allocations.
Science, knowledge and methods	Scientific method is reductionist and positivist, with a strong natural science bias; a complex world is split into independent variables and cause-effect relationships; scientists' categories and perceptions are central.	Peoples' knowledge and transdisciplinary approaches are key; scientific method is holistic and post-positivist; local categories of knowledge and perceptions are central; disruption of knowledge hierarchies; subject-object and method-data distinctions are blurred.
Strategy and context of intervention	Professionals know what they want; pre-specified project design or research plan; top-down approach. Information and results are extracted from controlled situations and communities; context is independent and controlled – blueprint-oriented.	While clear about the need for sustainable food systems and biocultural diversity, professionals working with farmers do not know where research projects will lead; they are engaged in a living, emergent and open-ended learning process that cannot be fully pre-determined. Understanding and focus emerges through interaction; context of inquiry and intervention is fundamental – process-oriented.

(*continued*)

TABLE 8.2 Continued

	From the old professionalism	To the new professionalism
Assumptions about reality	Assumption of singular, tangible reality.	Assumption of multiple realities that are socially constructed.
Relationship between actors involved in the process	Professionals control and motivate from a distance; they tend not to trust people (farmers, food workers, indigenous and rural people etc.) who are simply the object of inquiry or intervention.	Professionals engage in close dialogue; they attempt to build trust through joint analyses and negotiation; understanding arises through this engagement, resulting in more power-equalizing ways of knowing based on cognitive justice.
Mode of working	Single disciplinary – working alone.	Multidisciplinary and transdisciplinary – working in self-organizing groups that include scientists and other knowledge holders (peasant farmers, pastoralists, men and women etc.). External researchers and extension agents shift to new roles that facilitate and support local people's analysis, deliberations and production of knowledge.
Attitudes to food & agricultural policy, technology or services	Rejected policy, technology, knowledge or service assumed to be the fault of local people or local conditions; centrally designed policy and technology first.	Rejected policy, technology, knowledge or service is a failed innovation or the outcome of faulty research and inappropriate framing assumptions; people first.
Career development	Careers are inwards and upwards – as practitioners get better, they become promoted, take on more administration and spend less time in the field and with local communities.	Careers include outward and downward movement – professionals stay in touch with action at all levels and spend time with local communities and social movements.

Modified from Pretty and Chambers (1993) and Pimbert (2009).

This multifaceted literature on organizational change offers important insights for citizens, social movements and policy champions who seek to democratize, decolonize and re-orient public research and donor support for food sovereignty, agroecology and biocultural diversity. For example, mutually reinforcing and simultaneous actions are required to fundamentally change organizations that produce social, environmental, economic and technical knowledge. This transformation must be systemic and encompass academic cultures, the self-image of researchers and academics, teaching pedagogies, research agendas and methodologies, organizational cultures, operational procedures and the roles that universities and research institutes play in society (Bainbridge *et al.*, 2000; Pimbert, 2009). Some of the key levers for democracy and organizational transformation for transdisciplinarity in public research and education are listed below:

- Diversify the governance and the membership of budget allocation committees of public sector planning and research institutes to include representatives of diverse citizen groups and axes of difference (age, gender, age, race, ethnicity, disability, sexual orientation etc.). Establish procedures to ensure transparency, equity and accountability in the allocation of funds and dissemination of new knowledge.
- Encourage shifts from hierarchical and rigidly bureaucratic structures to 'flat', self-organizing, flexible, and responsive organizations.
- Redesign practical arrangements and the use of space and time within the workplace to meet the diverse needs of women, men and older staff and to help them fulfil their new professional obligations to work more closely with peasant farmers and other citizens (timetables, career paths, working hours, provision of paternity and maternity leave, childcare provisions, mini sabbaticals, promotion criteria etc.).
- Build the capacity of technical and scientific staff in the participatory skills, attitudes and behaviour needed to learn from citizens (mutual listening, respect, gender sensitivity, empathy etc.), decolonize research methodologies and engage in self-organizing horizontal processes.
- Provide capacity-building and experiential learning for staff to develop their ecological literacy and skills in agroecology as well as their political knowledge about cognitive justice, food sovereignty and biocultural diversity as a basis for self-determination.
- Reverse gender biases, colonial attitudes, racism and neo-Malthusian environmental crisis narratives in the ideologies and disciplines animating research organizations and their projects.
- Ensure that senior and middle-management positions are occupied by competent facilitators of organizational change with the vision, commitment and ability to reverse gender and other discriminatory biases in the ideologies, disciplines and practices of the organization.
- Promote and reward management that is consultative and participatory rather than hierarchical and efficiency-led, as well as command and control management styles based on a culture of blame.
- Establish incentive and accountability systems that are equitable for women and men, and do not discriminate based on race, ethnicity, age, disability or sexual orientation.
- Provide incentives and high rewards for staff and members of organizations to experiment, take initiatives and acknowledge errors as a way of learning-by-doing and engaging with the diverse local realities of citizens living in rural and urban areas.

- Encourage and reward the use of gender disaggregated and socially differentiated indicators (e.g. by class, age, race, ethnicity etc.) in monitoring and evaluation to enhance social justice, fairness and inclusion – both within organizations and in their external interventions.

In sum, far reaching and fundamental changes in organizations are necessary so that their *ethos*, policies, programmes, operational procedures, resource allocations and ways of working facilitate gender inclusive participation and transdisciplinarity in R&D; nurture attitudes grounded in empathy, respect and solidarity; and develop skills in ecological literacy that are needed for the local adaptive management of agroecosystems and diverse biocultural landscapes (Chapter 3 and 4; Borrini-Feyerabend *et al.*, 2007; Pimbert 2009).

Protecting public research

The following section briefly identifies some strategic actions that could help protect higher education and research from corporate capture, privatization and the commodification of knowledge.

Job security in university education and research

Despite the fact that they are among the most highly skilled and prestigious professions, university teachers and researchers are increasingly faced with the consequences of low-paid, insecure work. For example, *The Guardian* recently revealed the extent of casual labour and job insecurity among UK universities:

> Academics teaching or doing research in British universities will typically have spent years earning doctorates or other qualifications, yet more than half of them – 53% – manage on some form of insecure, non-permanent contract. They range from short-term contracts that typically elapse within nine months, to those paid by the hour to give classes or mark essays and exams.
>
> *(Chakrabortty and Weale, 2016)*

Strikingly, the richest British universities rely the most on insecure academic workers with fixed short-term contracts. For example, 70% and 68% of teaching staff are on insecure contracts at the universities of Birmingham and Warwick respectively. The universities of Birmingham, Edinburgh, Oxford and Warwick have the largest share of frontline teaching staff on short-term flexible and insecure contracts (Chakrabortty and Weale, 2016). In France, there are currently 60,000 doctoral students and most will be on casual contracts, low pay and periodically unemployed once they have obtained their PhDs and are on the job market (Trublet, 2016). This increasing casualization and spread of poverty-line pay in universities has also been the trend in the USA: 76% of academics are now on casual contracts with little job security, and growing numbers are even on food stamps (O'Hara, 2015).

Job insecurity is an integral part of the neoliberal university project, 'marked by the decline of the humanities and social sciences, cuts in public financing, enfeeblement of faculty and student roles in governance, increases in tuition fees, reductions in tenured faculty and increasing use of adjunct professors' (Heller, 2016b).

The casualization of the academic workforce is increasingly widespread and seriously undermines the quality of education and research in universities. Lack of job security militates against the changes in attitudes and behaviours needed for transdisciplinary co-inquiry – it promotes conformity to established research traditions and their cognitive routines (Trublet, 2016). Similarly, it is difficult to see how universities can re-invent and transform themselves for participatory and transdisciplinary ways of knowing when so many academic staff experience job insecurity, stress, low morale, lack of recognition and low pay (Weale, 2016). Chronic job insecurity in a climate of hypercompetition heightens the challenge of maintaining scientific integrity and makes it more difficult to 'incentivize altruistic and ethical outcomes, while de-emphasizing output' (Edward and Roy, 2017).

As both the products and victims of the capitalist division of labour, academic workers will probably need to engage in joint action with citizens and social movements to reverse these debilitating trends. However, the moral and political goal here

> is not the highest possible professional standards of a few specialists but, instead, the general progress and diffusion of knowledge within the community and the working class as a whole. Any progress in knowledge, technology and power that produces a lasting divorce between the experts and non-experts must be considered bad. Knowledge, like all the rest, is of value only if it can be shared.
>
> *(Gorz, 1976)*

Safeguards against the corruption of science by corporations

Intellectual suppression, competitive 'cognitive capitalism' and institutionalized bias in the halls of science significantly constrain the possibilities of an open and disinterested inquiry (Chapters 1, 2 and 5 in this volume; Heller, 2016b; Roger, 2013). This is particularly evident in the case of researchers assessing the risks of genetically modified organisms (GMOs) for public health and the environment, for example. When Berkeley plant geneticists Ignacio Chapela and David Quist uncovered the transgenic contamination of maize landraces growing in remote regions of Mexico and reported it in *Nature* in November 2001, they were subjected to vicious attacks and intimidation resembling the Pusztai[16] episode in Britain. Attacks and smear campaigns were orchestrated from within their own department, aided and abetted by Monsanto. Chapela was refused tenure by his university at the end of 2003 (Mantell, 2002; Rowell, 2003).[17]

More widely, industry uses its power to vilify, marginalize and reject scientists whose experimental results contradict the central dogma of molecular biology. As Barry Commoner says: 'The fact that one gene can give rise to multiple proteins ...

destroys the theoretical foundation of a multibillion-dollar industry' (Commoner, 2003). This corporate censorship of science is likely to worsen if a new wave of mega-mergers goes ahead in the seed industry: just three global seed corporations would be able to exert unprecedented control over what scientists can publicly say and write about their research in plant genetics and synthetic biology.[18]

Several actions can help insulate research from corporate abuse and capture. For example, the Union of Concerned Scientists (2012) has identified key areas where governments (national, regional and municipal) can act more to protect science against undue corporate influence and corruption, including protecting scientists from retaliation and intimidation; reforming the regulatory process; and strengthening monitoring and enforcement. Similarly, the work of the *Fondation Sciences Citoyennes* in France shows how researchers and citizens can organize and act against the enclosure of public research by corporations and systematic attacks on whistle blowers (see http://sciencescitoyennes.org).

More generally, increased government funding for public research is necessary to reverse the privatization and corporate capture of higher education and research. Additional public funds are also needed to generate the kinds of knowledge and liberatory technologies (*sensu* Bookchin, 1986) that can significantly expand the realm of freedom by reducing peoples' dependence on commodity markets controlled by corporations. For example, the potential of agroecological research to develop more autonomous food systems can only be realized if supported by much more public funding than it has received to date (see Chapters 1 and 2; Union of Concerned Scientists, 2015). After decades of neglect by government spending, whole areas of science urgently need new funds to recruit additional people with appropriate skills – from taxonomists who can identify the natural enemies of pests for use in biological control programmes, soil biologists able to develop knowledge-intensive agroecological methods for soil fertility management and carbon sequestration, to eco-linguists who can help understand how language encodes the stories that society is based on (ideologies, framings, metaphors, evaluations, identities etc.). The FAO's publication *The State of Food and Agriculture 2016* warns that 'achieving the transformation to sustainable agriculture is a major challenge ... available finance for investment in agriculture falls well short of needs ... The time to invest in agriculture and rural development is now' (FAO, 2016a).

Reclaiming universities as a commons for knowledge democracy

Ensuring that the cultural, intellectual and other resources of universities are accessible to all members of society – and are held in common[19] – is key for knowledge democracy. Stories of peoples' struggles to regain control over the commons and the production of knowledge can inspire and offer new models for the governance, re-structuring, organizational practices and roles of higher education and research. Past and present initiatives by peasant farmers, unemployed youth, casual labour and other citizens to recuperate factories, urban land for food production and abandoned workplaces in the Americas, Africa, Asia and Europe show how knowledge and wealth can be produced differently, with citizens in charge (Dion, 2016; Duchatel and Rochat, 2008; Sitrin, 2006; Sitrin and Azzelini, 2014; Zibechi, 2010; *Cooperativa Integral Catalana*, https://cooperativa.cat/en). For example, in war-torn Syria and south-east Turkey, Kurdish men and women are putting into practice their demands for autonomy and democratic confederalism (Öcalan, 2011) by creating a region-wide web of villages and municipal councils through which they can govern themselves. In this 'stateless democracy' Kurdish communities are formulating their own laws, creating their own parliament and building their own universities and capacity for research (New World Academy, 2015; TATORT Kurdistan, 2013). The Lucas Aerospace workers' plan for socially useful production is another emblematic example of how citizens can reclaim control over the production of knowledge (Box 8.8).

BOX 8.8 THE LUCAS AEROSPACE PLAN FOR SOCIALLY USEFUL PRODUCTION IN THE UK

Faced with the prospect of massive job redundancies, the employees of Lucas Aerospace developed their own plan for the re-structuring and re-conversion of their workplaces from arms manufacture to socially useful production. Published in 1976, the Alternative Corporate Plan of Lucas Aerospace proposed radical alternatives to closure in manufacturing – from the production of appropriate technology for community needs, the development of skill enhancing human-centred technology, to participatory design and industrial democracy (Cooley, 1982; Smith, 2014; Wainwright and Elliott, 2017). About half of Lucas' output supplied military contracts which depended on public funds, as did many of the company's civilian products. Workers argued for state support to go instead to more socially useful products. Lucas employees argued they had the *right* to socially useful production instead of redundancies. The Lucas plan was based on employees' careful collective assessment of the diversity of workers' knowledge, skills, work organization, machinery and economic options. Most notably, the workers' reconversion plans contested established hierarchies of knowledge and valued plural knowledge, including peoples' experiential and tacit knowledge. The Lucas workers wanted to 'demonstrate in a very practical and direct way the creative power of "ordinary people"' (Cooley, 1982).

Through their alliances with wider social movements (radical science, environmental, feminist and peace movements), the Lucas workers' concept of socially useful production increasingly emphasized not only jobs, but also democratic control and direct participation in R&D and the design of technology for social need and environmental sustainability (Cooley, 1982). Conversations with community development activists helped deepen awareness that socially usefully production had to be guided by the needs identified and defined by local communities, including the most economically deprived.

For example, a Coventry workshop brought together grassroots community groups and Lucas shop stewards committees 'to explore the links, in concept and in practice, between industry and the community, the economy and the state, production and consumption, home and work' (Coventry Workshop, 1978). The movement that emerged sought more direct control for workers, communities and citizens in R&D and production processes. With the support of progressive city councils (e.g. Greater London Council – GLC), a number of Technology Networks were created as part of an industrial strategy committed to socially useful production. The GLC created the Greater London Enterprise Board (GLEB), with Mike Cooley appointed as its Technology Director after he had been sacked by Lucas Aerospace for his activism. As Adrian Smith recalls,

> Technology Networks aimed to combine the 'untapped skill, creativity and sheer enthusiasm' in local communities with the 'reservoir of scientific and innovation knowledge' in London's polytechnics. Hundreds of designs and prototypes were developed, including electric bicycles, small-scale wind turbines, energy conservation services, disability devices, re-manufactured products, children's play equipment, community computer networks, and a women's IT co-operative. Designs were registered in an open access product bank. GLEB helped co-operatives and social enterprises develop these prototypes into businesses.
>
> (Smith, 2014)

The Lucas plan and the wider initiatives it inspired 'came up against trade union, government and management institutions stuck in the command and control mentalities of the 1950s, and the power of the movement was destroyed by Thatcher's onslaught against the unions and radical local government in the 1980s' (Wainwright, 2009). However, the underlying ideas of the Lucas plan for socially useful production are still relevant today.

Insights from these social experiments can help re-invent universities and democratize research for the common good. However, lessons learnt and underlying principles have to be carefully adapted to each specific historical context and its possibilities. They cannot simply be copied from one place to another. Re-inventing universities as the commons will require wide-ranging deliberations and dialogues between academic workers, researchers, grassroots peasant networks and other citizens. As stated by Robin Hahnel:

> The goal is clear enough: We must convince a majority of people that ordinary people are perfectly capable of managing our own economic affairs without capitalist employers or commissars to tell us what to do. We must convince a majority of people that groups of self-managing workers and consumers are capable of coordinating their own division of labor through participatory, democratic planning, rather than abdicating this task to the

market system or central planners. But how this goal will be achieved, and how people will be prepared to defend necessary changes from powerful, entrenched, minority interests who will predictably attempt to thwart the will of the majority, will vary greatly from place to place. All that can be said about it with any certainty is that in most places it will require a great deal of educational and organizing work of various kinds, given where we are today.

(Hahnel, 2016)

Conclusion

One of the clearest demands of the food sovereignty movement is for peasant farmers, indigenous peoples, pastoralists, fishers and other citizens to exercise their fundamental human right to decide their own food and agricultural policies (Nyéléni, 2007; Nyéléni, 2015). This implies that the construction of technical and policy related knowledge for food sovereignty, agroecology and biocultural diversity should be actively shaped by food producers and consumers. A two-pronged approach to democratizing the production of transdisciplinary knowledge has been proposed in this chapter: (1) strengthening horizontal networks of grassroots self-managed research and innovation; and (2) fundamentally transforming and democratizing public research institutions and universities. Depending on context and history, one approach may be favoured over another. However, when these two approaches are used in complementary and mutually reinforcing ways this can significantly expand democracy and the construction of knowledge for food sovereignty, agroecology and biocultural diversity.

In each pathway for transformation, contesting and constructing knowledge depends on subverting hierarchies of knowledge – erasing the boundaries between peoples' knowledge and the disciplinary knowledge of the natural and social sciences as well as the arts and humanities. This transdisciplinarity implies participatory ways of knowing that give the least powerful actors more significant roles than before in the production and validation of knowledge. Power-equalizing processes in the co-construction of transdisciplinary knowledge are indeed central to the two transformative pathways described here, and they include a politics of cognitive justice, reversals from normal professional practice, organizational change, *diálogos de saberes* and intercultural dialogues, the strengthening of local organizations for autonomous learning and action, and citizens' direct democratic control over research priorities and resource allocations for the construction of knowledge(s).

Given the inherently conservative nature of states and the professional-managerial class, it is perhaps wishful thinking that academics alone can contest and transform the dominant culture of cognitive capitalism in universities and research institutions. In practice, activist researchers and critical scholars, grassroots networks of peasant innovators, as well as citizens and wider social movements will have to work together to exert the countervailing power needed to democratize research and construct knowledge for food sovereignty, agroecology and biocultural diversity. Moreover, transforming knowledge depends on many different actors engaging in large-scale counter-hegemonic practices for at least two other reasons. First, knowledge broadly reflects and reinforces specific power relations and worldviews in any society. Deep social change is often needed for the emergence of new knowledge paradigms. Secondly, while clearly vitally important, new knowledge alone will not lead to the widespread adoption of food sovereignty, agroecology and biocultural diversity. Deeper-seated political and economic changes are necessary throughout society, including policies that can reverse the ongoing economic genocide of family farmers as well as provide the time and material security which food producers and other citizens need to fully engage in participatory democracy.

The approaches to knowledge construction described here must therefore be seen as part of a wider process of transformation that seeks to invent a new modernity based on plural definitions of human well-being and an active citizenship that can fundamentally democratize economic, political, ecological, social and cultural realms (Bookchin, 2005; Fotopoulos, 1997). For example, given the scale of today's democratic deficit, new political structures are required to combine localism with interdependence for co-ordinated action across large areas for food sovereignty, agroecology and biocultural diversity. One option is democratic confederalism, which involves a network of citizen-based (as opposed to government) bodies or councils with members or delegates elected from popular face-to-face democratic assemblies, in villages, towns and neighbourhoods of large cities. When combined with an education for active citizenship, these confederal bodies or councils become the means of interlinking villages, neighbourhoods, towns and agro-ecological regions into a confederation based on shared responsibilities, full accountability, firmly mandated representatives and the right to recall them if necessary (Bookchin, 2015; Öcalan, 2011). Citizens can thus participate in a direct and democratic way in the decentralized and distributed production of post-normal knowledge that is now needed for the local adaptive management of ecosystems and economies in today's context of rapid change and uncertainty. Mainstreaming the construction of knowledge for food sovereignty, agroecology and biocultural diversity ultimately depends on these deeper transformations for direct democracy, freedom and justice.

Notes

1   Transformation is the creative re-visioning and fundamental re-design of whole systems. It involves 'seeing things differently', 'doing better things' and re-thinking whole systems on a participative basis. It is a form of triple loop learning (Senge, 1990) which asks the deeper 'underlying why' questions and focuses on underlying paradigms, norms and values that frame and legitimize the purpose of knowledge, policies, organizations, technologies and practice. As such transformation sharply differs from reform (second loop learning) and from adaptation and maintenance of the *status quo* (accommodation, first loop learning).

2   In this chapter, I use the term 'farmer' interchangeably with '*campesino*' and 'peasant'. Small-scale food producers – farmers, artisanal fisherfolks, pastoralists, forest dwellers, hunters and gatherers – provide the food to the majority of the world population. They also constitute the largest group of 'economically active people'. About 40% of all

working people are small-scale farmers – peasants – and around 43% of the agricultural labour force in developing countries are women (FAO, 2016a).

3 *Dalit*, meaning 'oppressed' in Sanskrit, is the name of castes in India which are 'untouchable'. These are social groups confined to menial and despised jobs.

4 In Inca mythology, *Pacha* means the different spheres of the cosmos. In the Quechua language, *Pacha* is often translated as 'world', and it includes the sky, the sun, the moon, the stars, the planets and constellations (*Hanan pacha*); the tangible world where people, animals and plants all live (*Kay pacha*); and the inner world associated with the dead as well as with new life (*Ukhu pacha*).

5 Post-normal science is the sort of inquiry in which the facts are uncertain, values are often in dispute, stakes are high and decisions are urgent. Central to post-normal science is the idea of an 'extended peer community' and the recognition that there is a plurality of legitimate perspectives on every issue (see Ravetz, 1971; Funtowicz and Ravetz, 1994).

6 Strictly speaking, organizations are not the same as institutions. Institutions are 'the humanly devised constraints that shape human interaction … they structure incentives in human exchange, whether political, social or economic … Institutions reduce uncertainty by providing a structure to everyday life … Institutions include any form of constraint that human beings devise to shape interaction' (North, 1990). Land tenure rules and other rules regulating access, use and control over natural resources are examples of institutions. Although they embrace them, institutions are not organizations; they are best understood as a set of informal and formal rules that are administered by organizations. Organizations are thus 'groups of individuals bound by some common purpose to achieve objectives' (North, 1990). Organizations operate within the framework – the rules and constraints – set by institutions. Examples include government departments or local beekeeper associations which administer sets of formal and informal 'rules of the game'.

7 In India, women's '*sanghams*' seek to create a space where women can talk about their problems, share their worries and seek advice. As women's associations, *sangham* groups play a key role in building the confidence, respect and freedom of women for *dalit* and other marginalized people. In Buddhist teachings, the word '*Sangha*' means a group. Being part of the *Sangha* with '*Sangham saranam gacchami*' is about taking refuge in that which is good, virtuous, kind, compassionate and generous.

8 In the European Union, for example, there were 8 million farms in the 12 member states that made up the EU in 1990. Ten years later – after the accession of three additional member states (Austria, Finland and Sweden) – the EU had lost 1.4 million farms, reducing the total in 2000 to 6.6 million farms (Choplin, 2017). Overall, the number of European farmers is decreasing every year by about 2%, though falls of more than 8% were registered between 2002 and 2003 in the Czech Republic, Hungary, Poland, Slovenia, Slovakia and the UK. There is also a negative demographic trend in Europe: currently only 6% of farmers are under the age of 35 across the EU, and 34% of all farmers are over 65 years old (CEJA, 2011). In France, where the percentage of the active working population in agriculture decreased from 30% to 3% over a period of 50 years, 10,000 farmers per year leave farming before reaching retirement age – i.e. one third of the total number of farmers who quit farming every year, according to a recent inter-ministerial study. The reasons for leaving farming in France are the same as for many other countries worldwide, including banks refusing to give loans, lack of cash, inability to reimburse money borrowed for farm investments and farm enterprises being less and less able to absorb impacts of two consecutive years of crisis (ASP, 2016).

9 UNORCA is the National Union of Autonomous Regional Peasant Organisations. It is a national network of 1,400 Mexican *campesino* and indigenous farming organizations representing 200,000 producers in 27 Mexican states. UNORCA is also a member of La Via Campesina: http://unorcamexico.org/author/unorcaeditor.

10 For example, see Levidow (2008) for a discussion on shortcomings of government-controlled citizens' juries in four European countries.

11 For example, see the film *Imagining Research for Food Sovereignty* (http://www.excluded-voices.org/st-ulrich-workshop-democratising-agricultural-research-food-sovereignty-and-peasant-agrarian-culture), and the video films posted on www.excludedvoices.org/video and www.agroecologynow.org.

12 In April 2018, four West African farmers involved in the deliberative process to date will travel to London to ask the UK Government and British taxpayers to de-prioritize AGRA in its overseas aid and support instead research on agroecology for family farming in Africa.

13 A cosmovision is a particular way of viewing the world or of understanding the universe.

14 For example, the UK's Research Excellence Framework (http://www.ref.ac.uk/about), the German Universities Excellence Initiative (https://www.wissenschaftsrat.de/en/about.html) and the Excellence in Research for Australia (www.arc.gov.au/era).

15 The UN Food and Agriculture Organization (FAO) and the Government of Hungary hosted the *Regional Symposium on Agroecology for Sustainable Agriculture and Food Systems in Europe and Central Asia* from 23 to 25 November 2016 in Budapest (Hungary). Comments on donors' research support were made during the session on 'Research, innovation and knowledge sharing for agroecological transition' (Module 3). See: www.fao.org/europe/events/detail-events/en/c/429132.

16 Arpad Pusztai was a senior scientist from the Rowett Institute in Aberdeen, UK. In the summer of 1998, he told the British public that feeding young rats GM potatoes appeared to harm them. Dr Pusztai lost his job, his research group was disbanded, and a gag order was placed on him. The Royal Society – the top society of scientists in the UK – issued a hasty official report discrediting Pusztai's findings (Randerson, 2008).

17 Following widespread public protest, the University of Berkeley (California) reversed its decision and finally decided to grant tenure to Dr. Ignacio Chapela in 2005.

18 At the time of writing, three mega-mergers in the agri-chemical industry are simultaneously underway around the globe, namely (1) ChemChina's takeover of Syngenta; (2) Bayer CropScience's acquisition of Monsanto; and (3) Dow Chemical Company (Dow) and E.I. du Pont de Nemours and Company (DuPont). If these mergers are approved by EU and US regulators, only three corporations will control nearly 60% of the world's commercially marketed seeds, nearly 70% of the chemicals and pesticides used to grow food and nearly all of the world's GM crop genetic traits (Vidal, 2016).

19 Michael Hardt argues that our choices are not limited to businesses controlled privately (private property) or by the state (public property). The third option is to hold things *in common* – where resources and services are produced, distributed and controlled democratically and equitably according to peoples need (Hardt, 2011).

References

Adger, W.N., T.A. Benjaminsen, K. Brown and H. Svarstad (2001) 'Advancing a political ecology of global environmental discourses', *Development and Change* 32(4): 681–715.

Agarwal, B. (1995) *Gender, Environment and Poverty Interlinks in Rural India*, United Nations Research Institute for Social Development, Geneva.

Amsler, S.S. (2015) *The Education of Radical Democracy*, Routledge, London.

Apffel-Marglin, F. and S.A. Marglin (1996) *Decolonizing Knowledge: From Development to Dialogue*, Clarendon Press, Oxford.

Apffel Marglin F., S.A. Marglin and L. Jayawardena (1990) *Dominating Knowledge: Development, Culture, and Resistance*, Clarendon Press, Oxford.

Argumedo, A. and M.P. Pimbert (2005), *Traditional Resource Rights and Indigenous People in the Andes*, IIED and ANDES, International Institute for Environment and Development (IIED), London.

Argumedo A. and M.P. Pimbert (2008) 'Protecting farmers' rights with indigenous biocultural heritage territories: The experience of the Potato Park', Paper presented at the *9th Conference of the Parties of the Convention on Biological Diversity*, 9–30 May 2008, Bonn. Available at: http://pubs.iied.org/G03072/?k=pimbert+argumedo.

Argumedo A. and M.P. Pimbert (2010) 'Bypassing globalization: Barter markets as a new indigenous economy in Peru', *Development*, 53 (3): 343–349.

Arnold, P. and I. Cole (1987) 'The decentralisation of local services: rhetoric and reality', In P. Hoggett and R. Hambleton (eds.), *Decentralisation and democracy. Localising public services*, School for Advanced Urban Studies, University of Bristol, Bristol.

Arnstein, S. R. (1969) 'A ladder of citizen participation', *JAIP*, 35(4): 216–224.

ASP (2016) *Départs précoces en agriculture. Analyse d'une situation peu connue*, Agence de Service et de Paiement, Paris.

Bachmann, L., E. Cruzada and S. Wright (2009) *Food Security and Farmer Empowerment: A study of the impacts of farmer-led sustainable agriculture in the Philippines*, MASISPAG, Laguna.

Bainbridge, V. S. Foerster, K. Pasteur, M. P., Pimbert G. Pratt and I. Y. Arroyo (2000) *Transforming Bureaucracies: Institutionalising Participation and People Centred Processes in Natural Resource Management – an Annotated Bibliography*, International Institute for Environment and Development (IIED), London.

Barbosa, L.P. (2016) '*Educação do Campo* [Education for and by the countryside] as a political project in the context of the struggle for land in Brazil', *The Journal of Peasant Studies*, 44(1): 118–143.

Bawden, R.J. (2007) 'A paradigm for persistence: A vital challenge for the agricultural academy', *International Journal of Agricultural Sustainability*, 5(1): 17–24.

Beaverstock, J.V. and I. Hay (2016) 'They've never had it so good': The rise and rise of the super-rich and wealth inequality, In: Hay, I. and J.V. Beaverstock (Eds) *Handbook on Wealth and the Super-Rich*, Edward Elgar Publishing, Cheltenham.

Beckie, M. and R. Berezan (2017) 'Engaging with Cuba's permaculture movement through transformative learning', In: People's Knowledge Editorial Collective (ed.) *Food Justice: Participatory and Action Research Approaches*, Reclaiming Diversity and Citizenship Series, Coventry.

Beder, S. (2006a) *Free Market Missionaries: The Corporate Manipulation of Community Values*, Earthscan, London.

Beder, S. (2006b) *Suiting Themselves: How Corporations Drive the Global Agenda*, Earthscan, London.

Bellows, A. C., Valente, F. L. S., S. Lemke and M.D.N.B. de Lara (2016) *Gender, Nutrition, and the Human Right to Adequate Food*, Routledge, London.

Benessia, A., Funtowicz, S., Giampietro, M., Guimarães Pereira, Â., Ravetz, J., Saltelli, A., Strand, R. and van der Sluijs, J. P. (2016) *The Rightful Place of Science: Science on the Verge*, Consortium for Science, Policy & Outcomes, Tempe, Arizona.

Bergmann, M., T. Jahn, T. Knobloch, W. Krohn, C. Pohl and E. Schramm (2013) *Methods for Transdisciplinary Research: A Primer for Practice*, Campus Verlag, Munich.

Bernstein, J. H. (2015) 'Transdisciplinarity: A review of its origins, development, and current issues', *Journal of Research Practice,* 11(1), Article R1.

Bollier, D. and S. Helfrich (2012) *The Wealth of the Commons. A World beyond the Market and State*, Levellers Press, Amherst.

Bollier, D. and S. Helfrich (2015) *Patterns of Commoning*, Heinrich Böll Foundation and Levellers Press, Amherst.

Bookchin, M. (1986) 'Toward a liberatory technology', In: Murray Bookchin (ed.), *Post Scarcity Anarchism*, 2nd Edition, Black Rose Books, Montreal.

Bookchin, M. (2005) *The Ecology of Freedom. The Emergence and Dissolution of Hierarchy*, AK Press, Edinburgh.

Bookchin, M. (2015) *The Next Revolution. Popular assemblies and the promise of direct democracy*, Verso, London and New York.

Borrini-Feyerabend, G., M.P. Pimbert, M. Taghi Farvar, A. Kothari and Y. Renard (2007) *Sharing Power: A Global Guide to Collaborative Management of Natural Resources*, Routledge and Earthscan, London.

Bradbury, H. and P. Reason (2000) 'Broadening the bandwidth of validity: issues of choice points for improving the quality of action research', In: P. Reason and H. Bradbury (eds) *Handbook of Action Research: Participative Inquiry and Practice*, SAGE Publishing, London.

Brown, P. (1992) 'Popular epidemiology and toxic waste contamination: Lay and professional ways of knowing', *Journal of Health and Social Behavior*, 33 (3): 267–281.

Bunch, R. (1998) *Soil Conservation Protects Farms from Hurricane Mitch*, Cornell University's MULCH-L discussion group, Cornell University, Ithaca.

Calas, M.B. and L. Smircich (1997) 'The woman's point of view: Feminist approaches to organisation studies', In S.R. Clegg, C. Hardy *et al.* (eds), *Handbook of Organisation Studies*, Sage Publications, London.

Callon, M., P. Lascoumes and Y. Barthe (2001) *Agir dans un monde incertain. Essai sur la démocratie technique*, Le Seuil, Paris.

CEJA (2011) The European Council of Young Farmers, Press Release, 31st January 2011, Conseil Européen des Jeunes Agriculteurs, Brussels, www.ceja.eu/press-corner/

Chakrabortty, A. and S. Weale (2016) 'Universities accused of "importing Sports Direct model" for lecturers' pay', *The Guardian*, 16 November 2016.

Chambers, R. (1993) *Challenging the Professions. Frontiers for Rural Development*, Intermediate Technology, Publications, London.

Chambers, R. (1996) *Whose Reality Counts?* Intermediate Technology Publications, London.

Chambers, R. (2008) *Revolutions in Development Inquiry*, Earthscan, London

Chambers, R. (2014) 'Perverse payment by results: Frogs in a pot and straitjackets for obstacle courses', Participation, Power and Social Change (PPSC) blog, 3 September 2014, Institute of Development Studies, Falmer, https://participationpower.wordpress.com/tag/robert-chambers/.

Charvolin, F., A. Micoud, L.K. Nyhart (2007) *Des sciences citoyennes? La question de l'amateur dans les sciences naturalistes*, Éditions de l'Aube, La Tour d'Aigues.

Chevalier, J.M. and D.J. Buckles (2008) *SAS2 : A Guide to Collaborative Inquiry and Social Engagement.* Sage and IDRC, London.

Chilisa, B. (2012) *Indigenous Research Methodologies*, Sage Publishing, London.

Chomsky, N. (2017) *Requiem for the American Dream: The Principles of Concentrated Wealth and Power*, Seven Stories Press, New York.

Choplin, G. (2017) *Paysans mutins, paysans demain. Pour une autre politique agricole et alimentaire*, Editions Yves Michel, Gap.

Claeys, P. (2015) *Human Rights and the Food Sovereignty Movement. Reclaiming control*, Routledge, London.

Clegg, S.R., C. Hardy, T. Lawrence and W.R. Nord (2006) *The SAGE Handbook of Organization Studies*, 2nd Edition, Sage Publishing, London.

Cohn, A., J. Cook, M. Fernández, R. Reider and C. Steward (2006) *Agroecology and the Struggle for Food Sovereignty in the Americas*, University of Yale, IIED and IUCN, London and Yale.

Commoner, B. (2003) 'Unravelling the DNA myth', *Seedling*, July 2003: 6–12, GRAIN, Barcelona.

Community Media Trust, P.V. Satheesh and M.P. Pimbert (2008) *Affirming Life and Diversity: Rural images and voices on food sovereignty in South India*, International Institute for Environment and Development, London.

Cooley, M. (1982), *Architect or Bee? The Human/Technology Relationship*, South End Press, Boston.

Coote A. and J. Lenaghan (1997) *Citizens' Juries: Theory into Practice*, Institute for Public Policy Research, London.

Coventry Workshop (1978) *Progress Report 1976–77*, Coventry Workshop, Coventry.

Crozier, M. and E. Friedberg (1977) *L'acteur et le systéme. Les contraintes de l'action collective*, Seuil, Paris.

De Rosnay, J. (1975) *Le Macroscope. Vers une vision globale*, Editions du Seuil, Paris.

Desmarais, A. A. (2007) *Globalization and the Power of Peasants: La Vía Campesina*, Pluto Press, London.

DFID (2014) 'DFID's strategy for payment by results: Sharpening incentives to perform', UK Government's Department for International Development, London, https://www.gov.uk/government/publications/dfids-strategy-for-payment-by-results-sharpening-incentives-to-perform.

Dilts R. (1999) *Facilitating the Emergence of Local Institutions. Reflections from the Experience of the Community IPM Programme in Indonesia*, available at www.communityipm.org/downloads.html.

Dion, C. (2016) *Demain et après … Un nouveau monde en marche*, Actes Sud, Arles.

Dominguez Rubio, F. and P. Baert (2012) *The Politics of Knowledge*, Routledge, London.

Duchatel, J. and F. Rochat (2008) *Produire de la richesse autrement*, Centre Europe-Tiers Monde – CETIM, Geneva.

Edwards, M.A. and S. Roy (2017) 'Academic research in the 21st century: maintaining scientific integrity in a climate of perverse incentives and hypercompetition', *Environmental Engineering Science*, 34 (1): 51–61.

Ernst, M. (1998), *Sustainable Agriculture Protects Livelihoods from Impacts of Hurricane Mitch*, USAID, Office of Foreign Disaster Assistance, Tegucigalpa.

Esteva, G. and M.S. Prakash (2014) *Grassroots Postmodernism: Remaking the Soil of Cultures*, Zed Books, London.

Fakih, M., T. Rahardjo and M.P. Pimbert (2003) *Community Integrated Pest Management in Indonesia: Institutionalising Participation and People-Centered Approaches*, IIED and REaD, London and Yogyakarta.

Fals Borda, O. (1987) 'The application of participatory action-research in Latin America', *International Sociology*, 2(4): 329–347.

Fals Borda, O. (2001) 'Participatory (action) research in social theory: Origins and challenges', In: P. Reason and H. Bradbury (eds) *Handbook of Action Research: Participative Inquiry and Practice*, SAGE Publishers, London.

FAO (2010) *The Second Report on the State of the World's Plant Genetic Resources for Food and Agriculture*, Food and Agriculture Organization of the United Nations, Rome.

FAO (2015) *The Second Report on the State of the World's Animal Genetic Resources for Food and Agriculture*, FAO Commission on Genetic Resources for Food and Agriculture Assessments, Rome.

FAO (2016a) *The State of Food and Agriculture*, Food and Agriculture Organization of the United Nations, Rome.

FAO (2016b) *Zero Budget Natural Farming in India*, Family Farming Knowledge Platform, Food and Agriculture Organization of the United Nations (FAO), Rome.

Forsyth, T. (2004) *Critical Political Ecology*, Routledge, London.

Fotopoulos, T. (1997) *Towards an Inclusive Democracy*, Cassel, London.

Francis, C.A., N. Jordan, P. Porter, T.A. Breland, G. Lieblein, L. Salomonsson, N. Sriskandarajah and M. Wiedenhoeft (2011) 'Innovative education in agroecology: Experiential learning for a sustainable agriculture', *Critical Reviews in Plant Sciences* 30(1–2): 226–237.

Francis, C., T.A. Breland, E. Østergaard, G. Lieblein and S. Morse (2013) 'Phenomenon-based learning in agroecology: A prerequisite for transdisciplinarity and responsible action', *Agroecology and Sustainable Food Systems* 37(1): 60–75.

Freire, P. (1970) *Pedagogy of the Oppressed*, Seabury Press, New York.

Freire, P. (1973) *Extension or Communication?* McGraw, New York.

Fressoli, M., E. Arond, D. Abrol, A. Smith, A. Ely and R. Dias (2014) 'When grassroots innovation movements encounter mainstream institutions: implications for models of inclusive innovation', *Innovation and Development* 4 (2): 277–292.

Funtowicz, S.O. and J. R. Ravetz (1990) *Uncertainty and Quality in Science for Policy*, Kluwer Academic Publishers, Dordrecht.

Funtowicz, S. O. and J. R. Ravetz (1993), 'Science for the post-normal age', *Futures* 25(7): 739–755.

Gharajedaghi, J. (2011) *Systems Thinking: Managing Chaos and Complexity*, Morgan Kaufmann, Burlington.

Gibbs, P. (2015) *Transdisciplinary Professional Learning and Practice*, Springer, New York.

Gibson-Graham, J. K. (2006), *A Postcapitalist Politics*, University of Minnesota Press, Minnesota.

Goetz, A.M. (1997a) 'Introduction: Getting institutions right for women in development', In: A.M. Goetz (ed.), *Getting Institutions Right for Women in Development*, Zed Books, London.

Goetz, A.M. (1997b) 'Managing organisational change: the 'gendered' organisation of space and time', *Gender and Development* 5(1): 17–27.

Gorz, A. (1976) 'On the class character of science and scientists', In: H. Rose and S. Rose (Eds) *The Political Economy of Science. Ideology of/in the natural sciences*, Macmillan Press, London.

Gramsci, A. (1978) *Selections from the Prison Notebooks*, International Publishers, New York.

Guijt, I. and M.K. Shah, (eds.) (1998) *The Myth of Community: Gender issues in participatory development*, Intermediate Technology Publications, London.

Habermas, J. (1971) *Knowledge and Human Interests*, Beacon Press, Boston,

Hahnel, R. (2016) *Participatory Economics and the Next System*, The Next System Project, http://thenextsystem.org/wp-content/uploads/2016/03/NewSystems_Robin Hahnel.pdf.

Haire-Joshu, D. and T.D. McBride (2013) *Transdisciplinary Public Health: Research, Education, and Practice*, John Wiley & Sons, Hoboken, NJ.

Haklay, M. (2015) *Citizen Science and Policy: A European Perspective*, Woodrow Wilson International Center for Scholars, Washington DC.

Hall, B. (1992) 'From margins to center? The development and purpose of participatory research', *The American Sociologist*, 23(4): 15–28.

Hall, B. (1997) 'Looking back, looking forward: reflections on the origins of the International Participatory Research Network and the Participatory Research Group in Toronto, Canada', paper presented at the *Midwest Research to Practice Conference in Adult, Continuing and Community Education*, Michigan State University, East Lansing, Michigan, Available at: http://www.canr.msu.edu/dept/aee/research/hallpr.htm.

Hall, B. and R. Tandon (2015) 'Are we killing knowledge systems? Knowledge, democracy and transformation', paper presented at *Public Engagement and the Politics of Evidence*, 23–25 July, Available at: http://www.politicsofevidence.ca/349/.

Haraway, D. (1988) 'Situated knowledges: The science question in feminism and the privilege of partial perspective', *Feminist Studies*, 14(3): 575–599.

Harcourt, W. and I.L. Nelson (2015) *Practising Feminist Political Ecologies: Moving Beyond the 'Green Economy'*, Zed Books, London.

Harding, S. (1991) *Whose Science? Whose Knowledge? Thinking from Women's Lives*, Cornell University Press, Ithaca, NY.

Harding, S. (2006) *Science and Social Inequality: Feminist and Postcolonial Issues*, University of Illinois Press, Urbana.

Hardt, M. (2011) 'Reclaim the common in communism', *The Guardian*, 3 February 2011.

Harvey, D. (2004) 'The "new" imperialism: Accumulation by dispossession', *Socialist Register* 40: 63–87.

Harvey, D. (2014) *Seventeen Contradictions and the End of Capitalism*, Oxford University Press, Oxford.

Hay, I. and J. V. Beaverstock (2016) *Handbook on Wealth and the Super-Rich*, Edward Elgar Publishing, Cheltenham.

Heller, H. (2016a) *The Capitalist University: The Transformations of Higher Education in the United States, 1945–2016*, Pluto Press, London.

Heller, H. (2016b) 'How US universities became IP-based capitalists', *Times Higher Education*, 20 October 2016.

Hess, D.J. (2005) 'Technology- and product-oriented movements: Approximating social movement studies and science and technology studies', *Science, Technology and Human Values* 30(4): 515–535.

Hirsch Hadorn, G., Hoffmann-Riem, H., Biber-Klemm, S., Grossenbacher-Mansuy, W., Joye, D., Pohl, C., Wiesmann, U. and Zemp, E. (2008) *Handbook of Transdisciplinary Research*, Springer, the Netherlands.

Holdrege, C. (2013) *Thinking Like a Plant. A Living Science for Life*, Lindisfarne Books, Great Barrington, MA.

Holling, C.S. (1998) 'Two cultures of ecology', *Conservation Ecology* [online] 2(2): 4, www.consecol.org/vol2/iss2/art4.

Holt-Giménez, E. (2002) 'Measuring farmers' agroecological resistance after Hurricane Mitch in Nicaragua: A case study in participatory, sustainable land management impact monitoring', *Agriculture, Ecosystems and Environment* 93: 87–105.

Holt-Gimenez, E. (2006) *Campesino a Campesino: Voices from Latin America's Farmer to Farmer Movement for Sustainable Agriculture*, Food First Books, Oakland.

hooks, b. (1994). *Teaching to Transgress: Education as the Practice of Freedom*, Routledge, New York.

hooks, b. (2000) *Feminist Theory: From Margin to Center*, Pluto Press, London.

hooks, b. (2003) *Teaching Community: A Pedagogy of Hope*, Routledge, New York.

IIED, APPG Agroecology, CNOP, Kene Conseils, Centre Djoliba and IRPAD (2012) *High level policy dialogue between the Alliance for a Green Revolution in Africa (AGRA) and small scale farmers on the priorities and governance of agricultural research for development in West Africa*, International Institute for Environment and Development (IIED), the All Party Parliamentary Group on Agroecology and Sustainable Food Systems (APPG), The Coordination Nationale des Organisations Paysannes (CNOP), Kene Conseils, Centre Djoliba and Institut de Recherche et de Promotion des Alternatives de Développement en Afrique (IRPAD), London and Bamako.

IIRR (2000) *Going to Scale: Can We Bring More Benefits to More People More Quickly? Conference Highlights*, International Institute of Rural Reconstruction (IIRR), Philippines.

Ingram, D. (1987) *Habermas and the Dialectic of Reason*, Yale University Press, New Haven.

InPACT (2016) *Pour une souveraineté technologique des paysans*, Initiatives pour une Agriculture Citoyenne et Territorial, Bagnolet, Available at: http://www.latelierpaysan.org/IMG/pdf/impression_plaidoyer_long_janv_17.pdf.

Irwin, A. (1995) *Citizen Science : A Study of People, Expertise and Sustainable Development*, Routledge, London.

Kamminga, H. (1995) 'Science for the people?' In: Wakeford, T. and Walters, M. (Eds) *Science for the Earth*, Wiley, Chichester.

Keller, E.F. (1985) *Reflections on Gender and Science*, Yale University Press, Yale.

Kuhn, T.S. (1962) *The Structure of Scientific Revolutions*, University of Chicago Press, Chicago.

La Via Campesina (2009) *Declaration of Rights of Peasants – Women and Men, La Via Campesina*, Harare, https://viacampesina.net/downloads/PDF/EN-3.pdf.

Laloux, F. (2016) *Reinventing Organisations.* Nelson Parker.

Lamont M (2009) *How Professors Think: Inside the Curious World of Academic Judgment*, Harvard University Press, Cambridge, MA.

Lang, D.J., Wiek, A., Bergmann, M., Stauffacher, P. Martens, P. Moll, M. Swilling and C. J. Thomas (2012) 'Transdisciplinary research in sustainability science: practice, principles, and challenges', *Sustainability Science* 7(1): 25–43.

Latour, B. (1987) *Science in Action. How to Follow Scientists and Engineers through Society*, Harvard University Press, Cambridge MA.

Latour, B. (1993) *We Have Never Been Modern*, Harvard University Press, Cambridge, MA.

Lederman, M. and I. Bartsch (2001) *The Gender and Science Reader*, Routledge, London.

Leff, E. (2004) 'Racionalidad ambiental y diálogo de saberes. Significancia y sentido en la construcción de un futuro sustentable', *Polis. Revista Latinoamericana* (7), 1–29.

Levidow, L. (2008) 'Democratizing technology choices: European public participation in agbiotech assessments', *Gatekeeper Series*, No. 135, International Institute for Environment and Development, London.

Lorway, R. (2017) *AIDS Activism, Science and Community Across Three Continents*, Springer, New York.

Lowndes, V. and G. Stoker (1998) *Guidance on Enhancing Public Participation in Local Government*, Department of Environment, Transport and Regions, London.

Macdonald, M., E. Sprenger and I. Dubel (1997) *Gender and Organisational Change. Bridging the gap between policy and practice*, Kit Publications, Royal Tropical Institute, Amsterdam.

MacGregor, S. (2014) Transdisciplinarity and Conceptual Change, *World Futures: The Journal of New Paradigm Research* 70 (3–4): 200–232.

Machín Sosa, B., A.M. Roque, D.R. Ávila and P. Rosset (2010) *Revolución agroecológica: el movimiento de Campesino a Campesino de la ANAP en Cuba. Cuando el campesino ve, hace fe*, ANAP and La Vía Campesina, Havana and Jakarta, http://www.viacampesina.org/downloads/pdf/sp/2010-04-14-rev-agro.pdf.

Machín Sosa, B., A. M. Roque Jaime, D. R. Ávila Lozano and P.M. Rosset (2013) *Agroecological Revolution: The Farmer-to-Farmer Movement of the ANAP in Cuba*, La Via Campesina, Jakarta, Available at: http://viacampesina.org/downloads/pdf/en/Agroecological-revolution-ENGLISH.pdf.

Macilwain, C. (2016) 'The elephant in the room we can't ignore', *Nature* 531, 277 (17 March 2016).

Mae-Wan Ho (2013) 'The new genetics and natural *versus* artificial genetic modification', *Entropy* 15: 4748–4781.

Mantell, K. (2002) 'Mexico confirms GM maize contamination', *SciDev Net*, 19 April 2002.

Marshall, J. and Reason, P. (2007) 'Quality in research as "taking an attitude of inquiry"', *Management Research News* 30 (5): 368–380.

Marti, N. and M.P. Pimbert (2007) 'Barter markets for the conservation of agro-ecosystem multi-functionality: the case of the chalayplasa in the Peruvian Andes', *International Journal of Agricultural Sustainability* 5(1): 51–69.

Martínez-Torres, M.E. and P. M. Rosset (2014) 'Diálogo de saberes in La Vía Campesina: food sovereignty and agroecology' *The Journal of Peasant Studies*, 41(6): 979–997.

Mason, K. (2013) 'Academics and social movements: Knowing our place, making our space', *ACME: An International Journal for Critical Geographies* 12(1): 23–43.

Masson, D., A. Paulos and E. Beaulieu Bastien (2017) 'Struggling for food sovereignty in the World March of Women', *Journal of Peasant Studies* 44(1): 56–77.

Maus, M. (1990) *The Gift*, Routledge, London.

Mauser, W., G. Klepper, M. Rice, B. S. Schmalzbauer and H. Hackmann (2013) 'Transdisciplinary global change research: The co-creation of knowledge for sustainability', *Current Opinion in Environmental Sustainability* 5: 420–431.

Maxigas (2012) 'Hacklabs and hackerspaces – tracing two genealogies', *Journal of Peer Production*, 2: 1–10.

Mazhar, F., J.M. Chevalier, F. Akhter, D.J. Buckles and M. Bourassa (2006) 'Towards a Peasant World University: Food for thought, a concept note', mimeo, UBINIG and Carleton University.

Meek, D. (2015) 'Learning as territoriality: The political ecology of education in the Brazilian landless workers' movement' *Journal of Peasant Studies* 42 (6): 1179–1200.

Meek D. and R. Tarlau (2015) 'Critical Food Systems Education (CFSE): educating for food sovereignty', *Agroecology and Sustainable Food Systems* 40(3): 237–360.

Merchant, C. (1992) *The Death of Nature: Women, Ecology and the Scientific Revolution*, Reprint edition, Bravo Ltd., London.

Merchant, C. (1996) *Earthcare: Women and the Environment*, Routledge, London.

Meusburger, P., G. Derek and L. Suarsana (2015) *Geographies of Knowledge and Power*, Springer, New York.

Mollett, S. and C. Faria (2013) 'Messing with gender in feminist political ecology', *Geoforum* 45: 116–125.

Monsanto Tribunal (2017) *International Monsanto Tribunal Advisory Opinion*, The Hague, Available at: http://www.monsanto-ribunal.org/upload/asset_cache/189791450.pdf.

Moore, S., C. Neylon, M. P. Eve, D. P. O'Donnell and D. Pattinson (2017) '"Excellence R Us": University research and the fetishisation of excellence', *Palgrave Communications* 3, Article 16105, published online, 19 January 2017, 10.1057/palcomms.2016.105.

New World Academy (with the Kurdish Women's Movement) (2015) *Stateless Democracy*, New World Academy Reader No 5, Utrecht.

Nicolescu, B. (1994), *The Charter of Transdisciplinarity*, available at http://inters.org/Freitas-Morin-Nicolescu-Transdisciplinarity.

Nicolescu, B. (2008) *Transdisciplinarity – Theory and Practice*, Hampton Press, Cresskill, NJ.

Noble, D.F. (1995) *Progress without People: New Technology, Unemployment, and the Message of Resistance*, Between the Lines, Toronto.

North, D.C (1990) *Institutions, Institutional Change, and Economic Performance*, Cambridge University Press, Cambridge.

Nowotny, H. (2006) *The Potential of Transdisciplinarity*, available at www.helga-nowotny.eu/downloads/helga_nowotny_b59.pdf.

Nyéléni (2007) *Declaration of Nyéléni*, Forum for Food Sovereignty, http://nyeleni.org/spip.php?article290.

Nyéléni (2015) *Declaration of the International Forum for Agroecology*, www.foodsovereignty.org/forum- agroecology-nyeleni-2015.

Öcalan, A. (2011) *Democratic Confederalism*, Transmedia Publishing, London.

O'Hara, M. (2015) 'University lecturers on the breadline: Is the UK following in America's footsteps?' *The Guardian*, 17 November 2015.

O'Mahony S. and F. Ferraro (2007) 'The emergence of governance in an open source community', *Academy of Management Journal* 50(5): 1079–1106.

Ostrom, E. (2005) *Understanding Institutional Diversity*, Princeton University Press, Princeton.

Ostrom, E. (2010) 'Beyond markets and states: polycentric governance of complex economic systems, *American Economic Review* 100: 1–33.

Pachico, D. and S. Fujisaka (2004) 'Scaling up and out: Achieving widespread impact through agricultural research', *CIAT Economics and Impact Series* 3: 340, International Center for Tropical Agriculture, Cali.

Peet, R. and M. Watts (2004) *Liberation Ecologies*, 2nd Edition, Routledge, London.

People's Knowledge Editorial Collective (2016) *People's Knowledge and Participatory Action Research: Escaping the White-Walled Labyrinth*, Practical Action Publishing, Rugby.

People's Knowledge Editorial Collective (2017) '*Knowledge for food justice: Participatory and action research approaches*', Reclaiming Diversity and Citizenship Series, Coventry University.

Perfecto, I., J.Vandermeer and A.Wright (2009) *Nature's Matrix: Linking agriculture, conservation and food sovereignty*, Earthscan, London.

Piaget, J. (1972) 'The epistemology of interdisciplinary relationships', In: Apostel, L. and Centre for Educational Research and Innovation (eds), *Interdisciplinarity: Problems of Teaching and Research in Universities*, Organisation for Economic Co-operation and Development, Paris.

Pimbert, M.P. (2004) *Institutionalising Participation and People-Centered Processes in Natural Resource Management*, International Institute for Environment (IIED), London.

Pimbert, M.P. (2008) *Towards Food Sovereignty: Reclaiming Autonomous Food Systems*, International Institute for Environment and Devleopment (IIED), London.

Pimbert, M.P. (2009a) 'Local organizations at the heart of food sovereignty', In: M.P. Pimbert (ed.) *Towards Food Sovereignty: Reclaiming Autonomous Food Systems*, International Institute for Environment and Development (IIED), London.

Pimbert, M.P. (2009b) 'Transforming knowledge and ways of knowing', In: M.P. Pimbert (ed.). *Towards Food Sovereignty: Reclaiming Autonomous Food Systems*, International Institute for Environment and Development (IIED), London.

Pimbert, M.P. (2010) 'Reclaiming citizenship, empowering civil society in policy-making', In: M.P. Pimbert (ed.), *Towards Food Sovereignty: Reclaiming Autonomous Food Systems*, International Institute for Environment and Development (IIED), London.

Pimbert, M.P. (2011) *Participatory Research and On-Farm Management of Agricultural Biodiversity in Europe*, International Institute for Environment and Development (IIED), London.

Pimbert, M.P. (2012) 'Putting farmers first: Reshaping agricultural research in West Africa', IIED Policy Brief, International Institute for Environment and Development (IIED), London.

Pimbert, M.P., B. Barry, A. Berson and K. Tran-Thanh (2011) *Democratising Agricultural Research for Food Sovereignty in West Africa*, the International Institute for Environment and Development (IIED), the Coordination Nationale des Organizations Paysannes (CNOP), le Centre Djoliba, l'Institut de Recherche et de Promotion des Alternatives en Dévelppement (IRPAD), Kene Conseils, l'Union des Radios et Télévisions Libres du Mali (URTEL), Bamako and London.

Pimbert, M.P., P. V. Satheesh, A. Argumedo and T. M. Farvar (2017) 'Participatory action research transforming local food systems in India, Iran and Peru', In: People's Knowledge Editorial Collective (Ed) *Knowledge for Food Justice: Participatory and Action Research Approach*. Reclaiming Diversity and Citizenship Series, Coventry.

Pimbert, M.P. and T. Wakeford (2003) '*Prajateerpu*, power and knowledge. The politics of participatory action research in development', *Action Research*, 1(2): 184–207.

PLA Notes (2002) *PLA Notes CD Rom 1998–2001 – full set of back issues 1 to 40*, International Institute for Environment and Development, London.

Pohl, C. and G. Hirsch Hadorn (2007) *Principles for Designing Transdisciplinary Research*, The Swiss Academies of Arts and Sciences.Verlag, Munich.

Pontius, J., Dilts, R. and Bartlett, A. (2002) *From Farmer Field School to Community IPM. Ten years of IPM training in Asia*, FAO Community IPM Programme, FAO, Bangkok.

Posey, D.A. (1999) *Cultural and Spiritual Values of Biodiversity*, United Nations Environmental Programme, Nairobi and Intermediate Technology Publications, London.

Pouteau, S. (2007a) 'Emergence and auto-organisation: Revising our concepts of growth, development and evolution toward a science of sustainability', In: Zollitsch, W., Winckler, C., Waiblinger, S. and A. Haslberger (eds) *Sustainable Food Production and Ethics*, Wageningen Academic Publishers, Wageningen.

Pouteau, S. (2007b) *Génétiquement indéterminé – Le vivant auto-organisé*, Quae, Versailles.

Pretty, J.N. (1994), 'Alternative systems of inquiry for sustainable agriculture', *IDS Bulletin* 25(2): 37–48, Institute of Development Studies, University of Sussex, Brighton.

Pretty, J.N. and R. Chambers (1993), 'Towards a learning paradigm: New professionalism and institutions for sustainable agriculture', *IDS Discussion Paper DP 334*, Institute of Development Studies, University of Sussex, Brighton.

Randerson, J. (2008) 'Arpad Pusztai: Biological divide', *The Guardian*, 15 January 2008.

Rao, A. and R. Stuart (1997) 'Rethinking organisations: A feminist perspective', *Gender and Development* 5(1): 10–16.

Ravetz, J.R. (1971) *Scientific Knowledge and Its Social Problems*, Clarendon Press, London.

Ravetz, J.R. (2006) *The No-Nonsense Guide to Science*, New Internationalist Publications, Oxford.

Reason, P. and H. Bradbury (2008) *The SAGE Handbook of Action Research*, 2nd Edition, SAGE Publishing, London.

Rengifo Vasquez, G. (1998) 'The Ayllu', In: F. Apffel-Marglin (ed) *The Spirit of Regeneration: Andean Culture Confronting Western Notions of Development*, PRATEC and Zed Books, London.

Réseau Semences Paysannes (2017) *Semences Paysannes*, website, http://www.semences paysannes.org/definition_des_semences_paysannes_532.php.

Rittel, H.W.J. and Webber, M. M. (1973) 'Dilemmas in a general theory of planning', *Policy Sciences*, 4, 155–169.

Rocheleau, D. E., Thomas-Slayter, B. P. and E. Wangari (1996) *Feminist Political Ecology: Global Issues and Local Experiences*, Routledge, London.

Roger, A. (2013) 'Moissonner le champ scientifique. L'emprise des firmes multinationales de l'agrochimie sur la recherche académique roumaine, *Revue d'anthropologie des connaissances* 7(3): 717–745.

Rosset, P. and M. E. Martínez-Torres (2012), 'Rural social movements and agroecology: Context, theory and process', *Ecology and Society* 17(3): 17.

Rosset, P. M., Machín Sosa, B., Roque Jaime, A. M. and D.R. Ávila Lozano (2011) 'The Campesino-to-Campesino agroecology movement of ANAP in Cuba: Social process methodology in the construction of sustainable peasant agriculture and food sovereignty', *Journal of Peasant Studies* 38(1): 161–191.

Rowell, A. (2003) *Don't Worry, It's Safe to Eat*, Earthscan Ltd., London.

Salas, M. (2013) *Visualising Food Sovereignty in the Andes. Voices and Flavours from the Earth*, International Institute for Environment and Development, London.

Salas, M., H.J. Tillman, N. McKee and N. Shahzadi (2007) *Visualisation in Participatory Programmes. How to Facilitate and Visualise Participatory Group Processes*, Southbound, Penang in association with UNICEF Dhaka.

Saltelli, A. (2016) 'Science in crisis: From the sugar scam to Brexit, our faith in experts is fading', The Conversation, September 27, 2016, available at https://theconversation. com/science-in-crisis-from-the-sugar-scam-to-brexit-our-faith-in-experts-is-fading-65016.

Santos, Boaventura de Souza (2014) *Epistemologies of the South. Justice Against Epistemicide*, Paradigm Publishers, Boulder.

Saunders, K. (2002) *Feminist Post-Development Thought: Rethinking Modernity, Post-Colonialism and Representation*, Zed Books, London.

Scholz, R.W. and O. Tietje (2002) *Embedded Case Study Methods. Integrating Quantitative and Qualitative Knowledge*, Sage Publishers, London.

Scott, J. (1998) *Seeing Like a State: How Certain Schemes to Improve the Human Condition Have Failed*, Yale University Press, Yale.

Scott, J. (2009) *The Art of Not Being Governed: An Anarchist History of Upland Southeast Asia*, Yale University Press, Yale.

Seeley, C. (2011) 'Unchartered territory: Imagining and stronger relationship between the arts and action research', *Action Research* 9(1): 83–89.

Seeley, C. and Thornhill, E. (2014) *Artful Organisations*, Ashridge Executive Education, Berkhamsted.

Senge, P. (1990) *The Fifth Discipline: The Art and Science of the Learning Organization*, Currency Doubleday, New York.

Sitrin, M. (2006) *Horizontalism. Voices of Popular Power in Argentina*, AK Press, Edinburgh.

Sitrin, M. and D. Azzelini (2014) *They Can't Represent US: Reinventing Democracy from Greece to Occupy*, Verso Books, London.

Smith, A. (2014) 'The Lucas Plan: What can it tell us about democratizing technology today?' *The Guardian*, 22 January 2014.

Smith, A. (2014) 'Socially useful production', *STEPS Working Paper* 58, STEPS Centre, Brighton.

Smith, L.T. (2012) *Decolonizing Methodologies: Research and Indigenous Peoples*, 2nd Edition, Zed Books, London.

Soyini Madison, D. (2012) *Critical Ethnography: Method, Ethics, and Performance*, SAGE Publications, London.

Stirling, A. and S. Maher (1999) *Rethinking Risk. A pilot multi-criteria mapping of a genetically modified crop in agricultural systems in the UK*, Science Policy Research Unit, Brighton.

TATORT Kurdistan (2013) *Democratic Autonomy in North Kurdistan: The Council Movement, Gender Liberation, and Ecology*, New Compass Press, Porsgrunn.

Testart, J. (2015) *L'Humanitude au pouvoir, Comment les citoyens peuvent decider du bien commun*. Seuil, Paris.

Toledo, V.M. and N. Barrera-Bassols (2009) *La Memoria Biocultural: la importancia ecológica de las sabidurías tradicionales*, ICARIA Editorial, Barcelona.

Troxler, P. (2014) 'Fab labs forked: a grassroots insurgency inside the next industrial revolution', *Journal of Peer Production*, 5: 1–3.

Trublet, K. (2016) 'Quand la nouvelle génération de chercheurs français passe son temps à Pôle emploi et en contrats précaires', *Basta!* 19 Décembre 2016.

Union of Concerned Scientists (2012) *Heads They Win, Tails We Lose: How Corporations Corrupt Science at the Public's Expense*, Union of Concerned Scientists, Cambridge, MA.

Union of Concerned Scientists (2015) *Counting on Agroecology. Why we Should Invest More in the Transition to Sustainable Agriculture*, Union of Concerned Scientists, Cambridge, MA.

URGENCI (2016) *Overview of Community Supported Agriculture in Europe*, European CSA Research Group, http://urgenci.net/the-csa-research-group.

Van Breda, J. and M. Swilling (in press), 'Logics and principles for designing emergent transdisciplinary research processes: Learning experiences and reflections from a South African case study', *Sustainability Science*, 2017.

Vicente, P.R. (1993) 'The MASIPAG program: An integrated approach to genetic conservation and use', In: Nordic Development Cooperation Agencies (ed.), *Growing Diversity in Farmers Fields*, Nordic Development Cooperation Agencies, Lidingo, Sweden.

Vidal, J. (2016) 'Farming mega-mergers threaten food security, say campaigners', *The Guardian*, 26 September 2016.

Visvanathan, S. (1997) *A Carnival for Science: Essays on Science, Technology, and Development*. Oxford University Press, Oxford.

Visvanathan, S. (2005) 'Knowledge, justice and democracy', In: Leach, M., Scoones, I. and B. Wynne (eds.), *Science and Citizens: Globalisation and the challenge of engagement*, Zed Books, London.

Wadsworth, Y. (2005) 'Beloved Bangladesh': A western glimpse of participatory action research and the animator-resource work of Research Initiatives Bangladesh', *Action Research*, 3(4): 417–435.

Wainwright, H. (2009) 'A real green deal', *Red Pepper*, 7 October 2009, www.redpepper.org.uk/a-real-green-deal.

Wainwright, H. and D. Elliott (2017) *The Lucas Plan. A New Trade Unionism in the Making?* Spokesman Books, Nottingham.

Wakeford, T. and M.P. Pimbert (2004) '*Prajateerpu*, power and knowledge. The politics of participatory action research in development. Part 2. Analysis, reflections and implications', *Action Research*, 2(1): 25–46.

Walker, P.A. (2006) 'Political ecology: where is the policy?' *Progress in Human Geography* 30(3): 382–395.

Weale, S. (2016) 'Part-time lecturers on precarious work: "I don't make enough for rent", *The Guardian*, 16 November 2016.

Wheatley, M.J. (2006) *Leadership and the New Science: Discovering Order in a Chaotic World,* 3rd Edition. Berrett-Koehler Publishers, San Francisco.

Women Sanghams, P.V. Satheesh and M.P. Pimbert (1999) 'Reclaiming diversity, restoring livelihoods', *Seedling*, June 1999, GRAIN, Barcelona.

Xiang, W.N. (2013) 'Working with wicked problems in socio-ecological systems: Awareness, acceptance, and adaptation', *Landscape and Urban Planning* 110: 1–4.

Zavala, M. (2013) 'What do we mean by decolonizing research strategies? Lessons from decolonizing, indigenous research projects in New Zealand and Latin America', *Decolonization: Indigeneity, Education & Society*, 2 (1): 55–71.

Zibechi, R. (2010) *Dispersing Power: Social Movement as Anti State Forces,* AK Press, Edinburgh.

INDEX

Note: text within tables, number span in bold